A LEVEL CHEMISTRY

LONGMAN A-LEVEL AND AS-LEVEL REVISE GUIDES

Series editors
Geoff Black and Stuart Wall

Titles available
Biology
Chemistry
Economics
English
Geography
Mathematics
Physics
Art
Computer Science
French
History
Sociology

A-LEVEL
AND AS-LEVEL

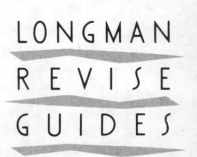

LONGMAN
REVISE
GUIDES

CHEMISTRY

Michael C. Cox

Longman

Longman Group UK Limited,
Longman House, Burnt Mill, Harlow,
Essex CM20 2JE, England
and Associated Companies throughout the world.

© Longman Group UK Limited 1990

First published 1990

Second impression 1990

British Library Cataloguing in Publication Data

Cox, Michael
  Chemistry.
  1. Chemistry
  I. Title
  540

  ISBN 0-582-05179-7

Set in 10/12pt Century Old Style

Produced by Longman Singapore Publishers (Pte) Ltd
Printed in Singapore.

EDITORS' PREFACE

Longman A-Level Revise Guides, written by experienced examiners and teachers, aim to give you the best possible foundation for success in your course. Each book in the series encourages thorough study and a full understanding of the concepts involved, and is designed as a subject companion and study aid to be used throughout the course.

Many candidates at A Level fail to achieve the grades which their ability deserves, owing to such problems as the lack of a structured revision strategy, or unsound examination technique. This series aims to remedy such deficiencies, by encouraging a realistic and disciplined approach in preparing for and taking exams.

The largely self-contained nature of the chapters gives the book a flexibility which you can use to your advantage. After starting with the background to the A, AS Level and Scottish Higher courses and details of the syllabus coverage, you can read all other chapters selectively, in any order appropriate to the stage you have reached in your course.

Geoff Black and Stuart Wall

ACKNOWLEDGEMENTS

The author is grateful to the Associated Examining Board (AEB), the University of Cambridge Local Examinations Syndicate (UCLES), the Joint Matriculation Board (JMB), the University of London Schools Examination Board (ULSEB) and Nuffield, the Northern Ireland Schools Examination Council (NISEC), the Oxford and Cambridge Schools Examination Board (OCSEB), the Oxford Delegacy of Local Examinations (ODLE), the Scottish Examination Board (SEB), the Welsh Joint Education Committee (WJEC) and the Wessex Advanced Level Modular Sciences Project for permission to reproduce questions from their examination papers. Any answers to these questions are solely my responsibility and have not been provided or approved by any Examination Board. The author is also grateful to BP Educational Service and to the Polytechnic of North London for permission to reproduce, in Chapter 1, the comprehension passage based upon extracts from *A Case Study of Ethanoic Acid*, R. Kelsey, BP Educational Service (1985) and *The Essential Chemical Industry*, Polytechnic of North London (1985).

I am especially grateful to the editors, Stuart Wall and Geoff Black, for their very helpful comments and guidance and to Professor Richard Kempa of the University of Keele for his extremely thought-provoking observations and constructive suggestions. His meticulous attention to detail has helped me to avoid errors and to call 'a spade a spade'. If there are mistakes or ambiguities in this book, it is entirely my fault.

Above all, I wish to thank my wife, Maureen, for her continuing love, patience and support yet again.

AUTHOR'S NOTE

Do you want to pass A-Level Chemistry with a good grade? Read on if your answer is yes.

This book is designed to guide you through your A-Level Chemistry course and examinations. It cannot replace your teacher or your textbook, but it can help you get the best out of both of them. If you do not want to study and be successful, then there is probably very little I can do or say that will make you want to study and succeed. No book can work miracles. However, if you want to be a successful student then I can help you with your study methods, your revision and your examination techniques.

Start by looking through the contents list. Chapter 1 is very important. You need to know right away which syllabus you are studying and how you will be examined. You may have to ask your teacher and write to the Examination Board for a copy of your syllabus. Do this today. You will find the names and addresses of the Examination Boards listed after the contents page.

Notice that Chapter 2 is about how to study. You should begin here, especially if you are just beginning your A-Level course. Successful students have good habits of study. Follow the advice in Chapter 2 and in the books which are recommended there. Make sure you acquire good study habits – they will help you with your other subjects and benefit you in many other ways.

The remaining chapters cover material essential to every A-Level Chemistry examination syllabus, and include the common core agreed by the GCE Boards of England, Wales and Northern Ireland. You can obtain a copy of the agreed common cores from any GCE Board. The Chemistry core is supposed to represent about 50 per cent of any A-Level syllabus. I believe that 75 per cent is a more realistic figure, and that A-Level syllabuses differ more in their methods of assessment than in the content of their chemistry.

The emphasis in this book is upon knowing and understanding the facts, patterns, principles, theories and applications which are central to A-Level Chemistry, and most likely to be assessed by any Examination Board. The chapters are as independent as possible, so you may study them in the order most suited to your course and textbook. They will give you extra help with the chemistry which students find particularly difficult and which candidates tend to get wrong in tests and examinations.

CONTENTS

NAMES AND ADDRESSES OF THE EXAM BOARDS

Associated Examining Board (AEB)
Stag Hill House
Guildford
Surrey GU2 5XJ

University of Cambridge Local Examinations Syndicate (UCLES)
Syndicate Buildings
1 Hills Road
Cambridge CB1 1YB

Joint Matriculation Board (JMB)
Devas St
Manchester M15 6EU

University of London Schools Examination Board (ULSEB)
Stewart House
32 Russell Square
London WC1B 5DN

Northern Ireland Schools Examination Council (NISEC)
Beechill House
42 Beechill Road
Belfast BT8 4RS

Oxford and Cambridge Schools Examination Board (OCSEB)
10 Trumpington Street
Cambridge CB2 1QB

Oxford Delegacy of Local Examinations (ODLE)
Ewert Place
Summertown
Oxford OX2 7BX

Scottish Examination Board (SEB)
Ironmills Road
Dalkeith
Midlothian EH22 1BR

Welsh Joint Education Committee (WJEC)
245 Western Avenue
Cardiff CF5 2YX

THE SYLLABUS AND EXAMINATION

GETTING STARTED

Do you know the difference between A-level, AS-level and H-grade chemistry? Do you know which syllabus you are following and which examination you will be taking? These are important questions. This chapter will help you to answer them. It will also help you to understand the aims, objectives and content of your chemistry syllabus. Do you know how to tackle OT and structured questions or comprehension and free response questions? These are the different types of question you will be set by your Examination Board. This chapter explains them to help you prepare for your examination.

ESSENTIAL PRINCIPLES

Are you doing A-level or AS-level chemistry? If you are not sure which, ask your teacher or college lecturer as soon as possible. The 'A' stands for 'Advanced' and in both cases means the same high standard that you must achieve beyond GCSE chemistry. The 'S' stands for 'Supplementary' and means the teaching and private study time you need is about half that required for an A-chemistry course. Think of S for semi-time but NOT semi-standard. AS-level chemistry is NOT easier than A-level! In Scotland you do H-grade chemistry.

> It really is important to have your own copy of the syllabus.

There are nine examination boards in the UK. You will find their names and addresses on page viii. Each board administers at least one approved A-level chemistry examination syllabus. AEB, JMB and ULSEB each administer two. Most boards also have at least one approved AS-level chemistry examination syllabus. Some syllabuses offer a choice of options or special studies. Discuss any choice with your teacher or lecturer and decide on your options or special studies at the earliest opportunity.

I strongly advise you to get a copy of your examination syllabus now and to find out your options as soon as possible. If you know where you are going you are more likely to get there.

An examination syllabus is a document published by an examination board for candidates and their teachers. In the strictest sense, it does not specify what you should learn or what your teacher should teach. The syllabus specifies what the board will examine, and how. So what will your board examine?

AIMS AND OBJECTIVES

In every A-level and AS-level syllabus you will find a concise statement of its aims. These tell you the general intentions behind the syllabus. For example, a typical aim will be to encourage you to see chemistry as an important practical science which is relevant to the world in which we live.

In every A-level and AS-level syllabus you will find a concise statement of its objectives. The objectives tell you the specific abilities that will be assessed. In 1983 the boards jointly published, as part of the A-level chemistry common core, the following assessment objectives to be included in every A-level chemistry syllabus:

Ability to:

- recognise and recall specific chemical facts, principles, methods and terminology;
- explain and interpret specific chemical facts in terms of principles;
- apply chemical knowledge to familiar and unfamiliar situations;
- use familiar and unfamiliar chemical data in diagrammatic, graphical, numerical, symbolic and verbal form, and to translate chemical data from one form into another;
- use chemical facts and principles to predict likely outcomes of events;
- organise chemical knowledge and present it clearly, and to communicate in an appropriate and logical manner;
- recognise and recall chemical laboratory techniques, to interpret observations and to plan simple laboratory investigations.

Since 1987 these assessment objectives have formed the basis of every A-level chemistry examination in the UK. You will find similar objectives in the AS-level syllabuses. Your board will examine your abilities by setting you written papers and subjecting you to either a 3 hr practical examination or a continuous internal assessment of your practical work by your teacher.

CONTENT

The content is a detailed statement that usually takes up most of the space in the published syllabuses. It sets out the facts, patterns, principles and theories that may be included in the examination to test your abilities. The contents page of this book lists the major interrelated topic areas you can expect to find represented in any A-level and AS-level chemistry syllabus. Each of the examination boards will develop these major topics in their own way and will specify the syllabus content as clearly, as concisely and as precisely as possible.

A = AEB;
B = UCLES;
C = JMB;
D = ULSEB;
E = NISEC;
F = NUFFIELD;
G = OCSEB;
H = ODLE;
I = SEB;
J = WJEC;
K = WESSEX;
L = RSC;
M = ALCC

	A	B	C	D	E	F	G	H	I	J	K	L	M
atomic structure – nuclear	✓	✓	✓	✓	✓	✓	✓	✓	✓	✓	✓	✓	✓
atomic structure – electronic	✓	✓	✓	✓	✓	✓	✓	✓	✓	✓	✓	✓	✓
structure & bonding – elements	✓	✓	✓	✓	✓	✓	✓	✓	✓	✓	✓	✓	✓
structure & bonding – compounds	✓	✓	✓	✓	✓	✓	✓	✓	✓	✓	✓	✓	✓
gases – kinetic theory & gas laws	✓	✓	✓		✓		✓	✓	✓	✓	✓	✓	✓
liquids – kinetic molecular picture	✓	✓	✓	✓	✓	✓	✓	✓		✓		✓	✓
equilibria – acid-base reactions	✓	✓	✓		✓		✓	✓	✓		✓	✓	✓
equilibria – redox reactions	✓	✓	✓	✓	✓	✓	✓	✓	✓	✓	✓	✓	✓
equilibria – gas reactions	✓	✓	✓	✓	✓	✓	✓	✓	✓	✓	✓	✓	✓
equilibria – liquid mixtures	✓	✓		✓	✓		✓	✓		✓			
equilibria – ionic precipitation			✓					✓		✓			
energetics – heat changes	✓	✓	✓	✓	✓	✓	✓	✓	✓	✓	✓	✓	✓
energetics – free energy changes	✓					✓							
energetics – entropy changes	✓						✓						
kinetics – rates of reaction	✓	✓	✓	✓	✓	✓	✓	✓	✓	✓	✓	✓	✓
kinetics – catalysis	✓	✓	✓	✓	✓	✓	✓	✓	✓	✓	✓	✓	✓
electrochemistry – electrolysis		✓						✓				✓	
electrochemistry – emf of cells	✓	✓	✓	✓	✓	✓	✓	✓	✓	✓	✓	✓	✓
periodic table – groups	✓	✓	✓	✓	✓	✓	✓	✓	✓	✓	✓	✓	✓
periodic table – periods	✓	✓	✓	✓	✓	✓	✓	✓	✓	✓	✓	✓	✓
periodic table – s-block elements	✓	✓	✓	✓	✓	✓	✓	✓	✓	✓	✓	✓	✓
periodic table – p-block elements	✓	✓	✓	✓	✓	✓	✓	✓	✓	✓	✓	✓	✓
periodic table – d-block elements	✓	✓	✓	✓	✓	✓	✓		✓		✓	✓	✓
structure in organic compounds	✓	✓	✓	✓	✓	✓	✓	✓	✓	✓	✓	✓	✓
mechanism in organic reactions	✓	✓	✓	✓	✓	✓	✓	✓		✓	✓	✓	✓
hydrocarbons	✓	✓	✓	✓	✓	✓	✓	✓	✓	✓	✓	✓	✓
halogeno hydrocarbons	✓	✓	✓	✓	✓	✓	✓	✓		✓	✓	✓	✓
alcohols	✓	✓	✓	✓	✓	✓	✓	✓	✓	✓	✓	✓	✓
aldehydes & ketones	✓	✓	✓	✓	✓	✓	✓	✓	✓	✓	✓	✓	✓
carboxylic acids	✓	✓	✓	✓	✓	✓	✓	✓	✓	✓	✓	✓	✓
amines	✓	✓	✓	✓	✓	✓	✓	✓	✓	✓	✓	✓	✓
esters & amides	✓	✓	✓	✓	✓	✓	✓	✓	✓	✓	✓	✓	✓
polymers	✓	✓	✓	✓	✓	✓	✓	✓			✓	✓	✓
social, environmental, technological	✓	✓	✓	✓	✓	✓	✓	✓	✓	✓	✓	✓	✓

Fig. 1.1 Analysis of A-level Chemistry Topics

In 1983 the boards jointly published a common core in A-level chemistry. You can obtain a copy from your board. The content of this common core represents at least 50% of the content of every A-level chemistry syllabus. As A-level syllabuses are revised to take account of the replacement of O-level by GCSE chemistry, the common core content may come to represent much more than 50% of the content of an individual syllabus.

In 1988 the Curriculum Subject Group of the Royal Society of Chemistry Education Division produced a report on post-16 chemistry. This report outlines the components which the Royal Society of Chemistry believes essential for all future A-level chemistry courses. The proposed conceptual and factual content matches closely the Inter-Board A-level Common Core. Moreover, these proposals were drawn up by four sub-groups of chemists working independently of each other and without reference to the published A-level common core.

This book covers not only the topics prescribed by the A-level common core but also where possible those other topics shared by all the syllabuses and examined by all the boards in England, Northern Ireland, Scotland and Wales: see Fig. 1.1. Many of these topics appear in the AS-level syllabus.

METHODS OF ASSESSMENT

In every chemistry syllabus you will find full details of the scheme of assessment. Figs. 1.2 and 1.3 show in outline how the different examination boards will test your A-level, Scottish Higher and AS-level chemistry. You may notice that most of the boards now use more than one type of question in their papers. The WJEC publish two versions of their examination papers, one set in English and one set in Welsh.

EXAM	PAPER 1	PAPER 2	PAPER 3	PRACTICAL
AEB	$2\frac{1}{4}$ hr comprehension and choice of 3 from 8 long answers	$1\frac{3}{4}$ hr 6–10 structured short answers	$1\frac{1}{4}$ hr 40 objective test questions	3 hr exam 3 exercises at least one quantitative
WESSEX	$1\frac{1}{2}$ hr structured short answers	2 hr 4 from 7 long answers	centre-based assessment of four modules	centre-based assessment
UCLES	$2\frac{1}{2}$ hr choice of 5 long answers minimum of 1 from 3 phys, 2 inorg, 3 org, 1 option	$1\frac{1}{2}$ hr 6 structured (5 on core & 1 on option)	1 hr 40 objective test questions	$3\frac{1}{4}$ hr exam or centre-based assessment
JMB	$2\frac{1}{2}$ hr A 40 objective test questions B & C structured short answers	$2\frac{1}{2}$ hr A structured short answers B choice of 2 long answers		3 hr exam or centre-based assessment
ULSEB	2 hr 7 structured short answers	$2\frac{1}{4}$ hr 4 compulsory questions with some choice	1 hr 40 objective test questions	3 hr exam or centre-based assessment
	Nuffield includes optional Paper 4 ($\frac{3}{4}$ hr) on special study but excludes the possibility of a practical examination			
NISEC	$2\frac{1}{2}$ hr 8 structured short answers	$1\frac{1}{4}$ hr 40 objective test questions	2 hr comprehension and choice of 2 from 6 long answers	centre-based assessment
OCSEB	$1\frac{1}{4}$ hr 40 objective test questions	$1\frac{1}{2}$ hr structured short answers	3 hr choice of questions from 3 sections	3 hr exam or centre-based assessment
ODLE	$2\frac{1}{4}$ hr physical	$2\frac{1}{4}$ hr inorganic	$2\frac{1}{4}$ hr organic	3 hr exam
	— two sections for each paper — A: compulsory structured questions – short answers B: choice of 3 from six questions – longer answers			

EXAM	PAPER 1	PAPER 2	PAPER 3	PRACTICAL
SEB	$1\frac{1}{2}$ hr 50 objective test questions	$2\frac{1}{2}$ hr A structured short answers B 3 compulsory 1 choice of long answers		
WJEC	3 hr structured short answers	3 hr 5 from 9 long answers		centre-based assessment

Fig. 1.2 A-level chemistry examinations

EXAM	PAPER 1	PAPER 2	PRACTICAL
AEB	$2\frac{1}{2}$ hr 3 sections: comprehension; structured; long questions, inc. compulsory on society & industry		centre-based assessment: minimum of five expts.; two must be quantitative and two qualitative
UCLES	$1\frac{1}{2}$ hr short answer and structured questions on core & modules	$1\frac{1}{2}$ hr longer open-ended questions on core & modules	centre-based assessment: minimum of 5 expts.; two must be quantitative
JMB	$1\frac{1}{2}$ hr multiple choice and compulsory questions on society & industry	$1\frac{1}{2}$ hr short answers and longer answers	centre-based assessment: each of four abilities tested at least once
ULSEB	1 hr multiple choice	$2\frac{1}{2}$ hr structured and longer essay-type questions	centre-based assessment: minimum of 5 expts.
Nuffield	$2\frac{1}{2}$ hr structured with one compulsory question on society & industry	$1\frac{1}{2}$ hr one question on each extension and special study	centre-based assessment which can include special study
ODLE	$1\frac{1}{2}$ hr structured questions	2 hr questions on options set to range of answers	centre-based assessment: about ten practical investigations

Fig. 1.3 AS-level chemistry examinations

TYPES OF EXAMINATION QUESTION

There are broadly four types of question. The AEB, London, London (Nuffield) and NISEC chemistry examinations use all four types.

OBJECTIVE TEST QUESTIONS

You may have to tackle several different kinds of objective test questions but you answer them all in the same way. You choose one letter (from A, B, C, D and sometimes E) that corresponds to your answer and you record your choice on a grid using a soft (HB) graphite pencil. The board's computer automatically scans the grid to mark and analyse your answers.

Buy past papers of objective tests and use them for revision.

You should follow the instructions carefully. An HB pencil is usually recommended. You could use a softer pencil (e.g. B) but never a harder one. Never use a pen. If you choose the correct answer the computer will give you one mark. If you choose the wrong answer you will score zero. The computer will not take off a mark for a wrong answer, so make sure you **attempt every question** in the test. Never mark more than one choice for a question. If you do, you will score zero for that question even if one of your choices is correct!

Multiple choice questions

Objective test (OT) questions are often wrongly called multiple choice (MC) questions. Here are the usual instructions and two examples (see Figs. 1.4a and 1.4b) typical of the multiple choice type of OT question. Notice the difference between the two. The first question gives you a choice of four and the second gives you a choice of five answers. In either case only one choice (the key) is correct. The other three or four choices (the distractors) are incorrect. In your objective test paper the MC questions will be of either one type (three distractors) or the other (four distractors): they will never be of both types.

What are the units of the rate constant for a first order reaction?

   A s^{-1}
   B $mol\,dm^{-3}$
   C $mol\,dm^{-3}\,s^{-1}$
   D $mol^{-1}\,dm^3\,s$

The electronic configuration of an element X is $1s^2 2s^2 2p^1$ and that of an element Y is $1s^2 2s^2 2p^5$. The molecular formula of a compound of X and Y could be

   A XY C X_2Y E XY_3
   B XY_2 D X_2Y_2

Fig. 1.4a) OT item with three distractors   **Fig. 1.4b) OT item with four distractors**

Multiple completion questions

These are quite different in style from multiple choice questions but they are still OT questions answered by marking a choice of A, B, C, D (and sometimes E) on a grid. Some candidates find the instructions difficult to follow at first so make sure you get some practice before you take the actual examination. The rubric (precise directions to be followed) may not be the same for every board. And your board may have adopted a new rubric for your exam. So be sure to follow the rubric carefully. Fig. 1.5 shows two sets of instructions applied to the same typical question.

In this section **one or more** of the options given may be correct. Select your answer by means of the following code:

A if 1, 2 and 3 are all correct	A if 1 and 2 only are correct
B if 1 and 2 only are correct	B if 2 and 3 only are correct
C if 2 and 3 only are correct	C if 1 only is correct
D if 3 only is correct	D if 3 only is correct
(AEB June 1988)	(AEB June 1989)

For the reaction $H_2(g) + I_2(g) \rightleftharpoons 2HI(g)$; $\Delta H = {}^-9.6\,kJ\,mol{-}1$. The factors that will decrease the amount of hydrogen iodide in the gas mixture at equilibrium include

  1 an increase in temperature
  2 an increase in pressure
  3 the use of a catalyst

AEB June 1988 rubric	AEB June 1989 rubric
gives answer = D	gives answer = C

Fig. 1.5 Two different multiple completion rubrics

Classification sets of questions

These are questions arranged in groups so that each one can refer to the same four (or five) choices of answer. They are essentially multiple choice questions in which the choice of answers comes before the question. Fig. 1.6 shows a typical rubric and set of questions.

Directions. Each group of questions below consists of five lettered headings followed by a list of numbered questions. For each numbered question select the one heading which is most closely related to it. Each heading may be used once, more than once, or not at all.

Questions 1–4 concern the following types of bonding or forces of interaction which can exist between particles

   A Dative covalent bonding
   B Dipole-dipole interactions
   C Hydrogen bonding
   D Ion-dipole interactions
   E Van der Waals' forces

Select from A to E the type of bonding or forces of interaction which best accounts for each of the following:

1   Water has a higher boiling point than hydrogen sulphide
2   Sodium chloride dissolves readily in water
3   Ammonia turns aqueous copper(II) sulphate dark blue in colour
4   The boiling points of the halogens increase with increasing atomic number

Fig. 1.6 Rubric and classification set

Situation sets of questions

These are groups of multiple choice or multiple completion questions related to some common data, information or laboratory situations. Fig. 1.7 shows a laboratory situation set of multiple choice questions taken from the June 1988 Nuffield A-level Chemistry paper 4.

SECTION III

Questions 31–39   **(nine questions)**

Directions. This group of questions deals with laboratory situations. Each situation is followed by a set of questions. Select the best answer for each question.

Questions 31–33. Cyclohexene may be produced in the laboratory by dehydrating cyclohexanol using phosphoric acid (85%) in the apparatus shown.

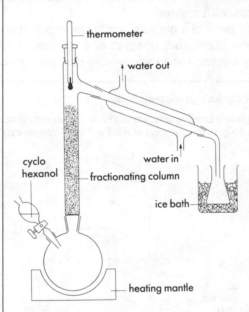

An excess of phosphoric acid (85%) was placed in the round-bottomed flask and 25 g of cyclohexanol was placed in the tap funnel. The cyclohexanol was slowly added, dropwise, ensuring that the temperature of the crude cyclohexene vapour produced during the reaction did not rise above 90° C.

The distillate was collected in an ice-cooled receiver and saturated with sodium chloride. The upper layer of crude cyclohexene was separated off and further purified.

The final yield of pure cyclohexene was 9 g.

Data:

	M /g mol^{-1}	ρ /g cm^{-3}	T_b /°C
Cyclohexanol	100	0.96	161
Cyclohexene	81	0.81	83

(M = mass of one mole; ρ = density; T_b = boiling point)

Fig. 1.7 Multiple choice laboratory situation set of questions

31   The most suitable practical method for measuring out the required amount of cyclohexanol is to

A   weigh the tap funnel empty and add cyclohexanol until the mass increases by 25 g
B   weigh 25 g of cyclohexanol on a watch-glass and transfer to the tap funnel
C   measure 26 cm^3 of cyclohexanol in a measuring cylinder and transfer to the tap funnel
D   run 24 cm^3 of cyclohexanol into the tap funnel from a burette
E   pipette 25 cm^3 of cyclohexanol into the tap funnel

32   The procedure by which the crude cyclohexene was further purified includes three steps (NOT shown in order)

  I Distil, collecting the fraction boiling at 81–83 °C
 II Wash with dilute sodium hydrogencarbonate solution
III Dry with anhydrous magnesium sulphate

Which is the correct order for the three steps?

A   I, III, II   D   III, I, II
B   II, I, III   E   III, II, I
C   II, III, I

33   The percentage of pure cyclohexene obtained in this procedure is

A $\dfrac{9 \times 0.96}{25 \times 0.81} \times 100$

B $\dfrac{9 \times 0.81}{25 \times 0.96} \times 100$

C $\dfrac{9}{0.81 \times 81} \times \dfrac{0.96 \times 100}{25} \times 100$

D $\dfrac{9 \times 0.81}{81} \times \dfrac{100}{25 \times 0.96} \times 100$

E $\dfrac{9}{81} \times \dfrac{100}{25} \times 100$

Assertion-reason questions

You may come across this type of question in some past examination papers. Each question consisted of two statements: an assertion and a reason, thus:

    1st statement (assertion):
        aqueous iron(III) chloride may be used as a test for phenol
    2nd statement (reason)
        phenol produces a distinctive mauve coloured complex with aqueous iron(III) cations

First you had to decide if each statement was true or false. If they were both true you then had to decide if the second statement (reason) was a valid reason for the first statement (assertion). For two correct statements and a valid reason you choose answer A. You would choose B for two correct but unrelated statements and C, D or E for combinations of one or other or both statements being false. These questions are useful for revision but they are no longer used in the examinations.

STRUCTURED QUESTIONS

You will be familiar with these questions from your GCSE examination. They consist of a number of parts, each part usually requiring a short answer. The marks allocated to each part are often shown on the question paper. This is an important and clear indication of how long to spend on that part of the question. Many of the boards now provide on the question paper spaces and lines for your answer. Sometimes the amount of space or the number of lines will help you to decide how much to write. But beware. You could be given a whole line for just a one word answer!

    Very often the parts of a question are linked together so that one part leads on to the next. This is most likely when one part of a question is itself further divided into parts. Look at the way the parts and their subdivisions are connected in Fig. 1.8.

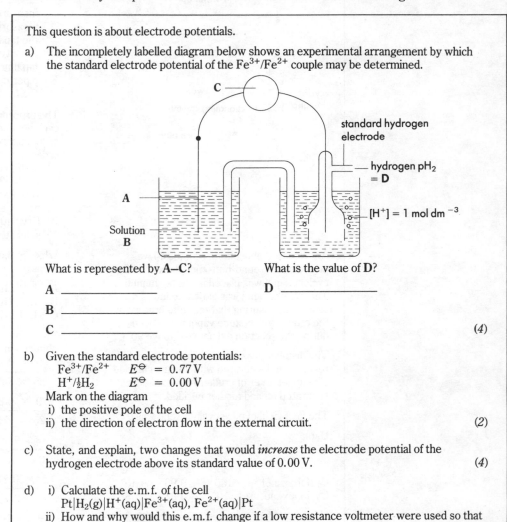

This question is about electrode potentials.

a) The incompletely labelled diagram below shows an experimental arrangement by which the standard electrode potential of the Fe^{3+}/Fe^{2+} couple may be determined.

What is represented by **A–C**?   What is the value of **D**?

A _____   D _____

B _____

C _____   *(4)*

b) Given the standard electrode potentials:
    Fe^{3+}/Fe^{2+} $E^{\ominus} = 0.77\,V$
    $H^{+}/\tfrac{1}{2}H_2$ $E^{\ominus} = 0.00\,V$
    Mark on the diagram
    i) the positive pole of the cell
    ii) the direction of electron flow in the external circuit.   *(2)*

c) State, and explain, two changes that would *increase* the electrode potential of the hydrogen electrode above its standard value of $0.00\,V$.   *(4)*

d) i) Calculate the e.m.f. of the cell
      $Pt|H_2(g)|H^{+}(aq)|Fe^{3+}(aq), Fe^{2+}(aq)|Pt$
   ii) How and why would this e.m.f. change if a low resistance voltmeter were used so that current was drawn from the cell?   *(4)*

Fig. 1.8 Parts of a question are often linked

> e) The electrode potentials given in b) suggest that a certain chemical reaction is possible. What is this reaction? Give two reasons why this reaction might not occur in practice. *(3)*
>
> *(Total 17 marks)*
> (ULSEB 1988)

Fig. 1.8 Parts of a question are often linked

You should always quickly read through the entire question before you tackle any of its parts. When my students read the parts towards the end of the question, they frequently find that they have a much clearer idea of the answers required by the parts at the beginning of the question. When a calculation is involved, the successive parts of the question often lead you through the working in easy stages. Sometimes, but not always, structured questions are stepped questions because within each question the parts become progressively more difficult.

COMPREHENSION OR DATA RESPONSE QUESTIONS

These questions are usually compulsory and, broadly speaking, are of two types. Both present you with a fairly substantial passage (about one or two pages) to read. The passage could contain diagrams, equations, graphs, tables of data or even a photograph.

In the AEB examination the passage is printed at the back of the paper on a sheet with a perforated margin. You should tear the passage out so you can easily refer to it when tackling the ten or twelve short answer questions based on the passage. The example in Fig. 1.9 is taken from the AEB June 1988 examination.

THE MANUFACTURE OF ETHANOIC ACID

There are two processes currently used in the U.K. for the large scale production of ethanoic acid.

An oxidation of naphtha

Light naphtha, consisting mainly of straight and branched chain C_5–C_7 alkanes with a boiling point range of 30–100 °C, can be oxidised directly to ethanoic acid. In this process, the hydrocarbons are oxidised by compressed air in a pressure vessel. The main products, methanoic, ethanoic and propanoic acids, are thought to arise by a free radical mechanism.

The oxidation is carried out in the liquid phase at a temperature of 150–200 °C. No catalyst is used. To maintain naphtha in the liquid state at these temperatures a pressure of 5 MPa is necessary. An important economic feature of the process, in addition to the cost of naphtha, is the energy requirement for providing air at high pressure.

The reaction is exothermic, as would be expected for a hydrocarbon-oxidation process. Two reactors are used. The reaction heat is removed by water-carrying cooling coils inside the reactors. Linked to the reactors and immediately above them is the overhead recovery system which is so placed to enable the condensed unreacted hydrocarbons to return to the reactor under gravity.

The reactor products are separated into three main fractions.

1 A low boiling point fraction. This contains unreacted and partially oxidised feedstock. This is recycled after propanone recovery.
2 A medium boiling point fraction. This contains an aqueous acid mixture. Water is removed by solvent extraction followed by azeotropic distillation. The acids are separated by fractional distillation and individually distilled.
3 A high boiling point fraction containing butanedioic acid. This can be crystallised out and purified but the crude mixture is often burnt as fuel.

For every 100 tonnes of ethanoic acid up 30 tonnes of propanone, over 25 tonnes of methanoic acid, and 15 tonnes of propanoic acid are produced, presenting considerable marketing problems.

The Monsanto process

National and international factors may have a powerful influence on the supply and cost of raw materials. This was the case in 1973–74 when the world price of petroleum rose sharply and, hence so did that of naphtha.

In the United States, the Monsanto company had been producing ethanoic acid from methanol and carbon monoxide.

$$CH_3OH + CO \rightarrow CH_3COOH$$

This process is now operated in the U.K. by B.P. Chemicals Ltd. The process operates at a pressure of 3 MPa and at a temperature of about 180 °C. A rhodium/iodine based catalyst is used. Yields of more than 99% ethanoic acid are obtained giving virtually no side reactions.

The source of the methanol feedstock is natural gas (methane). This is oxidised to synthesis gas (carbon monoxide and hydrogen) which is then converted into methanol.

However methanol can be made from any hydrocarbon source. As the petroleum and natural gas reserves diminish and become more expensive the synthesis gas can be produced from coal.

(Extracted from A CASE STUDY OF ETHANOIC ACID – R. KELSEY – B.P. Educational Service (1985) and THE ESSENTIAL CHEMICAL INDUSTRY – The Polytechnic of North London (1985).)

DO NOT SPEND MORE THAN 45 MINUTES ON QUESTION 1

1   Read the passage carefully and then answer questions a) to l). (The passage must be detached from the back of this question paper.)

a) Draw structural formulae to represent the **three** isomers of the C_5 alkane. Number the formulae 1, 2, 3 in order of increasing boiling point.   (4)

b) Give **two** characteristic properties of free radicals.
   Name **two** conditions under which they are formed.   (4)

c) For the air oxidation of naphtha manufacture of ethanoic acid give
   i) **one** process which requires energy;
   ii) **two** ways in which energy is recovered.   (5)

d) Explain why a high pressure (5 MPa) is necessary 'to maintain naphtha in the liquid state at these temperatures'.   (4)

e) Two reactor vessels are used in the air oxidation of naphtha process. These are constructed from stainless steel.
   i) Give **two** advantages of using two reactors rather than a single reactor.
   ii) Why are the reactors constructed from stainless steel?
   Give **one** disadvantage of using stainless steel.   (3)

f) What is the economic advantage of using an 'overhead recovery system'?   (1)

g) Give the structural formulae and names of **four** products of the air oxidation of naphtha apart from ethanoic acid.   (4)

h) i) What is an *azeotropic* mixture?   (3)
   ii) Explain why a pair of liquids which form an azeotropic mixture cannot be completely separated by fractional distillation. Include a temperature-composition diagram in your answer.   (4)

i) Explain why 'when the world price of petroleum rose sharply . . . so did that of naphtha'.   (1)

j) Write equations to represent the reactions in which
   i) methane is oxidised to synthesis gas;
   ii) synthesis gas is converted into methanol.   (3)

k) Give **three** advantages of the Monsanto process over the air oxidation of naphtha process.   (3)

l) For the Monsanto process give
   i) **two** fixed costs;
   ii) **two** variable costs.   (4)
   (AEB JUNE 1988)

Fig. 1.9 Comprehension questions

In the Nuffield examination you always have to write a summary of the most important points in the passage. It must be in your own words. You must not copy long sentences or phrases and you must not use more than a specified number of words (usually about 170). You will definitely lose marks if you use too many words. Sometimes there may be one or two short answer questions for you to tackle as well.

The NISEC examination will be somewhere between these two extremes. You will have to answer a number of structured questions based upon the passage and you will also have to write a summary of the main points.

ESSAY OR FREE RESPONSE QUESTIONS

Candidates generally find this type of question the most difficult. The question tends to be short, so little guidance is given, but long answers will be required. However, you will probably have a choice and you may find some of the questions will follow familiar patterns. You will certainly need to prepare for and practise answering these questions. This book contains a variety of examples. Revise your knowledge and understanding of chemistry by preparing outline answers for them.

WHAT DOES IT MEAN WHEN IT SAYS?

How many candidates (and their teachers) must have said at some time or other: 'I knew the answer but I didn't understand the question!' If you are studying H-grade chemistry in Scotland you can buy a book of past papers (with answers) from Robert Gibson (Publisher), 17 Fitzroy Place, Glasgow, G3 7SF. These are reprinted by special permission of the Scottish Examination Board but the answers do not emanate from the Board.

Unfortunately the other boards in the UK do not normally publish the detailed marks schemes and answers to their questions. If they did then you could learn the examiners' language just by studying their questions and answers. So far UCLES is the only board to include in its chemistry syllabus a glossary of some terms used in its science papers. I have compiled from the examination papers of the various boards a list of the most frequently used words and phrases: see Fig. 1.10.

" Study Fig. 1. 10 thoroughly to make sure you understand the question. "

Concise answers with the bare minimum of detail.

Classify each of the following oxides as acidic, basic or amphoteric.
Define the term molar first ionisation energy.
Give the oxidation number of uranium in the compound $UO_2(NO_3)_2$.
Indicate the conditions needed to increase the equilibrium yield.
Name the type of mechanism in the reaction of ammonia with bromoethane.
State Hess's law.
What is meant by a buffer solution?
Write a balanced equation for the complete combustion of ethanol.

Concise answers with essential but rather more detail.

Calculate the activation energy from the data provided.
Comment on the difference in physical properties of CO_2 and SiO_2.
Deduce the structure of the compound from the information provided.
Draw a labelled Born-Haber cycle for the formation of calcium oxide.
Identify the compounds X, Y and Z in the following observations.
Outline a laboratory method of measuring a named enthalpy change.
Show how you would detect the presence of sodium in a compound.
Sketch the unit cell of a body-centred cubic structure.

Longer reasoned answers supported by relevant facts and principles.

Explain why ammonia is basic and can form complexes with metal cations.
Explain what is meant by fractional distillation.
State and explain the effect of temperature upon reaction rates.
Suggest a way to distinguish between 1-bromobutane and 2-bromobutane.

Long well organised answers with all relevant facts and principles.

Compare the chemical reactions of ethanol and phenol.
Describe the principles of a mass spectrometer.
Discuss how properties of a solid depend on its structure and bonding.
Give an account of the use of catalysts in the chemical industry.

Fig. 1.10 Words and phrases in examination questions

The meaning of a word or phrase will depend upon the context of the question but you will see at a glance that some words or phrases require a short answer and some a long answer. You will find more help with the examiners' language in the questions and answers included in each chapter of this book.

GETTING STARTED

If you followed the advice given in Chapter 1, you should have a copy of your syllabus to tell you where you are going. Think of the syllabus as a reference map for routes to success in advanced chemistry. You now need to work out your best route to the top grade.

It is a definite advantage to have a good textbook, a good laboratory to work in and a good teacher to help you. However, success or failure is still in your own hands. So what can you do to help yourself? The answer is to develop good study habits, good revision programmes and good examination techniques. This chapter tells you how.

ESSENTIAL PRINCIPLES

Some people seem to like studying and find it easy. Others seem to dislike studying and find it difficult. What about you? Do you like studying or do you dislike it? Or is this question too general? Perhaps I should ask you more specific questions. Which subjects do you like to study and which do you dislike? Which subjects do you find easy and which do you find difficult? I guess this is what usually happens: (1) the more we like to study a subject, the easier we find it, and (2) the easier we find a subject, the more we like to study it. So, what can you do to increase your enjoyment of studying a subject? And what can you do to make the subject you are studying seem easier?

STUDY HOW TO STUDY

Use some of your time to study what other people have written about study methods. Here are two helpful paperbacks:

Title	Author	Publisher	Date	ISBN
Make the most of your mind	Tony Buzan	Pan Books	1988	0 330 30262 0
Double your learning power	G A Dudley	Thorsons	1986	0 7225 1211 2

In his book Tony Buzan describes the four major steps to studying any book: survey, preview, inview and review. At the back of his book Geoffrey Dudley gives a list of more than 70 other books that you could study to help you become a better student.

STUDYING IS A HABIT

A habit is something that doesn't require a great conscious effort or an act of will-power. You do it without thinking about it. This is how your studying should be. Roger Bacon wrote: 'Abeunt studia in mores.' Let studies become habits. I certainly agree that studying is a habit. Good students have been doing it for years and they never seem to have to think about it. So how have they acquired their study habits, and how can you build up your own good study habits? Here is a five-point plan on how to build the study habit:

❝ This is an excellent plan for developing good study habits. ❞

- Always do your private study in the same place at the same times each day. Always sit on the same fairly comfortable chair at the same table in the same quiet room on your own
- Warm up at the start of each study session by doing a simple, routine task left ready on purpose from your previous study session: e.g. underlining the headings and subheadings on a page of your notes
- Work up gradually to the study of a difficult part of a topic by tackling the easier parts of the topic first: e.g. draw a diagram of the apparatus before analysing your experimental results
- Always stop each session whilst you are still enjoying your studying. Never carry on for so long that you stop because you are fed up and no longer enjoying your work
- Clear your table at the end of each study session and just leave yourself a simple, routine task for the start of the next session

This five-point plan is simple and it works. Here is the psychology behind it. As soon as you sit at your table you have pleasant memories of your previous study sessions. There on your table is something simple for you to do and there is nothing to distract you. So as a matter of habit you find yourself doing something useful straightaway. In no time at all, without any conscious act of will power, you are automatically engrossed in your studies.

When I outline this plan to my new students, one or two (weaker students usually) raise difficulties and questions like these:

'I haven't got a room of my own. I share a tiny bedroom with my younger brother who goes to bed early. What should I do?'

If it really is impossible to find a corner of the bedroom for a chair and small table, you could fix a wide shelf on the wall and sit on your bed. Your studying shouldn't wake your brother and if he snores then you must either get used to it or, seriously, you could wear earplugs. Another possibility is to work somewhere else in the house (on the kitchen or dining table) when everybody else is in bed. If you must do this, then you should get up early in the morning. Do not work late into the night. I am making time to write this book by getting up at 5 am and doing two hours of work before breakfast as a matter of habit. If you really can't find a place at home, use the local library or find a neighbour who might let you study in a spare room. Whatever you do, don't make excuses for not building good study habits.

> 'I've got other subjects to study besides chemistry. I can't stop after 5 minutes because I've got too much to do already.'

If, when you first put the plan into action, you find that you have to stop your first study session even after 2 minutes, don't worry. You will soon catch up with your work. After all, there are times when we all have to catch up work missed because of absence or illness. The most important thing is to stop whilst you are still enjoying your work. It takes only a few study sessions for you to develop the habit of working longer and more efficiently than you ever did before because you now enjoy studying. One hour of effective studying that you enjoy is worth ten hours of ineffective studying that you dislike. Remember that your enjoyment of studying will breed success and your success will breed enjoyment of studying.

WHAT IS STUDY?

The five-point plan will lead you to enjoy studying but it does not say what you are doing when you are sitting there working at your table.

PHYSICAL ACTIVITIES

Studying consists of a variety of activities. It involves writing, drawing, calculating, computing and wordprocessing. These 'physical' activities leave you, at the end of your study session, with something that you can see: e.g. a neat set of notes, your account of an experiment, your solution to a problem, etc. This tangible evidence of your studying is something you can show to others. Your teacher can mark it. Your parents and friends can be made aware of it. You can see it piling up and it will remind you of how hard you have been working. All of this can give you a feeling of satisfaction and enhance your self-esteem. But there are other, perhaps more important, activities.

MENTAL ACTIVITIES

Studying also involves reading, listening, recognising, recalling, reflecting, remembering, knowing, thinking, understanding and learning. These 'mental' activities may leave you, at the end of your study session, with little or no tangible evidence of your efforts to show your teacher, your parents, your friends or even yourself. You have proof of your mental work probably only when you demonstrate to yourself and others what you have understood and learnt as a result of your studying. You do this when, for example, you get high marks in tests and exams. But you don't take a test or an exam every day. And what if it happens to be on something you haven't covered? You may not get high marks. It may look as though you haven't understood or learnt very much and it may seem as though you haven't done any studying. So there is a real danger that you will spend too much time on the 'physical' activities and not enough time on the 'mental' activities.

COMBINING MENTAL AND PHYSICAL ACTIVITIES

How do you strike a balance between the physical and mental activities of studying? You combine the two. Set yourself questions and problems. The act of setting yourself these tasks will involve the important mental activities. Put these questions and problems on paper (or your wordprocessor). The act of writing them out will involve the important physical activities. Answer the questions and solve the problems. This will involve not only the mental activities but also the physical activities as you jot down your thoughts and ideas. Put your final answers and solutions on paper (or your wordprocessor).

CHAPTER 2 ESSENTIAL PRINCIPLES 15

THE BENEFITS

You will enjoy studying when you combine your mental and physical activities by tackling specific tasks. If I ask my students simply to read some pages in their textbook, I rarely get an enthusiastic response or observe much purposeful activity. If, however, we have been tackling some problem and one of my students suggests that some vital piece of information or the answer to a specific question is in the textbook, the response is much more enthusiastic. The task is now different. We are not just reading generally and rather passively. We are now searching the text for some precise information. We are reading actively with a specific purpose. And the student who made the suggestion is usually the most enthusiastic and the most active.

You will always be ready to pass your tests and exams when you combine your mental and physical activities in tackling the right kind of questions and problems. This is where I come in. As a teacher I guide my students in the art of chemistry: drawing up the best questions to answer and framing the best problems to tackle. But I don't just mean exam questions. A good professional chemist usually knows what questions are worth asking and what problems are worth tackling to increase his knowledge and understanding. Fortunately for you, most of the A-level examination questions and problems are worth tackling to increase your knowledge and understanding of chemistry.

ORGANISE YOUR TIME

We are all different. Even identical twins are different. But we all have one thing in common – we each have 24 hours in the day. Some people achieve more than others in their 24 hour day. Why is this? The most important reason is that some people use their time more efficiently than others. Use your time wisely. Don't waste it, give it away or let others steal it.

USE YOUR OWN TEXTBOOKS

Buy your own chemistry textbook so that you can do more than just read it. You really need to use it by, for example, underlining or highlighting key points and writing notes in the margins.

STUDY FROM THE TOP DOWN

Start with the title and contents page of your textbook to get an overview of the main divisions of chemistry. You may see that A-level chemistry broadly divides into three parts: physical, inorganic and organic. Now skim quickly through the contents of each part. You may see that physical chemistry subdivides into sections on energetics, kinetics, equilibria, structure, bonding, etc. A quick glance through these sections may show you that each of them consists of subsections. By this process you can acquire a broad view of the subject and a familiarity with the language. Your aim is to appreciate the framework into which the facts, patterns, principles and theories of chemistry are fitted. In this way you are better able to pick out the essentials. This technique is well described by Tony Buzan.

PROCESS ACTIVELY

When studying, you should actively process the information to produce your notes in a different format. For example, condense text into the form of a diagram, graph or picture. Summarise the important features of a graph, diagram or picture by a few key words. Best of all, use Tony Buzan's techniques to produce a mind map. Whatever you do, always be active. Never be passive. Studying is an activity you should enjoy.

USE THIS BOOK

The art of chemistry lies in asking the right questions and posing the right problems. This book will help you to set yourself the most useful questions and problems and to judge the quality of your own answers and solutions. Turn the section and subsection headings into questions and think of the text as helping you study the chemistry and find the answers to your questions.

Revision is an essential component of good study habits. Good students make it part of their daily routine and use a planned programme of spaced revision. Unsuccessful students leave it too late and resort to last minute cramming.

REVISING IS A HABIT

Our brains are probably constantly engaged in revision without our being consciously aware of it. Every second our senses are sending new information to the brain to process. Much of this information will be compared with the information we have stored from past experiences. What you need to do as a student is to acquire the habit of consciously recalling and comparing your stored information with the new material. The mind mapping described by Tony Buzan is a way of helping you do this.

SPACING YOUR REVISION

Don't make the common mistake of starting your revision too late.

Various investigations and research have established the following features of the most effective revision:

- Periods of revision should be frequent and short.
- Periods of revision should be carefully spaced.
- New material should be revised as soon as possible.

Needless to say, many students tend to leave their revision to the last minute so that it becomes a long cramming session just before a test or an exam. Those students who revise more regularly don't always use carefully spaced revision. At the end of a lesson, lecture or study period, many students are too quick to 'switch off' and relax instead of reviewing the main points just covered.

In his 'programmed pattern of review' Tony Buzan recommends the following approach. Ten minutes after a one hour learning period spend 10 minutes in reviewing what you have learnt. This revision will prevent the very high percentage loss of memory that normally occurs in even a very short time after your learning period. The next day spend 5 minutes revising the same material. One week later, take 3 minutes, and then one month after that, take 2 minutes to revise this material again. You should find that this spaced revision (20 minutes in total) will be enough to lodge the material in your long term memory.

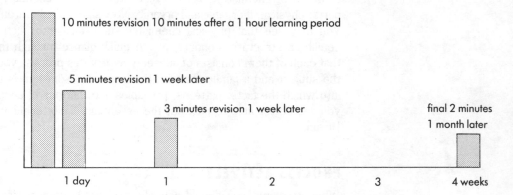

10 minutes revision 10 minutes after a 1 hour learning period

5 minutes revision 1 week later

3 minutes revision 1 week later

final 2 minutes 1 month later

Fig. 2.1 Spaced revision

1 day   1   2   3   4 weeks

CONDENSING YOUR NOTES

As an A-level chemistry student, you can expect to make notes throughout your course, so it seems inevitable that you will have more written notes at the end than at the beginning of the course. However, if you follow a programmed pattern of revision and acquire the habit of regularly reviewing and condensing your notes, you can keep their volume to a minimum.

Don't aim to acquire and retain a huge pile of notes. Aim to acquire and retain in your long term memory a firm understanding of the facts, patterns, principles and theories of chemistry. Make this your aim even if you are preparing for the Nuffield A-level Chemistry Paper 2 where you take your textbooks into the exam room.

HOW TO TAKE EXAMS

Don't lose valuable marks by wasting time.

The first thing to do is prepare yourself thoroughly. Check the syllabus to see what you will need to take into the exam room. And check the regulations to see what you are allowed to take in. For example, different boards have different regulations about calculators. Some do not allow you to use a calculator in the OT paper. If you are allowed a calculator, it must be silent, cordless and unable to store and display text or graphics. If the calculator is programmable, you must clear the memory of any programs – and you cannot take the calculator instruction manual into the exam room.

Tackle at least one set of past papers. Use the number of marks and the time allowed for each paper to calculate the *mark rate*: it is often about 1 mark per minute. Take off your wristwatch and put it in front of you where you can see it clearly. Get into the habit of checking your watch to keep to time. Do not spend too long on one question and not enough time on another. Time lost can rarely be recovered. Misjudging the time is one of the common mistakes you must avoid. If you are running out of time towards the end of an exam, abandon sentences and write your answers in note form.

Allow yourself at least one minute before the end of an OT paper to check through your answer grid to make sure you have attempted every question. You will score zero for any question you omit but you will NOT score −1 for a wrong choice. So a guess is better than nothing. In an OT paper, you must follow the precise instructions for correcting any mistake you may make. If you mark more than one choice you will score zero for that particular question. In all other written papers if you think you have made a mistake, cross it out neatly with one ruled line. Do not use an 'erasable pen' and do not use white correcting fluid (boards forbid it). Examiners look at your 'mistakes' and are sometimes allowed to award you marks for what you have crossed out.

ATOMIC STRUCTURE AND BONDING IN ELEMENTS

GETTING STARTED

An atom is pictured as a nucleus of protons and neutrons surrounded by electrons. The nuclei of radioactive elements may disintegrate spontaneously to emit either α-particles or β-particles and sometimes γ-radiation. The electrons surrounding the nucleus are arranged in orbitals on the basis of their energies. The electronic configuration represents, by a sequence of numbers, letters and superscripts, the grouping of orbitals into shells and subshells. Periodic patterns in the ionization energies, physical properties and structures of the elements are interpreted in terms of periodicity in their electronic configurations.

ESSENTIAL PRINCIPLES

ATOMIC STRUCTURE

There was a time when scientists believed that the elements were composed of submicroscopic particles which could not be split. They called these particles atoms (Greek: atomos – indivisible). Most recent research in theoretical and high-energy physics has indicated that, far from being indivisible, atoms seem to be composed of many different subatomic particles. However, in 1909, Geiger and Marsden performed their now famous experiments on the scattering of α-particles and in 1911, Ernest Rutherford proposed his theory of the nuclear atom to account for their results. In 1913, Moseley's work on X-rays related the atomic number to the number of protons in the nucleus and in 1932, Chadwick discovered the neutron.

Experimental work supports the theory that the atom consists of an extremely small nucleus composed of protons and neutrons surrounded by a large volume of space which, apart from the electrons, is empty: see Fig. 3.1. The nucleus is too small compared to the space occupied by the electrons for us to draw our diagrams of atoms to scale. You should be familiar with this simple picture (a nucleus surrounded by electrons) from your GCSE work. We use a similar but more detailed picture for A-level chemistry.

carbon isotope $^{13}_{6}C$: atomic number 6 and mass number 13

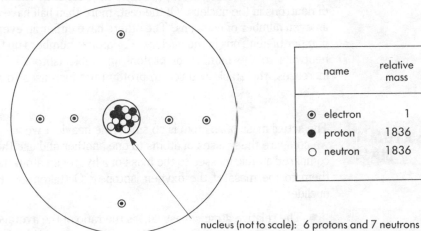

name	relative mass	relative charge
⊙ electron	1	−1
● proton	1836	+1
○ neutron	1836	0

Fig. 3.1 An isotope of carbon

nucleus (not to scale): 6 protons and 7 neutrons

ISOTOPES

The nucleus constitutes most of the mass of the atom because the mass of a proton is approximately the same as the mass of a neutron but about 2000 times the mass of an electron.

- Atomic number (Z) is the number of protons in the nucleus of an atom
- Mass number (A) is the number of protons and neutrons in a nucleus

All the atoms of a particular element must have the same number of protons in the nucleus but the number of neutrons may be different.

- Isotopes are atoms with the same atomic number but different mass numbers

RADIOACTIVE ISOTOPES

The nuclei of some isotopes are unstable and spontaneously disintegrate, radiating energy in the process. The radiation given out may or may not include very high energy gamma (γ-) rays but it always includes either alpha (α-) or beta (β-) particles (but never both) when a nucleus disintegrates. The isotope of a different element is always formed. These nuclear reactions may be represented by nuclear equations. You could be asked to balance and interpret equations for nuclear reactions.

- When a nucleus emits an α-particle (= He nucleus of 2n + 2p) the atomic number decreases by 2 and the mass number decreases by 4

$$^{235}_{92}U \rightarrow {}^{231}_{90}Th + \alpha^{2+}$$ uranium decaying to thorium

Think of two protons and two neutrons being taken from the nucleus.

- When a nucleus emits a β-particle (= an electron e⁻) the atomic number increases by 1 but the mass number stays the same

$$^{225}_{88}\text{Ra} \rightarrow {}^{225}_{89}\text{Ac} + \text{e}^-$$ radium decaying to actinium

Think of a neutron turning into a proton by losing an electron.

HALF-LIFE

> Calculations on half-life are quite popular with examiners.

- The half-life of a radioisotope is the time taken for its radioactivity to fall to half of its initial value
- The radioactive decay of an isotope is a first order rate of reaction so the half-life is independent of the mass of the radioisotope
- The half-life is characteristic of each radioisotope and is unaffected by catalysts or by changes in temperature
- The shorter the half-life, the more unstable is the radioisotope

STABLE ISOTOPES

Out of about 300 stable isotopes only 8 have an odd number of protons and an odd number of neutrons in the nucleus. Of the rest, more than half have an even number of protons and an even number of neutrons. The others have either an even number of protons or an even number of neutrons in the nucleus. For atomic numbers up to 20, the ratio of the number of neutrons to the number of protons in stable isotopes is 1:1. As the atomic number increases, the stable neutron to proton ratio increases to about 1.6:1.

RELATIVE ATOMIC MASS

The actual mass of an atom is so small (the heaviest weighs only about 4×10^{-25} kg) that we compare the masses of atoms to one another and use these relative values. At first we compared atomic masses to the mass of a hydrogen atom (taken as 1). Later we compared them to the mass of the oxygen isotope ^{16}O (taken as 16). Nowadays we use the ^{12}C nuclide:

- The relative atomic mass (A_r) is the ratio of the average mass per atom of the natural isotopic composition of an element to one-twelfth of the mass of an atom of nuclide ^{12}C

CALCULATING A RELATIVE ATOMIC MASS

The natural isotopic abundances of elements have been very accurately determined from mass spectra: see Fig. 3.2. You could be asked to read these abundances from a given

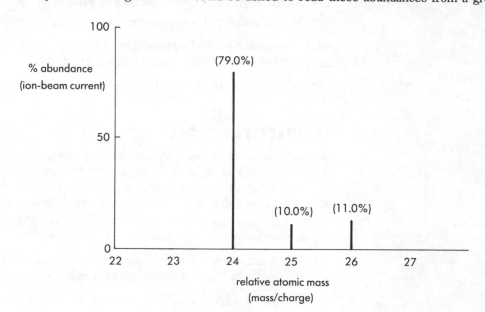

Fig. 3.2 Mass spectrum of magnesium

mass spectrum and use them to calculate the relative atomic mass from the mass numbers of the isotopes as follows:

A_r(Mg) is $(79.0 \times 24 + 10.0 \times 25 + 11.0 \times 26)/100 = 24.32$

The relative atomic mass of carbon itself is 12.011 because the element consists of two stable isotopes ^{12}C (98.9%) and ^{13}C (1.1%). Notice that the relative atomic mass (sometimes called the **atomic weight**) is a pure number: it does not have units. You will not be expected to remember A_r values and for most purposes in A-level chemistry they will be given to the first (or sometimes second) decimal place. For example, the AEB will print them on the periodic table provided with your examination paper.

AMOUNT OF SUBSTANCE AND THE MOLE

In chemistry we measure amounts of substances in moles. What is a mole? You may know it from your GCSE work as the mass (in grams) of a substance as indicated by its formula. How do you measure this mass? One way is to use the formula of the substance and the A_r values of the constituent elements to work out the relative formula mass and then measure out this value in grams: e.g.

formula of water is H_2O $A_r(H) = 1.0$ and $A_r(O) = 16.0$
relative formula mass of H_2O would be $(2 \times 1.0 + 16.0) = 18$
$\Rightarrow$ mass of one mole of H_2O would be 18 g.

Another way is to look up the formula and molar mass (M in g mol^{-1}) of a substance in a data book: see Fig. 3.3.

Formula	H	C	CH_4	H_2O	Na^+Cl^-	$Ca^{2+}CO_3^{2-}$
Molar mass/g mol^{-1}	1.0	12.0	16.0	18.0	58.5	100.0
6.02×10^{23}	$\longleftarrow$ atoms $\longrightarrow$		$\longleftarrow$ molecules $\longrightarrow$		$\longleftarrow$ ion-pairs $\longrightarrow$	

Fig. 3.3 Some molar masses

Why do we measure amounts of substances in moles? The main reason is that chemists want to deal with and compare similar numbers of particles of substances:

- The mole is the amount of substance which contains as many entities (atoms, molecules, ions) specified by the formula as there are carbon atoms in 0.012 kg of carbon nuclide ^{12}C

THE AVOGADRO CONSTANT

In 0.012 kg of carbon nuclide ^{12}C there are approximately 6.02×10^{23} atoms. So in 16 g of methane and 18 g of water there will be 6.02×10^{23} of CH_4 and H_2O molecules respectively. And in 58.5 g of sodium chloride and 100.0 g calcium carbonate there will be 6.02×10^{23} of Na^+Cl^- and $Ca^{2+}CO_3^{2-}$ ion-pairs respectively. How would we calculate the number of H_2O molecules in 45 g of water?

molar mass of H_2O is 18 g mol^{-1}
$\Rightarrow$ amount of H_2O in 45 g of water is $(45 \text{ g})/(18 \text{ g mol}^{-1}) = 2.5$ mol
$\Rightarrow$ number of H_2O molecules is $(6.02 \times 10^{23} \text{ mol}^{-1}) \times (2.5 \text{ mol})$
$= 1.505 \times 10^{24}$

Here is how we calculate the number of entities in any given amount of substance:

number of entities $= 6.02 \times 10^{23} \times$ amount of substance
or   number of entities $= L \times$ amount of substance

- The Avogadro constant is the proportionality constant connecting the number of entities (specified by a formula) to the amount of substance (expressed in moles)
- The value of the Avogadro constant, L, is 6.02×10^{23} mol^{-1}

ELECTRONIC STRUCTURE OF ATOMS

ATOMIC SPECTROSCOPY

If you look through a spectroscope at the colour produced by heating an s-block metal compound in a bunsen flame, you will see not one continuous band of varying colours but series of separate coloured lines (look in the Revised Nuffield Advanced Science Book of

energy to move electron completely out of H-atom from ground state level
is the ionisation energy $= 1312 \text{ kJ mol}^{-1}$

Balmer series:
frequency: 0.46 to 0.79 × 10¹⁵ Hz
lines in the visible
region of the spectrum

Lyman series:
frequency: 2.47 to 3.29 × 10¹⁵ Hz
lines in the 'invisible'
ultra violet region of the spectrum

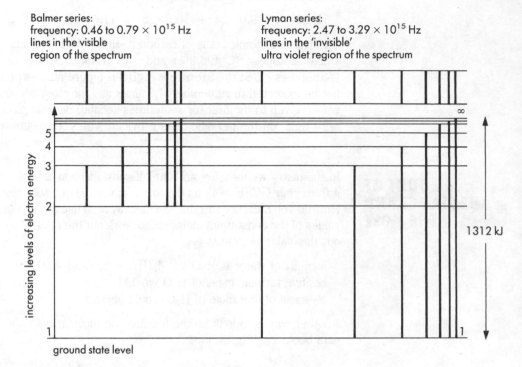

Fig. 3.4 Levels of electron energy in a hydrogen atom

Data between pages 54 and 55): see Fig. 3.4. We may explain the colour by saying that the atoms take in energy from the flame in the form of heat but give out energy in the form of light. It is the electrons in the atoms that are gaining and losing the energy. The lines in these atomic emission line spectra mean that the electrons can absorb and emit only definite values of energy because the electrons in an atom can have only certain levels of energy. Atomic spectroscopy is used to determine these electron energy levels.

ELECTRON ENERGY LEVELS

If we supply enough energy (electrical or thermal) to gaseous atoms they can lose electrons and become ionized.

- The molar **first ionization energy**, E_{m1}, of an element is the energy required to remove one mole of electrons from one mole of its gaseous atoms:
  $X(g) \rightarrow X^+(g) + e^-$

- The molar **second ionization energy**, E_{m2}, of an element is the energy required to remove one mole of electrons from one mole of its gaseous ions:
  $X^+(g) \rightarrow X^{2+}(g) + e^-$

You will find a table of successive ionization energies for the elements in data books. You should relate the pattern of these values to the principal levels of energy of the electrons: see Fig. 3.5.

logarithm of successive
ionisation energies (Emj)

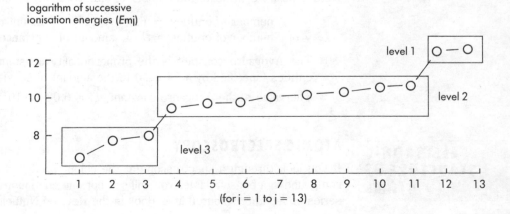

Fig. 3.5 Electron energy levels in an aluminium atom

(for j = 1 to j = 13)

We may suppose that electrons in an atom have kinetic energy (because they move) and potential energy (because they repel one another and are attracted by the nucleus). Notice that instead of saying an electron 'has a certain amount or level of energy', we often say that an electron 'is on a certain energy level' or 'is in a certain energy shell'. Unfortunately this can easily create misleading pictures in your mind, especially when these 'levels of energy' are represented on paper by 'platforms, boxes and concentric circles'. It is better to think of the electrons as being in sets according to their energies.

On the basis of the thirteen successive ionization energies, we arrange the levels of energy of the electrons in aluminium atoms into three sets (Al 2.8.3) and call this arrangement the **electronic configuration** of aluminium. The last number (3) in the sequence (Al 2.8.3) refers to the so-called 'valence electrons in the outer shell'. These are the electrons with the lowest ionization energies and the ones that are involved in bonding. The preceding numbers (2.8.) in the sequence (Al 2.8.3) refer to the so-called 'inner shell electrons'. These are the electrons with the higher ionization energies and the ones that you may assume are not usually involved in bonding. These inner shell electrons partially shield the valence electrons (by repulsion between similar charges) from the attraction of the opposite nuclear charge.

- Inner shell electrons partially shield the outer (valence) shell electrons from the attraction of the nucleus

We can use the successive ionization energies of the other elements to derive their electronic configurations in terms of their principal levels of energy: see Fig. 3.6.

atomic number	name of element	symbol and no. of electrons in the main energy levels
3	Lithium	Li 2. 1
11	Sodium	Na 2. 8. 1
19	Potassium	K 2. 8. 8. 1
37	Rubidium	Rb 2. 8. 18. 8. 1
55	Caesium	Cs 2. 8. 18. 18. 8. 1
87	Francium	Fr 2. 8. 18. 32. 18. 8. 1

Fig. 3.6 Electronic configurations of the alkali metal atoms

When we do this, two important patterns emerge:

- Atoms of elements in the same group of the periodic table have the same number of valence electrons in the outer shell

You should relate this pattern to the similarity of properties of the elements in each group of the periodic table.

- The maximum number of electrons at each of the first five levels of energy would be 2.8.18.32.32

You should relate these numbers to the number of elements in each period across the periodic table:

2(H-He); 8(Li-Ne); 8(Na-Ar); 18(K-Kr); 18(Rb-Xe) and 32(Cs-Rn).

SUBSHELLS AND ORBITALS

The periodic pattern of molar **first ionization energies** against atomic number indicates that we may arrange the principal sets of electron energies into subsets: see Fig. 3.7. The 'large' numbers 1, 2, 3 (up to 7) are called the principal quantum numbers. They still refer to the principal energy sets whilst the letters s, p, d and f refer to their subsets. The superscripts show the number of electrons whose levels of energy make them members of the subset. Now instead of saying an electron 'has a certain precise amount or level of energy', we tend to say that an electron 'is in a certain energy subshell'. We even use phrases such as 'two 1s electrons' and 'five 3d electrons'!

- Electrons in the same subshell do little to shield each other from the attraction of the nucleus

The pattern of **first ionization energies** also indicates that there are sub-subsets (called orbitals) within the subsets (called subshells) of the levels of energy of electrons: see Fig. 3.8. This is confirmed by atomic spectroscopy. The s-subshell has one orbital, the p-subshell has three, the d-subshell has five and the f-subshell has seven orbitals. When we wish to say that a specific electron 'has a very precise amount or level of energy' we say the electron 'is in or occupies a specific orbital'.

■ No more than two electrons may be in the same orbital

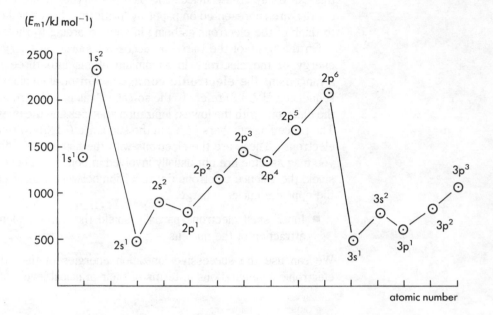

Fig. 3.7 Periodic variation of molar first ionisation energies

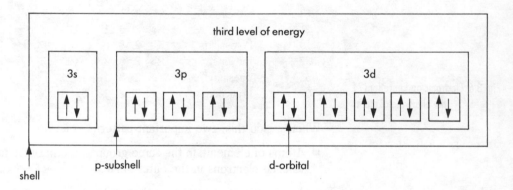

Fig. 3.8 Subdivision of electron energies

BUILDING UP ELECTRONIC CONFIGURATIONS

The ground state electronic configurations of the elements are listed in data books. You should be able to work out the configuration of an element from its atomic number using a plan of the periodic table and some simple rules: see Fig. 3.9.

Problem: Write the electronic configuration of an atom in the ground state of an element whose atomic number is 43.

Solution: From the plan of the periodic table the subshells would build up in the following order:

This is very important. Study these simple rules.

	1s	2s	2p	3s	3p	4s	3d	4p	5s	4d
	↑	↑	↑	↑	↑	↑	↑	↑	↑	
maximum number electrons:	2	2	6	2	6	2	10	6	2	10

	1s	2s	2p	3s	3p	4s	3d	4p	5s	4d
filling the orbitals and										5
keeping a check on the	↓	↓	↓	↓	↓	↓	↓	↓	↓	↑
cumulative total number:	2	4	10	12	18	20	30	36	38	43
shows five electrons										
left to go in the 4d:	$1s^2$	$2s^2$	$2p^6$	$3s^2$	$3p^6$	$4s^2$	$3d^{10}$	$4p^6$	$5s^2$	$4d^5$

$\Rightarrow$ electronic configuration is $1s^2 2s^2 2p^6 3s^2 3p^6 3d^{10} 4s^2 4p^6 4d^5 5s^2$

Notice that when you have worked out the number of electrons in each subshell you must write adjacent to each other the subshells with the same principal quantum number.

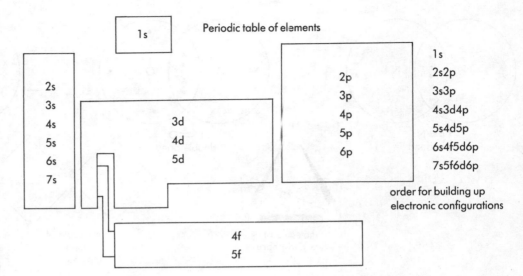

Fig. 3.9 Order for building up
electronic configurations

ELECTRONIC CONFIGURATIONS AND BONDING IN ELEMENTS

A minority of the elements are non-metals. Some of these are gases even at quite low temperatures and high pressures. One or two are very hard solids with very high melting points. The majority of the elements are malleable metals with good electrical and thermal conductivity. You need to explain the properties of these different types of structure in terms of the bonding in the elements. And you should relate the bonding to the composition of the atoms and their electronic configurations.

NOBLE GASES AND VAN DER WAALS FORCES

At very low temperatures and high pressures the noble gases will condense. The very weak short-range forces holding the atoms together in the liquid are called van der Waals forces. The movement of the electrons in relation to the nuclei produce weak instantaneous dipoles that attract one another. The more electrons there are in an atom the more van der Waals forces there will be. This explains the increase in the boiling point of the noble gases with increasing atomic number.

- van der Waals forces are weak instantaneous dipole-dipole attractions operating between atoms in all substances

Chemists have not yet made any compounds of helium, neon or argon. So the electronic configurations of their atoms and those of the other noble gases are considered to be particularly stable arrangements of electrons. Many (but not all) chemically reactive elements combine to form compounds so that the electrons in their atoms rearrange to match the electronic configurations of noble gases.

- Atoms have a tendency to achieve a noble gas electronic configuration

SIMPLE MOLECULAR STRUCTURES AND THE COVALENT BOND

- A covalent bond is the result of two nuclei sharing electrons
- A covalent bond forms because the attraction of the nuclei for the electrons is greater than the repulsions between the nuclei and between the electrons of the two atoms
- The covalent bond between the atoms is strong but the van der Waals forces between the molecules are weak

Hydrogen, the halogens, oxygen and nitrogen all exist as diatomic molecules. If we represent the outer valence set of electrons by Venn diagrams, we can represent the sharing of electrons by the intersection of the sets: see Fig. 3.10. The number of pairs of electrons being shared and therefore the number of covalent bonds formed between two atoms will usually be 8 – the non-metal's group number in the periodic table. For example, nitrogen is in group 5 so in N_2 the two atoms are joined by a triple bond. Sulphur (8−6 = 2 bonds) and white phosphorus (8−5 = 3 bonds) form polyatomic molecules for which dot and cross diagrams would be unsuitable. Remember that we represent electrons by 'dots and crosses' only as a convenience to show their origin and

> " Don't confuse electron sharing with electron transfer. "

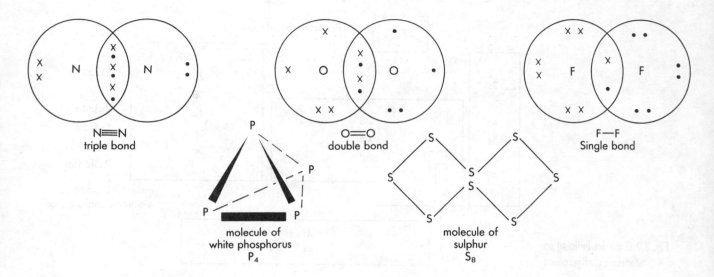

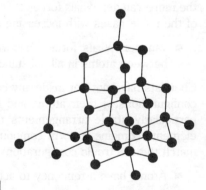

Fig. 3.10 Simple covalent molecular structures

help us keep count of the number of valence electrons in each atom. We cannot actually distinguish between them once the bond is formed.

- Simple molecular structures are usually poor electrical and thermal conductors that melt and boil at low temperatures and consume little heat in the process

GIANT COVALENT STRUCTURES

Diamond (an allotrope of carbon) and silicon form similar, covalently bonded, giant structures (sometimes called giant molecules). Your teacher may show you a model. You are unlikely to be asked to draw the structure but you should appreciate that each atom is covalently bonded to four other atoms located at the corners of a tetrahedron: see Fig. 3.11 (structure of diamond). The result is an extremely strongly bonded interconnected network of six-membered rings of atoms.

Fig. 3.11 Structure of diamond

- Giant molecular structures are usually poor electrical and thermal conductors that melt at very high temperatures and consume a great deal of heat in the process

Graphite (another allotrope of carbon) forms a giant structure that consists of carbon atoms covalently bonded closely together into large flat sheets of flat interconnected hexagons: see Fig. 3.12. These giant molecular sheets give the structure its high melting point. Within each sheet some of the valence electrons (not used in the covalent bonding) are free to move. These mobile electrons give the structure its good electrical and thermal conductivity. Only weak non-directional van der Waals forces may attract the sheets to each other. The ability of the sheets to slide over each other gives the structure its soft, slippery lubricating property.

METALLIC STRUCTURES

Metals are elements or mixtures of elements. You should be familiar with their characteristic properties: lustre, high electrical and thermal conductivity, hardness, toughness, ductility and malleability. Metals usually melt and boil at high temperatures,

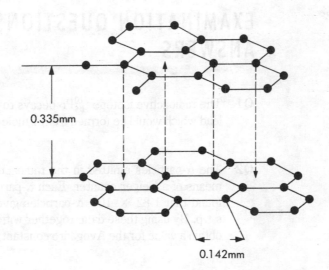

0.335mm

0.142mm

Fig. 3.12 **Structure of graphite**

consuming a great deal of heat in the process. You may account for these general features by the following simplified description of metallic bonding:

- A metal consists of a lattice of mutually repelling positive ions held together by their attraction for a 'sea' of mobile delocalised electrons

The mobile electrons account for the lustre and conductivity. The strong attraction between the positive ions and the electrons accounts for the hardness and toughness. The non-directional nature of the attractive forces accounts for the ductility and malleability.

You may think of the atoms of the s-block elements and the metals in group 3 of the periodic table as forming their stable cations (e.g. Na^+, Mg^{2+}, Al^{3+}) and a sea of their valence electrons. The d-block metals may also contribute electrons from their d-orbitals. You may regard the cations in all the metal structures as spheres that pack together to form one or other of three types of crystal lattice.

- There are only three kinds of metal structure: hexagonal close-packed (hcp), cubic close-packed (ccp) and body-centred: see Fig. 3.13.

Coordination number is an important idea when thinking about crystal structures. In metallic crystals the coordination number of an ion is the number of its nearest equidistant neighbouring ions.

- The coordination number in the close-packed structures is 12
- The coordination number in the body-centred structures is 8
- Most metals have close-packed structures but alkali metals have body-centred structures

The body-centred structure of the alkali metals accounts for their softness and low density when compared to the other metals that are harder, denser and have close-packed structures.

Fig. 3.13 **Metal crystal structures**

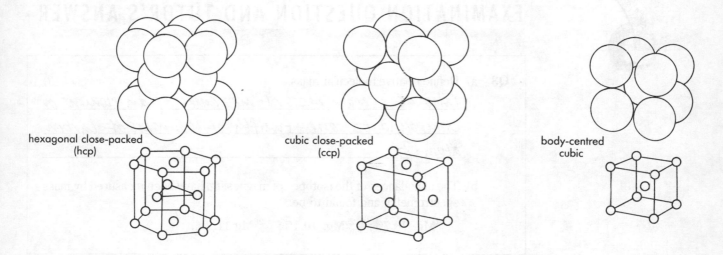

hexagonal close-packed (hcp)

cubic close-packed (ccp)

body-centred cubic

EXAMINATION QUESTIONS AND OUTLINE ANSWERS

Q1 The radioactive isotope $^{210}_{84}$Po decays to $^{206}_{82}$Pb, which is stable. Calculate the mass of lead which would be formed from 1 mole of $^{210}_{84}$Po after **two** half-lives. (2)

(SEB Higher Grade 1988 (part))

Q2 The α-particles emitted in the radioactive decay of radium-226 can be counted by means of a Geiger counter. Each α-particle gains electrons to form helium gas. It is found that 1.82×10^{17} α-particles give 6.75×10^{-3} cm³ of helium, measured at s.t.p. By using these data, together with any other data required in the *Data Booklet*, obtain a value for the Avogadro constant, *L*. (5)

(UCLES 1988 (part))

Outline answers

Q1 1 mol → ½ mol → ¼ mol → ¾ mol ^{206}Pb formed
⇒ ¾ × 206 = 154.5 g lead

Q2 $^{226}_{88}Ra \longrightarrow ^{222}_{86}Rn + ^{4}_{2}He$

Data book gives molar gas volume as 22400 cm³ mol⁻¹ at s.t.p

⇒ $\dfrac{22400}{6.75 \times 10^{-3}} \times 1.82 \times 10^{17}$ gives $L = 6.04 \times 10^{23}$ mol⁻¹

EXAMINATION QUESTION AND TUTOR'S ANSWER

Q3 a) Define relative molecular mass. (3)

ratio of average mass of molecules of an element or compound to one-twelfth of the mass of carbon atom $^{12}_{6}C$

b) The abundances of the isotopes of magnesium have been measured by mass spectrometry and found to be:

^{24}Mg, 78.7%; ^{25}Mg, 10.1%; ^{26}Mg 11.2%.

Assuming the relative atomic masses of the isotopes are exactly 24, 25 and 26, calculate the relative atomic mass of magnesium.   (3)

$$\frac{(78 \cdot 7 \times 24) + (10 \cdot 1 \times 25) + (11 \cdot 2 \times 26)}{100}$$

$$= 24 \cdot 3$$

c) i) Give **two** characteristic properties of neutrons.   (2)

Same mass as proton

not changed.

ii) Write a balanced nuclear equation for the production of a neutron from an atom of the beryllium isotope ^9_4Be.   (1)

$$^2_1H + {}^9_4Be \rightarrow {}^{10}_5B + {}^1_0n$$

(ODLE 1988)

STUDENT'S ANSWER WITH EXAMINER'S COMMENTS

Q4 a) Define the *mole*.

> Learn your definitions!
> See page 21.

The formula in grams

b) Identify J, X and Z in the following nuclear reactions by giving their chemical symbols, mass numbers and atomic numbers.

i) $^2_1H + {}^3_1H = J + {}^1_0n$; $\quad$ ^4_2He ✓

ii) $^{14}_7N + {}^4_2He = {}^{17}_8O + X$; $\quad$ ^1_1P ✓

iii) $^1_0n + {}^7_3Li = {}^4_2He + {}^1_0n + Z$. $\quad$ ^3_1H ✓

c) The soil in Welsh hills is contaminated with the caesium-137 isotope as a result of the Chernobyl nuclear accident. $^{137}_{55}Cs$ decays with a half-life of 30 years into the barium isotope, $^{137}_{56}Ba$.

i) State which particle is emitted during this decay.

β - particle ✓

ii) State briefly why there is concern over the caesium contamination.

Caesium is radioactive and the sheep could get radiation poisoning. ✗

> c)i) is correct but think about the chemistry in ii). Barium is very poisonous!

d) Fig. 1 is an incomplete sketch showing α, β and γ rays passing into an evacuated vessel containing metal foils or sheets, and a zone of magnetic field, acting perpendicularly to the plane of the paper.

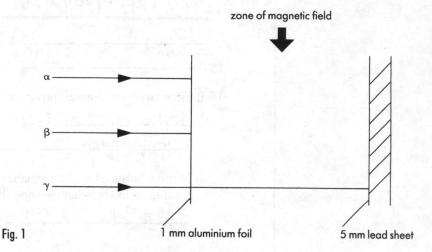

Fig. 1

Complete the sketch to show what will happen to the three rays.

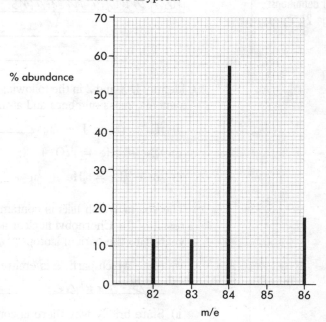

e) Fig. 2 shows a mass spectrum of krypton gas. Use this figure to calculate the relative atomic mass of krypton.

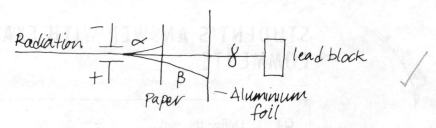

% abundance

Fig. 2

m/e

$$\frac{(12 \times 82) + (12 \times 83) + (58 \times 84) + (18 \times 86)}{100} = 84.0 \checkmark$$

f) Define the term *molar first ionisation energy*.

Energy needed to remove one mole of electrons from an element : $x \rightarrow x^+ + e^-$

a mole of gaseous atoms of

add state symbols (g)

g) Fig. 3 shows a plot of first ionisation energy against atomic number for the lighter elements.

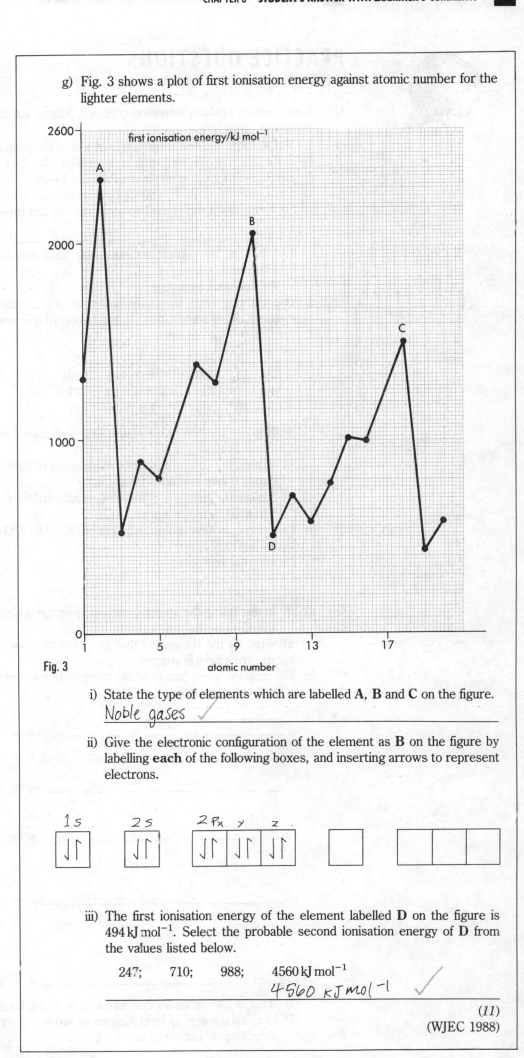

Fig. 3

i) State the type of elements which are labelled **A**, **B** and **C** on the figure.

Noble gases ✓

ii) Give the electronic configuration of the element as **B** on the figure by labelling **each** of the following boxes, and inserting arrows to represent electrons.

iii) The first ionisation energy of the element labelled **D** on the figure is 494 kJ mol^{-1}. Select the probable second ionisation energy of **D** from the values listed below.

247;   710;   988;   4560 kJ mol^{-1}

4560 KJ mol^{-1} ✓

(11)

(WJEC 1988)

PRACTICE QUESTIONS

Q5 This question is about ionisation energies, atomic structure and their importance in chemistry.

a) Define the term, *first ionisation energy* of an element: (1)

b) State any two factors which determine the first ionisation energies of the elements in a given group of the Periodic Table: (2)

c) Three elements, A, B, and C, have atomic numbers 8, 10 and 11 respectively.

i) Write down the electronic structures of the three elements:

A _____ B _____ C _____ (2)

ii) What is the order of increasing first ionisation energy of the three elements? (1)

Explain your reasoning. (2)

d) The following Table shows the first ionisation energies, I, in kJ mol^{-1} and atomic numbers, Z, of some Group II and Group III elements:

	Group II			Group III	
	Z	I		Z	I
Be	4	900	B	5	799
Mg	12	736	Al	13	577
Ca	20	590	Ga	31	577

i) Explain why the value of the first ionisation energy of Be is greater than that of B. (1)

ii) Explain the change in relative magnitudes of the first ionisation energies of the Group II elements from Be to Ca. (1)

iii) Explain the change in relative magnitudes of the first ionisation energies of the Group III elements from B to Ga. (2)

e) Two elements have atomic numbers of 19 and 9. What type of compound will they form when they react? (1)

Explain your reasoning. (2)

(OCSEB 1988)

Q6 a) i) In discussing the atomic emission spectrum of hydrogen, what is meant by a *series*? (1)

ii) What do the transitions that give rise to a specified series in the emission spectrum have in common? (1)

b) The diagram below (not to scale) represents some of the electronic energy levels in the hydrogen atom.

$$\begin{aligned}
&\underline{\hspace{6cm}}\ n = \infty \\
&\underline{\hspace{6cm}}\ n = 6 \\
&\underline{\hspace{6cm}}\ n = 5 \\[0.5em]
&\underline{\hspace{6cm}}\ n = 4 \\[1em]
&\underline{\hspace{6cm}}\ n = 3 \\[1.5em]
&\underline{\hspace{6cm}}\ n = 2 \\[2em]
&\underline{\hspace{6cm}}\ n = 1
\end{aligned}$$

i) What is the significance of an electron in the level marked $n = \infty$?

ii) Mark on the energy level diagram an arrow to represent the ionisation energy of hydrogen. Label this arrow 'A'.

iii) Mark on the energy level diagram an arrow to represent the lowest energy transition in the Balmer emission spectrum (the series found in the visible region). Label this arrow 'B'. (3)

c) Of the elements in the Periodic Table from hydrogen to neon, state which has the highest first ionisation energy and which has the lowest. Give reasons for your answers. (6)

(JMB 1988)

4

STRUCTURE AND BONDING IN COMPOUNDS

PREDICTING THE SHAPE AND POLARITY OF A MOLECULE: THE VSEPR THEORY

GIANT COVALENT STRUCTURES

IONIC COMPOUNDS

POLARISATION OF ANIONS

IONIC CRYSTAL STRUCTURES

GETTING STARTED

The bonding in binary compounds may be covalent, ionic or intermediate in character. The shapes of simple covalent molecules may be predicted by the valence shell electron pair repulsion (VSEPR) theory. Crystals of caesium chloride and sodium chloride are important examples of ionic structures. Intermediate bonding may be seen as polarisation of covalent bonds resulting from electronegativity differences and as distortion of anions resulting from their polarisability and the polarising power of cations. Ion-dipole attractions, hydrogen bonding, dipole-dipole interactions and van der Waals forces also influence the structure and properties of substances.

ESSENTIAL PRINCIPLES

COVALENT COMPOUNDS

Chemists have not yet made any compounds of helium, neon or argon. So the electronic configurations of their atoms and those of the other noble gases are considered to be particularly stable arrangements of electrons. In many compounds of the s- and p-block elements the arrangement of electrons in the atoms match the electronic configuration of the noble gases. This is the basis of the octet rule:

- Atoms have a tendency to achieve a noble gas electronic configuration

You should be able to use this rule to predict the formula of many simple compounds of the s- and p-block elements but not the d-block elements.

SIMPLE MOLECULES

A covalent bond exists between two atoms when they share electrons because the two nuclei attract the electrons and this attraction outweighs the repulsion between the nuclei and between the electrons. The balance of these forces will depend upon the atoms involved and will govern the length and strength of the bond. Bond strength is the energy needed to separate completely the atoms in the molecules of one mole of compound. You will find more about bond strengths in the chapter on energetics.

- Bond lengths can be measured by microwave spectroscopy
- Bond length is the average distance between the nuclei of two atoms

You will not be expected to know values of bond lengths or strengths but you should remember the following pattern:

- For any two given atoms the bond gets shorter and stronger as it changes from a single to a double to a triple covalent bond
- The number of covalent bonds formed by an atom is often 8 – the number of valence electrons in the outer shell of the non-metal

Although in Venn diagrams of simple molecules you may represent the outer valence electrons by 'dots and crosses' in order to show their origin and count them, you must remember that once the bond is formed electrons are not distinguishable by their origin.

- A dative (or coordinate) bond is a covalent bond in which one of the atoms has supplied both electrons being shared.

A dot and a cross (or two dots or two crosses) in the intersection, or, a dash '–' between the symbols of the elements represents a shared (bonded) pair of electrons. This is a single covalent bond. Dots and crosses *not* in the intersection represent unshared (non-bonded) electrons belonging to one atom or the other: see Fig. 4.1.

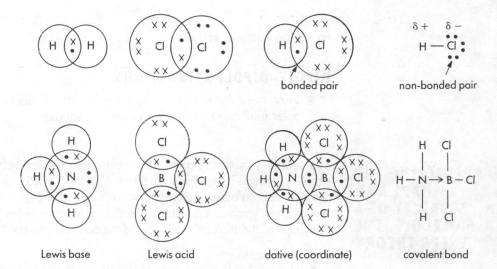

Lewis base   Lewis acid   dative (coordinate)   covalent bond

Fig. 4.1 Dot and cross diagrams for simple molecules

POLAR BONDS

In a diatomic molecule of hydrogen or chlorine the two identical nuclei must attract the pair of electrons equally strongly. In a molecule of HCl, the chlorine nucleus (albeit shielded by two inner shells of electrons) attracts the electrons more strongly than does the hydrogen nucleus.

■ The Pauling electronegativity index (N_p) is a measure of how strongly an atom in a compound attracts electrons in a bond

Fluorine is the most electronegative of all the elements. So in the H-F molecule the fluorine takes more than its fair share of the bonded electron-pair. Consequently both atoms become slightly (δ) charged and this makes the hydrogen fluoride molecule ($^{\delta+}H{-}F^{\delta-}$) polar. One of these slight charges multiplied by the distance between them is called the *dipole moment* (p). If $\delta+$ (or $\delta-$) is measured in coulombs and the distance is in metres, the units of p would be coulomb metres (C m). We usually use a unit called the debye (D) where $1\,D \simeq 3.34 \times 10^{-30}\,C\,m$.

■ A dipole moment gives a measure of the polarity of a molecule

Oxygen ($N_p = 3.5$) is more electronegative than carbon ($N_p = 2.5$) but the overall dipole moment of carbon dioxide is 0. Why is this so? The reason is that both the carbon–oxygen bonds are equally polar but act in exactly opposite directions: see Fig. 4.2.

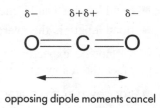

opposing dipole moments cancel

Fig. 4.2 Linear non-polar molecule

Oxygen is also more electronegative than hydrogen ($N_p = 2.1$). However, the dipole moment of water is not zero but is 1.84 D (a high value). What does this suggest? It suggests that the water molecule cannot be linear, but must be bent so that both the equally polar hydrogen–oxygen bonds are pointing in similar directions: see Fig. 4.3.

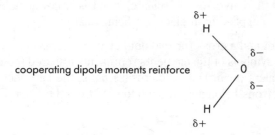

cooperating dipole moments reinforce

Fig. 4.3 Non-linear polar molecule

DIPOLE–DIPOLE ATTRACTIONS

■ Permanent dipole–dipole forces are the attractions between the negative end of one polar molecule and the positive end of another

PREDICTING THE SHAPE AND POLARITY OF A MOLECULE: THE VSEPR THEORY

You should be able to predict the approximate angles between bonds and the shape of simple molecules by applying rules based on the VSEPR theory that valence shell electron pairs repel each other. Follow the steps outlined in Fig. 4.4. Remember that the VSEPR theory gives you only a qualitative prediction of bond angles but it does allow you to predict the shape of the molecule well enough to decide if any polar bonds will make the molecule polar.

■ The greater the difference in the electronegativities of two atoms, the more polar is the covalent bond between the two atoms

1 Write the formula to display all bonded electron pairs by dashes (—) and all non-bonded pairs by two dots (:)

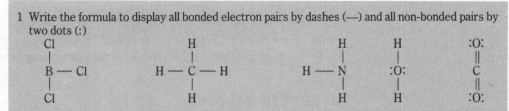

2 Assume the electron pairs (bonded and non-bonded) move equally as far apart as possible from each other but treat a double bonded pair (=) as a single bonded pair

trigonal 120°	tetrahedral 109.5°	pyramidal	non-linear	linear 180°

3 Adjust any bond angles affected by the rule that repulsion between a non-bonded and a bonded pair is greater than that between two bonded pairs and that two non-bonded pairs repel each other even more strongly

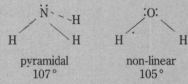

pyramidal 107°	non-linear 105°

H—N—H and H—O—H angles are 'squeezed together' by stronger repulsions from non-bonded electron pairs to less than 109.5° (tetrahedral angle)

Fig. 4.4 Predicting the shape of a simple molecule

You will not be expected to remember electronegativity index values but you should remember the trends that exist in the periodic table: see Fig. 4.5.

	increasing electronegativity			
H (2.1)				
B (2.0)	C (2.5)	N (3.0)	O (3.5)	F (4.0)
	Si (1.8)	P (2.1)	S (2.5)	Cl (3.0)
		As (2.0)	Se (2.4)	Br (2.8)
			Te (2.1)	I (2.5)

Fig. 4.5 Trends in the Pauling electronegativity index N_p

Here is an approximate test to distinguish between polar and non-polar liquids in the laboratory: comb your hair and hold the electrostatically charged plastic comb close to the stream of test liquid running from a burette: see Fig. 4.6. In the case of a polar liquid, the charged comb will attract polar molecules and you should see the stream of liquid make a

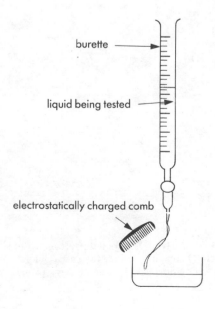

burette

liquid being tested

electrostatically charged comb

Fig. 4.6 Distinguishing between polar and non-polar liquids

significant change of course. Non-polar liquids will only be marginally affected. Practise with water first to check that you are charging your comb enough. Make sure you can explain the results in terms of the polarised bonds and the shape of the molecule: e.g. why is *cis*-1,2-dichloroethene polar but the *trans*-isomer non-polar?

HYDROGEN BONDING

> **Hydrogen bonding is an important topic at A-level.**

The trend of increasing boiling point with increasing molar mass of the noble gases is matched by the trend for the hydrides of groups 5, 6 and 7 in the periodic table excepting the first member of each group: see Figs. 12.19, 12.28, 12.31. You should be able to explain the exceptionally high boiling points of ammonia, water and hydrogen fluoride (compared to the values for the other hydrides in each group) by the increased intermolecular forces of attraction caused by hydrogen bonding.

- A hydrogen bond ($\cdots$) is a weak bond between a very electronegative atom (X = N, O or F) and a hydrogen atom bonded to a very electronegative atom (Y = N, O or F), thus $-X \cdots H-Y$.

You could explain the bond by saying the highly electronegative atom Y polarizes the H–Y bond so much that the nucleus of the hydrogen atom is exposed to attraction by a lone pair on another electronegative atom X.

- Hydrogen bonding is stronger than van der Waals forces and permanent dipole–dipole attractions but weaker than covalent bonding
- Hydrogen bonding is a major factor in determining the structure and properties of water, hydrated salts, carbohydrates and proteins

DELOCALISED BONDING

When you use the octet and VSEPR theories to predict the shapes of certain molecules you may find that you can write more than one structure: see Fig. 4.7. When you can find

Write down the arrangement of the symbols and the number of valence electrons of each atom then alter the numbers as you draw in the bonds between the symbols to give H-atoms 2 and other atoms 8 electrons to complete their outer shells:

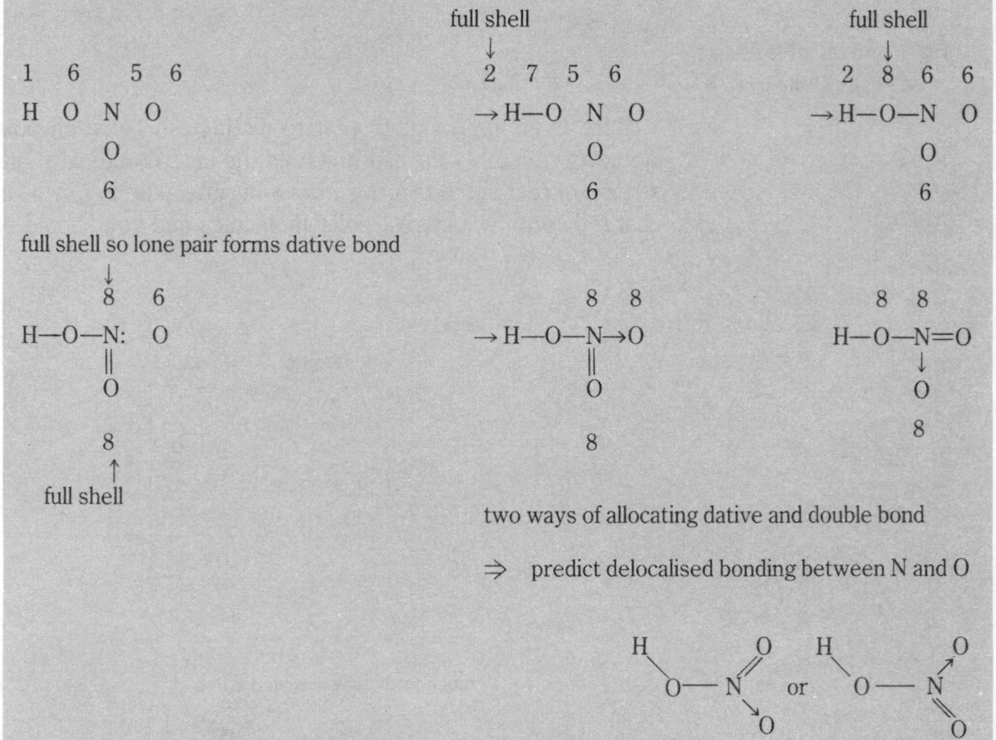

$$
\begin{array}{llll}
 & & & \text{full shell} \\
 & & & \downarrow \\
1 \quad 6 \quad\; 5 \quad 6 & \quad & 2 \quad 7 \quad 5 \quad 6 & \quad 2 \quad 8 \quad 6 \quad 6 \\
\text{H} \;\; \text{O} \;\; \text{N} \;\; \text{O} & & \rightarrow \text{H}\!-\!\text{O} \;\; \text{N} \;\; \text{O} & \rightarrow \text{H}\!-\!\text{O}\!-\!\text{N} \;\; \text{O} \\
\qquad\;\; \text{O} & & \qquad\quad \text{O} & \qquad\qquad \text{O} \\
\qquad\;\; 6 & & \qquad\quad 6 & \qquad\qquad 6
\end{array}
$$

full shell so lone pair forms dative bond

↓
8   6

H—O—N: O
‖
O

8

↑
full shell

8   8

→H—O—N→O
‖
O

8

8   8

H—O—N=O
↓
O

8

two ways of allocating dative and double bond

⇒   predict delocalised bonding between N and O

You predict the bond between the nitrogen atom and oxygen atom to be identical and to be longer than a N=O bond but shorter than a N–O bond. And since the nitrogen has no lone pair, the N-atom and three O-atoms will all be in the same plane with an ONO bond angle of 120°.

Fig. 4.7 Predicting delocalised bonding in nitric acid

more than one way of allocating some electrons to the bonding of certain atoms represented by symbols in fixed positions on your paper, you may assume delocalised bonding exists. For instance, when the nitric acid molecule ionises to NO_3^-, all three NO bonds become equal and the negative charge is delocalised across all three oxygen atoms.

■ The benzene molecule is an important example of a delocalised molecule

You will find more examples of delocalisation elsewhere in this book. You could consider a metal to be one giant molecule with all the valence electrons delocalised. But, remember that in simple molecular structures the delocalised electrons are confined within the simple molecule.

■ Compounds with simple molecular structures are usually poor electrical and thermal conductors. They melt and boil at low temperatures and consume little heat in the processes of melting and evaporating because there are no mobile electrons and the forces of attraction between the molecules are weak.

GIANT COVALENT STRUCTURES

Silicon has an allotrope with the same structure as diamond – silicon carbide. Silicon carbide (SiC) is formed when carbon and silicon are heated together above 1750 °C. This compound is used in cutting and grinding wheels. It sublimes at about 2700 °C. On the Mohs scale of hardness which ranges from 1 (talc) to 10 (diamond), the hardness of silicon is 7 and the hardness of silicon carbide is 9. You should relate these properties and the use of silicon carbide to its giant covalently bonded structure which is that of diamond where alternate carbon atoms have been replaced by silicon atoms.

Silicon oxide (silica), in sharp contrast to carbon dioxide, has a giant covalent structure of alternating Si and O atoms strongly bonded into a three-dimensional network of six-membered rings: see Fig. 4.8.

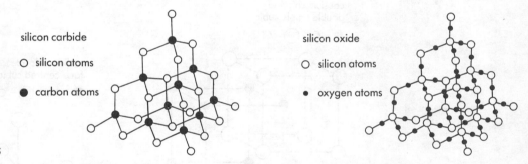

silicon carbide
O silicon atoms
● carbon atoms

silicon oxide
O silicon atoms
● oxygen atoms

Fig. 4.8 Giant covalent structures

■ Compounds with giant molecular structures are usually poor electrical and thermal conductors that melt at very high temperatures. They consume a great deal of heat in the process because there are no mobile electrons and the forces of attraction between the atoms are strong

IONIC COMPOUNDS

■ Ionic bonding is the result of electrons being transferred from one atom to another and the ions packing together into a crystal lattice

You can use the octet rule and the electronic configurations to predict the formula of simple binary ionic compounds formed by the s-block metals and some non-metallic elements: see Fig. 4.9.

Fig. 4.9 Predicting the formula of rubidium oxide

atomic number of rubidium is 37

⇒ electronic configuration Rb 2.8.18.8.1

⇒ predict loss of 1 valence electron to leave Rb⁺ 2.8.18.8

atomic number of oxygen is 8

⇒ electronic configuration is O 2.6

⇒ predict gain of 2 electrons to valence shell to give O^{2-} 2.8

⇒ predict $(Rb^+)_2O^{2-}$ as the formula

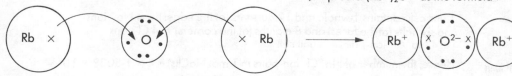

POLARISATION OF ANIONS

When cations and anions come together to form a crystal lattice the net positive charge on the cations can attract the electrons in the outer shell of the anions.

■ Anion polarisation is the distortion of the shape of a polarisable anion caused by the attraction of a polarising cation

If the charge density (amount of positive charge in a given volume) of the cation increases, the distortion or polarisation of the anion increases and the compound becomes less ionic (more covalent) in character.

> *Intermediate bonding is very important.*

■ Polarising power of a cation increases as its charge increases and as the ionic radius decreases

■ Polarisability of an anion increases as its charge increases and as its radius increases

■ Covalent character of an ionic compound is high if the cation is small, the anion is large and the charges on the ions are high

You should compare the previous point with Fajans' rules for predicting the nature of the bonding between two elements. They state that covalent bonding is more probable and ionic bonding less probable if

1   the ions possess multiple charges
2   the atoms produce small cations or large anions.

IONIC CRYSTAL STRUCTURES

You should be able to draw diagrams and describe the unit cells of the caesium chloride and sodium chloride crystal lattices. You should also be able to work out the number of ions in each unit cell and to deduce the empirical formula of the compound from the ratio of the coordination numbers of each ion: see Fig. 4.10. Take care not to confuse the unit cell of

caesium chloride
double simple cubic

● = Cs^+ ○ = Cl^-

Coordination number Cs^+ = 8
Coordination number Cl^- = 8
⇒ ratio of Cs^+ : Cl^- is 1 : 1
Hence formula of compound is CsCl

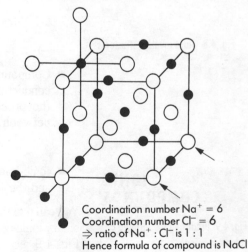

sodium chloride
face-centred cubic

● = Na^+ ○ = Cl^-

Coordination number Na^+ = 6
Coordination number Cl^- = 6
⇒ ratio of Na^+ : Cl^- is 1 : 1
Hence formula of compound is NaCl

To calculate the Avogadro constant from the dimensions and composition of a unit cell (found from X-ray analysis) of sodium chloride (density 2.165 g cm^{-3} and molar mass 58.44 g mol^{-1}):

volume of 1 mol NaCl(s) is 58.44 ÷ 2.165 = 26.993 cm^3 or 2.6993 × 10^{-5} m^3

volume of 1 unit cell is (0.5641)^3 = 0.17950 nm^3 or 1.7950 × 10^{-28} m^3

⇒ 1 mol NaCl(s) contains 2.6993 × 10^{-5} ÷ 1.7950 × 10^{-28} = 1.5038 × 10^{23} unit cells

1 unit cell contains 1 whole and 12 quarters (at the edges) = 4 Na^+ ions
and six halfs (at the faces) and 8 eighths (at the corners) = 4 Cl^- ions:
i.e. 4 Na^+Cl^- ion-pairs in a unit cell

Hence, the number of Na^+Cl^- ion-pairs in 1 mol NaCl(s) is 4 × 1.5038 × 10^{23}

Hence the Avogadro constant = 6.015 × 10^{23} mol^{-1}.

Fig. 4.10 Crystal structures of ionic compounds

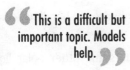

the caesium chloride with that of an alkali metal. Caesium chloride is NOT a body-centred cube because the ion at the centre of the cube is different from the ions at the corners. Notice also that in the drawings of the unit cells, the lines between the ions do NOT represent bonds; they merely give the drawing perspective.

- The dimensions of the unit cell, and the arrangement of ions in it, may be determined by X-ray (diffraction) crystallography and the data used to determine a value for the Avogadro constant

COORDINATION NUMBER AND RADIUS RATIO

There are only seven different crystal systems: cubic, hexagonal, monoclinic, ortho-rhombic, rhombohedral, tetragonal and triclinic; based upon the seven regular flat-faced shapes that can tesselate, i.e. be fitted together to fill a space completely without any gaps. Caesium chloride and sodium chloride belong to the cubic crystal system. The type of crystal structure adopted by an ionic compound is governed by the relative numbers, shapes and sizes of the cations and anions.

- The radius ratio is the radius of the smaller ion divided by the radius of the larger ion

In a metal all the ions are identical; therefore the radius ratio is 1 and the number of nearest equidistant neighbouring ions (the coordination number) is either 12 (in the close-packed structures) or 8 (in the body-centred structure).

- The coordination number of an ion in a crystal of an ionic compound is the number of nearest, equidistant, oppositely charged ions

In an ionic crystalline compound the coordination number of the smaller ion is usually, but not always, related to the radius ratio as follows:

$\dfrac{\text{radius (small ion)}}{\text{radius (large ion)}}$	< 0.41	> 0.41 and < 0.73	> 0.73
coordination number	4	6	8

- Anhydrous compounds with ionic crystal structures are poor electrical and thermal conductors because the ions are not free to move
- Anhydrous ionic compounds usually melt at high temperatures, consuming a great deal of heat in the process, because the electrostatic forces of attraction between the oppositely charged ions are very strong
- Crystals of ionic compounds are brittle and cleaved when struck because planes of ions move relative to each other resulting in repulsion between the cations and between the anions facing each other
- Molten ionic compounds conduct electrolytically because the ions are no longer fixed in a lattice but are free to move

ION-DIPOLE FORCES AND HYDRATED CRYSTALS

An electrostatically charged comb has the same effect on a stream of water irrespective of the sign of its charge. If the comb is positively charged then it will attract the negative (oxygen) end of the polar water molecules. If the comb is negatively charged it attracts the positive (hydrogen) end of the water molecules. The charges on cations and anions also attract water molecules.

- Many ionic compounds form hydrated crystals in which the water molecules are held in the lattice by ion-dipole forces
- Hydrated ionic compounds melt at much lower temperatures than the corresponding anhydrous ionic compounds. They require less heat for the process which often involves decomposition by loss of water
- Hydrated ionic crystals are usually much softer and easier to cleave than their anhydrous counterparts

The ions are further apart when surrounded by water molecules so the electrostatic forces of attraction between the oppositely charged ions will be less. In some cases the crystal structure may consist of hydrated layers held together by weak hydrogen bonding between the water molecules of hydration.

- Many ionic compounds dissolve well in water as well as some other polar liquids because the energy released by the formation of the ion-dipole forces compensates for the lattice energy consumed in overcoming the electrostatic forces of attraction between the cations and anions

You may picture the way an ionic substance dissolves in water like this:

1   cations and anions attract polar water molecules
2   water molecules get between ions (thereby lessening mutual attraction of the ions)
3   water molecules completely surround cations and anions
4   hydrated ions leave the crystal and mix with the water

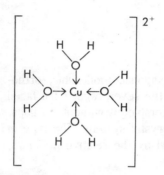

Fig. 4.11 Tetraaquacopper (II) cation

COMPLEX CATIONS

Many cations form hydrated ions in which the water molecules are attached to the central metal ion by dative bonds: see Fig. 4.11.

POLYATOMIC NON-METAL IONS

Many molecules may form anions (or cations) by acting as acids (or bases) and losing (or gaining) protons: see Fig. 4.12.

Fig. 4.12 Polyatomic anions and cations

$$H-O-H \quad + \quad :N-H \quad \longrightarrow \quad H-O^- \quad + \quad H-N-H$$

bent   pyramidal   linear   tetrahedral
molecule   molecule   anion   cation

You should be able to draw 'dot and cross' diagrams and predict the shape of some simple anions and cations: see Fig. 4.13.

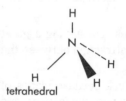

ammonium ion: NH_4^+

tetrahedral

mutual repulsion of four bonded electron pairs

Fig. 4.13 Predicting the shape of some anions and cations

carbonate ion: CO_3^{2-}

trigonal planar

no lone pair on the carbon atom and all three bonds mutually repel

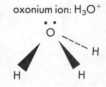

oxonium ion: H_3O^+

trigonal pyramidal

one lone pair on the oxygen atom and repulsion between four electron pairs

ANISOTROPIC CRYSTALS

The *refractive index* (consult a physics book for more details) is an important physical property that measures the effect of substances upon the speed of light passing through them. You will find the refractive index of a substance listed in data books alongside other important properties such as its molar mass, density, melting point, boiling point and solubility.

- An anisotropic substance is a crystalline solid whose refractive index varies with the direction of the light passing through its crystal

Most substances are anisotropic because their crystal structures are different when viewed from different directions. In ionic compounds these differences often arise because the crystal structure contains non-spherical ions: e.g. the trigonal planar carbonate ion in calcite, $CaCO_3$. Other properties of anisotropic crystals (e.g. conductivity in graphite) may also depend upon the direction in which they are measured. Anisotropic solids do not belong to the cubic crystal system.

- Isotropic ionic crystalline compounds belong to the cubic system and their properties do not depend upon the direction of their measurement
- Caesium chloride and sodium chloride form isotropic cubic crystals

EXAMINATION QUESTION AND OUTLINE ANSWER

Q1 a) Aluminium crystallises in a face-centred cubic lattice and polonium in a simple cubic lattice. Draw and label clear diagrams which show how the atoms are arranged in these two structures. *(3)*

b) i) The lattice energy of sodium chloride is $-772\,kJ\,mol^{-1}$. Explain the meaning of this statement. *(2)*

ii) Explain why, although the lattice energy of sodium chloride is numerically large, the enthalpy of solution in water is small numerically. *(3)*

iii) Use the following lattice energies to discuss and illustrate the factors controlling the magnitudes of lattice energies

LiF	$-1030\,kJ\,mol^{-1}$
LiI	$-744\,kJ\,mol^{-1}$
CaF_2	$-2600\,kJ\,mol^{-1}$

(4)

c) Illustrate the correlation of physical properties with crystal structure and/or bond type by discussing the cases of:
  i) silica; *(3)*
  ii) a molecular crystal; *(2)*
  iii) a metal. *(3)*

(ODLE 1988)

Outline Answer

Q1

(a) see text page 27

(b) (i) Energy given out when one mole of crystalline sodium chloride is formed from its **gaseous ions**

$$Na^+(g) + Cl^-(g) \rightarrow NaCl(s); \quad \Delta H = -772\,kJ\,mol^{-1}$$

(ii) The hydration energy is also large: i.e.

$$(H_2O(l)+)Na^+(g) + Cl^-(g) \rightarrow Na^+(aq) + Cl^-(aq);$$
$$\Delta H = -large\ value$$

(iii) F^- is a smaller ion than I^- so F^- ion change density is greater ⇒ stronger attracting force.

Ca^{2+} has twice the charge of Li^+ so Ca^{2+} charge density is greater ⇒ stronger attracting force.

(c) (i) see text page 39
  (ii) iodine or carbon dioxide are good examples — van der Waals forces between molecules are weak
  (iii) Positive ions strongly held by a 'sea' of mobile electrons.

STUDENT'S ANSWER WITH EXAMINER'S COMMENTS

Q2 a) Draw clear diagrams to show the shape of a molecule of each of the following compounds:

i) CH_4 ii) NH_3 iii) H_2O iv) HF   *(4)*

Explain the shape of the NH_3 molecule as shown in a) ii).   *(3)*

b) The graph shows the variation in boiling point of the hydrides in groups IV to VII of the Periodic Table.

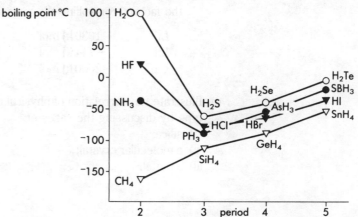

i) Suggest a reason for the regular increase in boiling point from CH_4 to SnH_4.   *(1)*

ii) What is responsible for the anomalously high boiling points of NH_3, HF and H_2O?   *(1)*

iii) Carefully explain, with the aid of a diagram, the nature of the forces *between* molecules of water.   *(3)*

c) Copper is a good conductor of electricity in the solid state while sodium chloride will not conduct in the solid state, but does so when molten.

Briefly compare the bonding in copper and sodium chloride and hence explain this fact.   *(4)*

Total 16 marks
(WESSEX (specimen))

Student Answer

Q2

(a)(i)   [diagram of CH₄] ✓

(ii)   [diagram of NH₃] ✗

> " The N-atom has a lone pair of electrons. "

(iii)   [diagram of H₂O] ✓

(iv)   H — F ✓

The electrons repel each other, so the bonds spread out as much as possible. ✓

> " For full marks this repulsion between the lone pair and the bonded pairs must be mentioned. "

(b) (i) Van der Waals forces increase because the relative molecular mass increases ✓

   (ii) Molecules have hydrogen bonds between them ✓

   (iii)
   H—
       O:————H—O—H
   H⁄   hydrogen bond

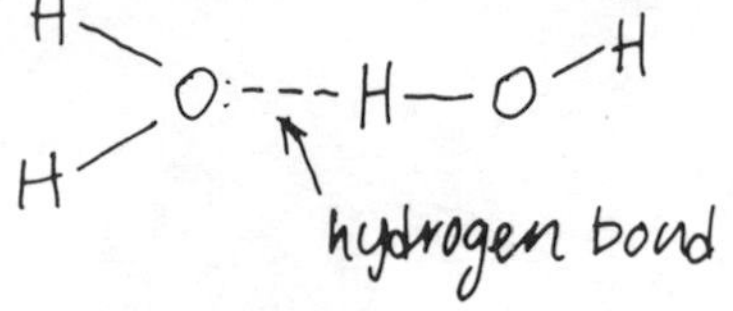

(c) Copper has metallic bonding with mobile electrons, sodium chloride has ionic bonding and no mobile electrons.

Too brief.

PRACTICE QUESTION

Q3 For the substances sodium, iodine, diamond and caesium chloride:

a) Tabulate the structure of **each** solid, and the type of bonding in the solid, using the headings

Name of substance	Description of the structure of the solid	Type of bonding in the solid.

(8)

b) Discuss the relationship between structure and bonding in **each** solid. (4)

c) State the physical properties – hardness, volatility (i.e. melting point and boiling point), electrical conductance – of **each** solid. (4)

d) Discuss how the physical properties depend on the crystal structure and bonding of **each** solid (4)

(WJEC 1988)

GASES, LIQUIDS, MIXTURES AND SOLUTIONS

GETTING STARTED

The physical properties of gases are described by various laws and explained by the kinetic molecular theory. The ideal gas equation and Graham's law of diffusion are related to methods of determining relative molecular mass. Steam distillation may be understood in terms of Dalton's law of partial pressures and fractional distillation in terms of Raoult's law. Intermolecular forces can account for deviations from ideal gas behaviour and from Raoult's law.

GASES

THE GAS LAWS

IDEAL GAS EQUATION

KINETIC THEORY OF GASES

DETERMINING MOLAR MASSES

MASS SPECTROMETRY

LIQUIDS

TROUTON'S RULE

DISTILLATION

RAOULT'S LAW

IDEAL SOLUTIONS

NON-IDEAL SOLUTIONS

ESSENTIAL PRINCIPLES

GASES

In principle the three states of matter are interconvertible. In practice it may be almost impossible to turn some solids into a liquid let alone into the gaseous state. Some gases only liquefy at extremely low temperatures and high pressures. Gases fill their containers and are easily compressed, unlike liquids and solids which have their own surfaces and are difficult to compress.

- Gases are substances that are entirely in the gaseous state under standard conditions of temperature and pressure

- Vapours are substances that are normally liquid or solid under standard conditions of temperature and pressure

Substances such as helium, neon, argon, hydrogen, oxygen, carbon dioxide and ammonia should be called gases but water, ethanol and trichloromethane in the gaseous state should be called vapours. The angles, lengths and strengths of bonds of molecules in the gaseous state can be determined by electron and neutron diffraction techniques.

THE GAS LAWS

“You will meet this topic if you study physics.”

In physical chemistry, your experiments with gases and vapours will often require you to measure the temperature, pressure and volume of a gas.. Sometimes you may have to measure its mass as well. Remember to convert your temperature reading from centrigrade (°C) to Kelvin (K) by adding 273. You can conveniently measure volume in cubic centimetres (cm^3) or millilitres (ml) ($1\,cm^3 = 1\,ml$). Pressure is the force (exerted by the gas) per unit area (of its container). In SI (système international) units, force is measured in newtons (N) and area in square metres (m^2) so pressure would be in newtons per square metre ($N\,m^{-2}$). This unit is usually called a pascal (Pa). You are likely to encounter the following units of pressure:

a) kilopascals (kPa) ($1\,kPa = 1000$ Pa)
b) bars (bar) (1 bar $= 100{,}000$ Pa or 100 kPa)
c) atmospheres (atm) (1 atm $= 101{,}325$ Pa)
d) millimetres of mercury (mm Hg) ($760\,mm$ Hg $= 1$ atm)

Atmospheric pressure is around one bar (one thousand millibars), so weather forecasters find it convenient to record pressure in mbar on their charts. In the physics laboratory we often use a mercury (Fortin) barometer to measure accurately the pressure of the atmosphere in mm Hg. In the chemical industry, processes usually take place at pressures anywhere from slightly above atmospheric pressure, to pump the gases through the pipes, up to very high values; so we use pressure gauges to measure in atm.

The gas laws are the result of experiments to investigate how the volume of gas changes with, say, pressure and temperature. You may already have met these laws in your GCSE work. You will come across them again if you study physics. They are important. Make a mental picture of an everyday illustration to help you understand and remember them.

- Boyle's law: $V \propto 1/p$
  The volume of a fixed mass of gas, at a constant temperature, is inversely proportional to its pressure.

Think of putting your finger over the hole of a bicycle pump and pushing on the pump handle. You squeeze the air into a smaller volume and feel the pressure rise: V gets smaller as p gets higher: see Fig. 5.1.

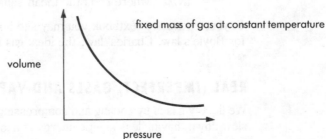

Fig. 5.1 Boyle's law

- Charles' law: $V \propto T$
  The volume of a fixed mass of gas, at a constant pressure, is directly proportional to its temperature on the Kelvin scale.

Think of the air in a balloon being heated and the hot-air balloon rising: V gets bigger as T gets higher: see Fig. 5.2.

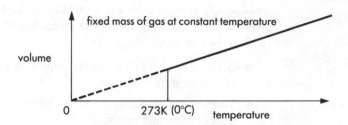

Fig. 5.2 Charles' law

- Avogadro's law: $V \propto n$
  The volume of gas, at a constant pressure and temperature, is directly proportional to its amount in moles.

THE IDEAL GAS EQUATION

> pV = nRT is the most important equation to understand and learn.

The above laws of Boyle, Charles and Avogadro may be combined into one law.

- Ideal gas law: $pV = nRT$
  p is the pressure, V the volume, T the temperature in Kelvin, n the amount of gas in moles and R is called the gas constant.

R is a fundamental constant. You will find eighty-four different values listed in the CRC Handbook of Chemistry and Physics. These values correspond to combinations of the different units chosen for p, V, T and mass. You will not be expected to remember any value of R but you must realise that pV is an energy term and R is energy per degree per mole.

KINETIC THEORY OF GASES

Strictly speaking, all these gas laws apply only to an ideal or perfect gas. But, what is an ideal or perfect gas? Or, under what conditions does a real or imperfect gas obey these laws? You should consult a physics textbook for a more detailed and mathematically rigorous answer to these questions. For A-level chemistry it is sufficient for you to understand the basic assumptions (and their limitations) for ideal gas behaviour:

- The pressure (p) exerted by a gas is the result of its molecules (mass m) which are in continuous random motion, colliding with the walls of the vessel containing the gas
- The collisions with the walls and between molecules are perfectly elastic so the total kinetic energy of all N molecules of gas is constant
- The average kinetic energy of the molecules is directly proportional to the temperature of the gas measured on the Kelvin scale
- The forces of attraction between the molecules as well as between the molecules and the walls of the containers are negligible
- The actual volume of a gas molecule and, therefore, of all the gas molecules is negligible compared to the volume (V) of the container

In a physics textbook you can see how the above assumptions are used to derive the following equation:

$pV = \frac{1}{3}Nm\bar{c}^2$ where $\bar{c}^2$ is the mean square speed of the molecules

In your chemistry textbook you may see how this equation can be interpreted to account for Boyle's law, Charles' law, the ideal gas law and Graham's law of diffusion (see later).

REAL (IMPERFECT) GASES AND VAPOURS

We liquefy gases by cooling and compressing them: i.e. we remove heat so the molecules slow down, having less kinetic energy at a lower temperature and we decrease the size of the container so the molecules come closer together. As the temperature goes down and the pressure goes up, it becomes more and more unreasonable to neglect the actual volume of the molecules and the forces between them.

- Deviation from ideal behaviour increases as temperature decreases, pressure increases and gases approach their point of liquefaction

VAN DER WAALS' EQUATION

In 1873, a Dutch chemist called J C van der Waals proposed the following gas equation:

$$(p + a/V_m^2)(V_m - b) = RT$$

where V_m is the volume of one mole of gas and a and b are constants. The constant a allows for the attractive forces between the molecules; b is the effective volume in one mole of gas of the molecules themselves. For A-level chemistry you will not have to use van der Waals' equation or know any more about it. However, you may have to use the ideal gas equation.

DETERMINING MOLAR MASSES

In the equation $pV = nRT$, n is the amount of gas measured in moles. If the mass of gas is m and its molar mass is M then the amount of gas is m/M and we can rewrite the ideal gas equation as:

$$\blacksquare \quad pV = \frac{m}{M}RT \quad \text{or} \quad M = \frac{mRT}{pV}$$

It follows that we can determine a value for M if we can measure the values of m, T, p and V. We also need to find a value for R and to assume that our gas or vapour does not deviate too far from ideal behaviour.

In your practical work you could determine M for, say, carbon dioxide by:

i) weighing a 250 ml volumetric flask full of air (w_1) and then carbon dioxide (w_2) at atmospheric pressure (p_a) and temperature (T_a),

ii) finding the volume of gas: by finding the volume of the neck from the graduation mark to the bottom of the stopper and adding it to the 250 ml: see Fig. 5.3

iii) calculating the buoyancy correction (w_b) for this volume (V_a) of carbon dioxide by multiplying V_a by the density of air at pressure p_a and temperature T_a to find the upthrust on the gas by the air it has displaced – look up Archimedes' principle in a physics book,

iv) looking up a value for R in a data book and substituting your values into the ideal gas equation to calculate M:

$$M = (w_2 - w_1 + w_b)RT_a/p_aV_a$$

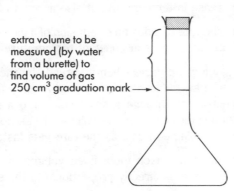

extra volume to be measured (by water from a burette) to find volume of gas
250 cm³ graduation mark ⟶

Fig. 5.3 Measuring the molar mass of a gas

You could also determine M for a volatile liquid such as hexane by:

i) weighing a sample of the liquid in a hypodermic syringe (w_x),

ii) injecting some of it into a steam-heated glass gas-syringe where it vaporises at atmospheric pressure (p_a) and steam temperature (T_a): see Fig. 5.4

iii) reweighing the hypodermic syringe (w_y) to determine the mass of liquid injected into the gas-syringe and noting the increase in volume (V_a) of gas (due to hexane vapour) in the glass syringe,

iv) looking up a value for R in a data book and again substituting your values into the ideal gas equation to calculate M:

$$M = (w_x - w_y)RT_a/p_aV_a$$

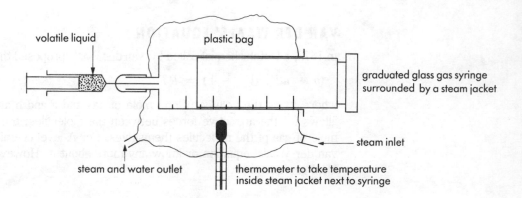

volatile liquid

plastic bag

graduated glass gas syringe surrounded by a steam jacket

steam inlet

steam and water outlet

thermometer to take temperature inside steam jacket next to syringe

Fig. 5.4 Measuring the molar mass of a vapour

In these experiments the measurement of the mass of gas is subject to the greatest error because the mass is so small. Notice that you do not have a buoyancy correction to make with the second experiment because you find the mass of the vapour whilst it is still a liquid displacing a negligible volume of air. Notice also that if you refer to M as the molar mass you must record your experimental result with the units g mol^{-1} but if you refer to M as the relative molecular mass it will not have units.

GRAHAM'S LAW OF DIFFUSION

In the equation $pV = \frac{1}{3}Nm\bar{c}^2$ the product of Nm (number of molecules × mass of each molecule) is the mass of gas. So if M is the molar mass, then the amount of gas (in moles) is $Nm/M = n$. Hence, $Nm = nM$ and the equation can be rewritten as

$$pV = \frac{1}{3}nM\bar{c}^2$$

But $pV = nRT$ for an ideal gas

so $\frac{1}{3}M\bar{c}^2 = RT$

$$\Rightarrow \sqrt{\bar{c}^2} = \sqrt{\left(\frac{3RT}{M}\right)}$$

where $\sqrt{\bar{c}^2}$ is called the root mean square speed of the molecules. This derivation from the kinetic theory shows that the root mean square speed of gas molecules increases with an increase in temperature. It also accounts for Graham's law of diffusion:

- The rate of diffusion or effusion of a gas is inversely proportional to the square root of its molar mass at constant temperature and pressure

We say a gas *diffuses* when it spreads by the random movement of its molecules from a high to a low concentration region. A gas *effuses* when it escapes from a region of high concentration, through a small hole, into a region of low concentration. If the rate of effusion is measured by the volume (V) escaping in a measured time (t) as V/t, then for a constant volume of gas, as the rate gets faster the time gets shorter:

- The time taken for a fixed volume of gas to effuse at constant temperature and pressure is directly proportional to the square root of its molar mass

You could measure the time (t_1) for a definite volume (V at a fixed pressure and temperature) of gas (molar mass M_1) to effuse from an apparatus: see Fig. 5.5. If you repeat the experiment with the same volume of a different gas (molar mass M_2) in the same apparatus under the same conditions and measure the new time (t_2), you can calculate the ratio of the molar masses of the two gases using this equation:

$$M_1/M_2 = (t_1/t_2)^2$$

If you know the molar mass of one of the gases, you can calculate the molar mass of the other.

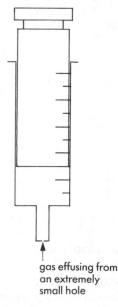

graduated glass syringe

gas effusing from an extremely small hole

Fig. 5.5 Graham's law of diffusion

GAY-LUSSAC'S LAW OF COMBINING VOLUMES

In 1809 the French chemist Louis J. Gay-Lussac summarised the results of his experiments on reactions involving gases:

- The volumes of gaseous reactants and products of a reaction measured at the same temperature and pressure will be in a simple ratio to each other

The Italian chemist Amadeo Avogadro provided an explanation for this law when he stated his principle in 1811:

- Equal volumes of gases at the same temperature and pressure contain the same number of molecules

You probably already know a consequence of Avogadro's principle:

- One mole of any gas occupies $24\,dm^3$ at $25\,°C$ and $1\,atm$

An important consequence of Gay-Lussac's law and Avogadro's principle is that chemists could determine experimentally the equation for a gaseous reaction by measuring the volumes of reactants and products. Moreover, they could determine the molecular formula of gaseous hydrocarbons by volumetric combustion analysis: see Fig. 5.6.

Problem: $20.0\,cm^3$ of a gaseous hydrocarbon C_xH_y were exploded with $200.0\,cm^3$ of oxygen to produce $140\,cm^3$ of gaseous products which contracted to $40\,cm^3$ of oxygen after absorption with concentrated alkali. All volumes were measured at a temperature of 298K and a pressure of 1 atm. Find the value of x and y.

Solution:
$$C_xH_y(g) \quad + \quad (x + y/4)O_2(g) \quad \rightarrow \quad xCO_2(g) \quad + \quad (y/2)H_2O(l)$$
$$20\,cm^3 \qquad (200 - 40)\,cm^3 \qquad (140 - 40)\,cm^3 \quad \simeq 0\ (liquid)$$
$$\Rightarrow 20\,cm^3 \qquad\quad 160\,cm^3 \qquad\qquad 100\,cm^3$$

applying Avogadro's principle to these volumes of gases gives the ratio of the amounts of gases (in moles):

$$1C_xH_y(g) + 8O_2(g) \rightarrow 5CO_2(g) + (y/2)H_2O(l)$$
$$\Rightarrow x = 5\ and\ 5O_2(g)\ is\ used\ to\ produce\ 5CO_2(g)$$
leaving $3O_2\ (= 6O)$ to produce $(y/2)H_2O(l)$, so $y/2 = 6$
$$\Rightarrow y = 12\ and\ the\ hydrocarbon\ is\ pentane,\ C_5H_{12}.$$

Fig. 5.6 Combustion analysis of a hydrocarbon

MASS SPECTROMETRY

Nowadays chemists use a mass spectrometer to determine the relative molecular mass (M_r), the molecular formula and, very often, the structure of compounds: see Fig. 5.7.

The substance is vaporised in the instrument at extremely low pressure. The gaseous molecules are ionised and usually fragmented into smaller gaseous ions. A fine beam of

> Make sure you understand the principles of this technique.

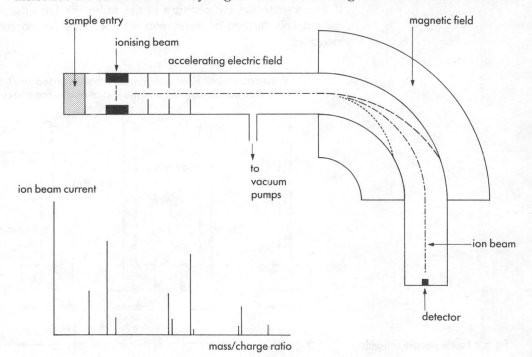

Fig. 5.7 Mass spectrometer

these ions (having different masses) is accelerated by an electric field and deflected by a magnetic field onto a (positive ion) current detector. The magnitude of this ion-beam current registered by the detector depends upon the number of ions in the beam focused on the detector. The mass of the ions being detected at any time depends upon the settings of the electric and magnetic fields. Increasing the magnetic field strength enables us to focus and detect ions of higher mass. A mass spectrum may be a chart recording of ion-beam current against magnetic field strength: see Fig. 5.7. You may think of it as a plot of the abundance of the ions against their relative masses.

To the nearest whole number $M_r(CH_3CO_2H) = 60$ and $M_r(NH_2CONH_2) = 60$. But modern high resolution spectroscopy can give the relative molecular mass of the parent molecule ion to four decimal places so that $M_r(CH_3CO_2H) = 60.0210$ and $M_r(NH_2CONH_2) = 60.0323$. Consequently, by using very accurate values for $A_r(H)$, $A_r(C)$, $A_r(N)$ and $A_r(O)$, together with the very accurate value for M_r(parent molecule), we can determine the formula of the compound. If we identify the ions into which the molecule fragments, we can often determine the structure of the compound: see Fig. 5.8.

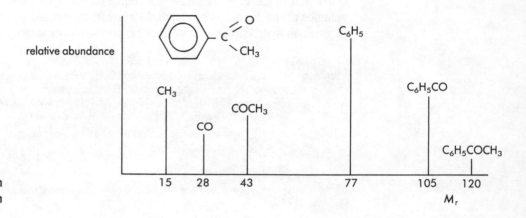

Fig. 5.8 Deducing a structure from a mass spectrum

LIQUIDS

If we put some water into a barometer tube containing only mercury, the height of the mercury column drops: see Fig. 5.9. This change in height of the mercury column is caused by the pressure of the water vapour in the tube above the mercury. If we use ethanol instead of water the mercury column drops even more because the liquid is more volatile than water. The vapour presure of ethanol is greater than that of water at the same temperature.

■ The saturated vapour pressure of a liquid is the pressure exerted by its vapour in equilibrium with the liquid in a closed container

If we increase the temperature of the water, or the ethanol, the level of the mercury column falls further because more of the liquid evaporates and its vapour pressure increases.

Barometer tubes filled with mercury then inverted into trough of mercury
Different liquids injected into the space above the mercury in the tube

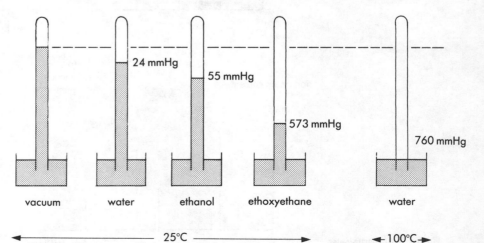

Fig. 5.9 Vapour pressure of liquids

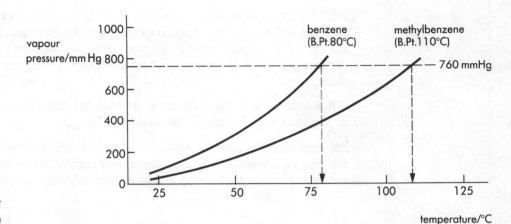

Fig. 5.10 Change of vapour pressure with temperature

- The vapour pressure of a liquid increases with increasing temperature: see Fig. 5.10. If we heat a liquid in a closed container, the pressure of the vapour inside the container will increase. Eventually, it will exceed the pressure of the atmosphere outside the container. If the container is not closed, the liquid will boil when its vapour pressure reaches the atmospheric pressure.

- The boiling point of a liquid is the temperature at which its saturated vapour pressure equals the pressure of the atmosphere

- A pure liquid at constant pressure boils at a constant temperature

The normal boiling point of a liquid is recorded in data books and is the temperature at which the vapour pressure is 1 atm. The boiling point is an important characteristic property which helps us to identify substances: e.g. ethoxyethane boils at 35 °C and ethanol at 78 °C.

- The more volatile the liquid, the lower the boiling point and the less energy required to evaporate one mole

TROUTON'S RULE

If you plot a graph of the standard molar enthalpy change of evaporation ($\Delta H^{\ominus}_{\text{evap}}$) against the boiling temperature (T_b) most, but not all, of the points lie on or close to a straight line: see Fig. 5.11. So for most liquids $\Delta H^{\ominus}_{\text{evap}}/T_b = $ constant ($\simeq 88\,\text{J K}^{-1}\,\text{mol}^{-1}$). This was observed by Frederick Trouton in 1884 and expressed in his rule:

- The molar enthalpy change of vaporisation of a liquid, at its normal boiling point, divided by its normal boiling point, in Kelvin, is constant

When a substance turns from the liquid into the gaseous state, its matter and energy become more widely and randomly dispersed. For one mole of substance the change in

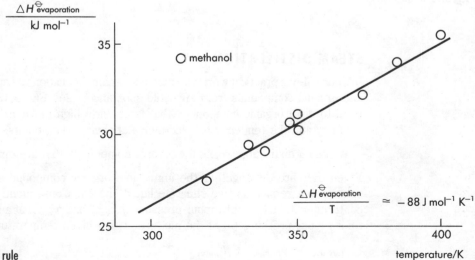

Fig. 5.11 Trouton's rule

volume from liquid to gas is similar for most liquids; the molar volume is also similar for most gases. So you might expect the dispersion of matter and energy to be about the same for one mole of any gas. $\Delta S_{evap}^{\ominus} = \Delta H_{evap}^{\ominus}/T_b$ and is the approximately constant increase in entropy per mole of substance changing from the liquid state to the gaseous state.

■ Liquids with hydrogen bonding between the molecules have $\Delta S_{evap}^{\ominus}$ values that are greater than Trouton's rule would predict

For example, the value of $\Delta S_{evap}^{\ominus}$(methanol) is $104\,\mathrm{J\,K^{-1}\,mol^{-1}}$. The hydrogen bonding causes these liquids to be more ordered than 'normal' liquids so the change to more widely and randomly dispersed energy and matter on evaporation is greater than for other liquids.

DISTILLATION

This is an important technique in which a liquid evaporates and its vapour condenses and collects elsewhere. In your GCSE coursework you will have separated salt and water by distillation: see Fig. 5.12. The technique works because the salt is involatile and the water volatile. The salt is left behind in the flask and the water is collected from the condenser as the distillate. The temperature recorded on the thermometer is the boiling point of the distillate.

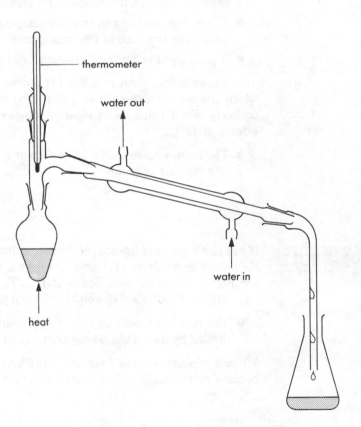

Fig. 5.12 Distillation

STEAM DISTILLATION

In your A-level practical work you may use steam distillation to separate a mixture of water and organic compounds from involatile substances: see Fig. 5.13. This technique is also used to distil organic compounds that have fairly high molar masses, and therefore low volatility and the tendency to decompose at their boiling points.

■ Steam distillation works for organic compounds that are immiscible with water

When they are put together the immiscible organic compound and water do not form a solution but remain as two separate liquid phases. Consequently, each liquid separately contributes its full (pure) vapour pressure ($p_{organic}$ and p_{water}) at a given temperature to the total vapour pressure (p) of the mixture at that given temperature: i.e.

$$p_{mixture} = p_{organic} + p_{water}.$$

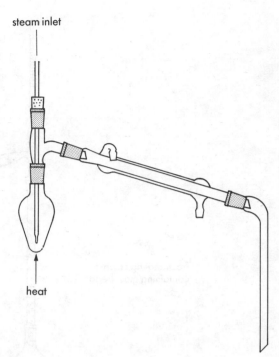

steam inlet

heat

Fig. 5.13 Steam distillation

If, say, the atmospheric pressure is 760 mm Hg, then the mixture will boil when $p_{\text{mixture}} = 760$ mm Hg and therefore when $p_{\text{organic}} + p_{\text{water}} = 760$ mm Hg. Under these conditions a mixture of an immiscible organic compound and water will distil at 97 °C if the compound has a vapour pressure of 78 mm Hg at 97 °C because the vapour pressure of water is 682 mm Hg at 97 °C. The distillate would contain more water than organic compound because the water is almost nine times more volatile than the organic compound but how can we estimate the relative amounts? We can treat the vapours as perfect gases and apply the ideal gas equation and the following law:

■ Dalton's law of partial pressures
  In a mixture of gases each gas exerts its pressure independently as if the other gases were not present.

The partial pressures of the organic compound and water in the vapour (in equilibrium with the liquid mixture) are

$$p_{\text{organic}} = n_{\text{organic}}RT/V \text{ and } p_{\text{water}} = n_{\text{water}}RT/V$$

$$\Rightarrow n_{\text{water}} : n_{\text{organic}} = p_{\text{water}} : p_{\text{organic}}$$

$$\Rightarrow \frac{\text{mass water}/18}{\text{mass compound}/M} = \frac{p_{\text{water}}}{p_{\text{organic}}} \quad \text{where } M \text{ is the molar mass of the compound.}$$

FRACTIONAL DISTILLATION

In your A-level practical work you may use fractional distillation to separate a mixture of two volatile miscible substances: see Fig. 5.14. When the two liquids are put together they mix to form a solution. When this solution is heated in a distillation flask, the vapour and, therefore, the distillate will contain a higher proportion of the more volatile component than the solution being heated. If you distil a fraction, say, for example, one half of the solution and then distil a fraction of that distillate, you may gradually separate the more volatile component from the less volatile one.

■ Fractional distillation may be used to separate the components of a solution of two or more volatile miscible liquids

The technique may not work very well if the boiling points of the components are too close together. For example, 2-methylnonane (b.pt. 166.8 °C) and 3-methylnonane (b.pt. 167.8 °C) are difficult to separate by fractional distillation. Sometimes fractional distillation does not work at all: e.g. a liquid consisting of 96% by mass of ethanol (b.pt. 78.32 °C) and 4% by mass of water (b.pt. 100.0 °C) boils at 78.15 °C and a liquid consisting of 67.4% by

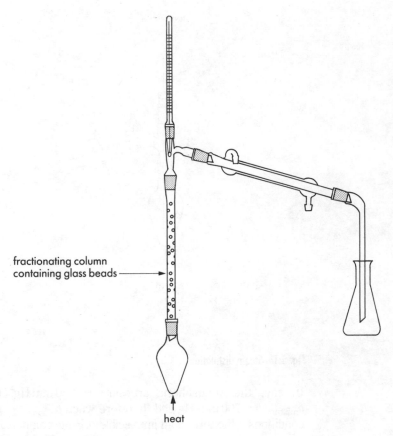

fractionating column
containing glass beads

heat

Fig. 5.14 Fractional distillation

mass of nitric acid (b.pt. 86 °C) and 32.6% by mass of water boils at 120.7 °C. We call such solutions constant boiling mixtures or azeotropes: see later.

RAOULT'S LAW

❝ This law is important for understanding fractional distillation. ❞

At 60 °C the vapour pressure of benzene (b.pt. 80 °C) is about 400 mm Hg and that of methylbenzene (b.pt. 111 °C) is about 150 mm Hg. If 0.5 mol C_6H_6 and 0.5 mol $C_6H_5CH_3$ are mixed they form a solution with a total vapour pressure of about 275 mm Hg. The partial pressure of the benzene in the vapour is about 200 mm Hg and that of methylbenzene is about 75 mm Hg. So the ratio of the amount of the more volatile benzene to the amount of the less volatile methylbenzene is 1 : 1 in the liquid but 8 : 3 in the vapour. Dalton's law of partial pressures accounts for the total vapour pressure (275 mm Hg) being equal to the sum of the partial vapour pressures (200 mm Hg + 75 mm Hg). Raoult's law accounts for the partial pressure of benzene being 200 mm Hg and that of methylbenzene being 75 mm Hg:

- Raoult's law
  The partial vapour pressure (p_A) of component (A) in a solution of two miscible volatile liquids is equal to the vapour pressure (p_A^0) of the pure component multiplied by its mole fraction (x_A) in the solution:
  $$p_A = x_A \times p_A^0 \quad \text{and} \quad p_B = x_B \times p_B^0$$
- The mole fraction (x_A) of component (A) is its amount (n_A) divided by the total amount ($n_A + n_B$) of components in the two-component solution
  $$x_A = n_A/(n_A + n_B) \quad \text{and} \quad x_B = n_B/(n_A + n_B) \quad \text{therefore } x_A + x_B = 1$$

According to Raoult's law, the partial pressure of benzene, in the vapour above its solution with methylbenzene, will be 0 mm Hg if its mole fraction (x_b) is 0, i.e. no benzene in the solution, and will increase linearly up to 400 mm Hg when $x_b = 1$, i.e. no methylbenzene in the solution. In the same way the partial pressure of methylbenzene will increase linearly from 0 to 150 mm Hg as its mole fraction ($x_m = 1 - x_b$) increases from 0 to 1. Consequently the total vapour pressure of the solution will vary from 150 mm Hg, only methylbenzene, to 275 mm Hg, equal amounts of benzene and methylbenzene in the solution, to 400 mm Hg, only benzene: see Fig. 5.15. The information provided by the four separate graphs in Fig. 5.15 can be displayed in one vapour pressure/composition diagram: see Fig. 5.16.

Variation of partial (p_B, p_M) and total (p_{TOTAL}) vapour pressure with composition (mole fractions x_B and x_M) for a mixture of benzene (B) and methylbenzene (M) at a constant temperature of 60°C

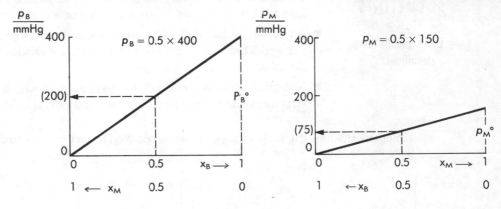

ALL AT THE SAME CONSTANT TEMPERATURE = 60°C

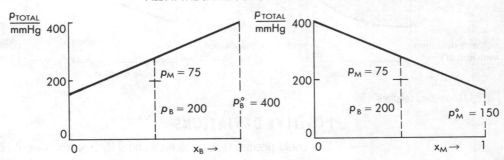

Fig. 5.15 Raoult's law

Variation of vapour pressure (p_{TOTAL}) with composition (mole fractions x_B and x_M) for a mixture of benzene (B) and methylbenzene (M) regarded as an ideal solution at a constant temperature of 60°C

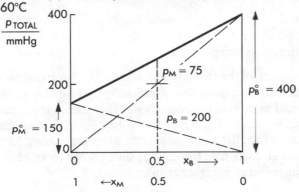

**Fig. 5.16 Vapour pressure/
composition diagram for an ideal
solution**

IDEAL SOLUTIONS

■ An ideal solution is a solution of two volatile liquids whose total vapour pressure at constant T varies linearly with composition

■ Ideal solutions are solutions that obey Raoult's law

■ Ideal solutions are formed by liquids whose molecules have very similar structures and intermolecular forces; and $\Delta H_{\text{mixing}} \simeq 0$

A solution of benzene and methylbenzene is a good example of an ideal solution. You could predict that ideal solutions obeying Raoult's law would be formed by two chain isomers of a saturated hydrocarbon (e.g. 2- and 3-methylpentane) or by two successive members of an homologous series (e.g. pentanol and hexanol). You could also predict no heat change on mixing the two liquids because the energy consumed in overcoming the forces holding the molecules together in the separate pure liquids should be provided by the same forces reforming between the very similar molecules in the solution: e.g.

| benzene molecules | + | methylbenzene molecules | $\xrightarrow{\text{mix}}$ | solution of benzene and methylbenzene molecules |

| energy taken in when van der Waals' forces are broken down | | | | energy given out when van der Waals' forces are set up |

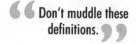

NON-IDEAL SOLUTIONS

Don't muddle these definitions.

- A non-ideal solution is a solution of two volatile liquids whose total vapour pressure at constant T does not vary linearly with composition: i.e. the solution does not obey Raoult's law but shows either a positive or a negative deviation from the law
- A non-ideal solution shows a positive deviation if its vapour pressure is higher than that predicted by Raoult's law and a negative deviation if its vapour pressure is lower than that predicted by the law
- Non-ideal solutions are usually formed by liquids whose molecules have very different structures and intermolecular forces; and $\Delta H_{mixing} \neq 0$: see Fig. 5.17.

Variation of total vapour pressure (p_{TOTAL}) with composition (mole fraction) at a constant temperature

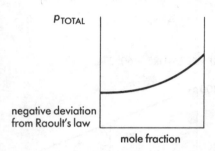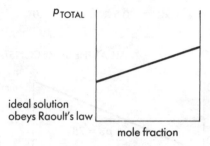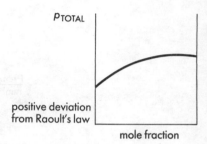

Fig. 5.17 Non-ideal solutions

POSITIVE DEVIATIONS

Positive deviation means high vapour pressure and low boiling point.

You could predict that an alcohol and a hydrocarbon would cool down when you mix them together to form a solution if you compare the intermolecular forces in the separate liquids with those in the solution: e.g.

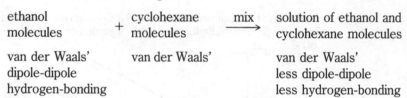

In a solution of the two different liquids, the hydrocarbon molecules would come between the alcohol molecules and significantly reduce the dipole-dipole and hydrogen-bonding forces of attraction between them. This means that energy is used to overcome these attractive forces. The mixing of the two liquids is therefore an endothermic process (ΔH is positive) and the temperature falls when the solution forms. Furthermore, the molecules are less strongly held together so they escape more easily into the gaseous state and the vapour pressure is higher than expected.

NEGATIVE DEVIATIONS

Only a small minority of non-ideal solutions show negative deviations. One example is ethyl ethanoate and trichloromethane. These two liquids warm up when you mix them to form a solution:

ethyl ethanoate molecules	+	trichloromethane molecules	mix →	solution of ethyl ethanoate and trichloromethane molecules
van der Waals' dipole-dipole		van der Waals' dipole-dipole		van der Waals' dipole-dipole hydrogen-bonding

In the trichloromethane molecule the three chlorine atoms attract the electronic charge in the valence shell of the carbon atom to which they are bonded. This increases the effective electronegativity of the carbon atom beyond its normal value so that the H-atom attached to it is capable of hydrogen bonding. But note that $CHCl_3$ molecules are not hydrogen bonded to each other because a) the C-atom does not have a lone pair of electrons and b) the Cl-atoms have lone pairs of electrons but the electronegativity of chlorine is too low. In ethyl ethanoate the molecules are not hydrogen bonded because there are no hydrogen atoms attached to the electronegative oxygen atoms. In the solution the H-atom, attached

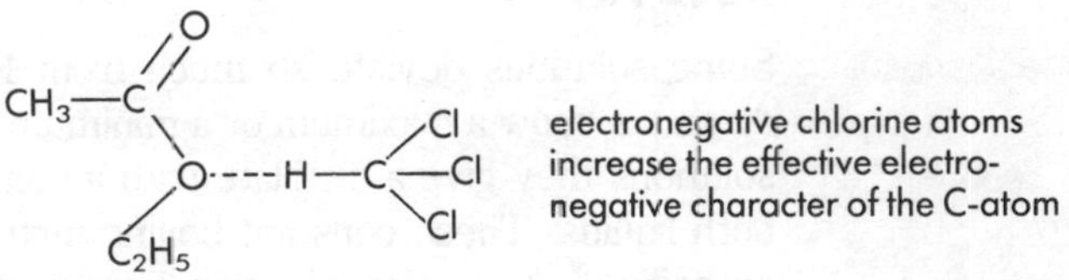

Fig. 5.18 Unusual case of hydrogen
bonding

electronegative chlorine atoms
increase the effective electronegative character of the C-atom

to the carbon atom with its unusually enhanced electronegativity, can form a hydrogen
bond with the electronegative oxygen atoms of the ethyl ethanoate: see Fig. 5.18. The
formation of hydrogen bonds is an exothermic process (ΔH is negative) so the liquids
warm up on mixing. The hydrogen-bonding between the molecules makes the mixture less
volatile than expected; vapour pressure is lower.

BOILING POINT/COMPOSITION DIAGRAMS

Diagrams of vapour pressure against composition at a constant temperature can be
converted into diagrams of boiling point against composition at a constant pressure of
1 atm. You need not be able to do the conversion but you could be asked to interpret the
result: see Fig. 5.19. Notice that for an ideal solution obeying Raoult's law, the straight
line, of p vs. x_A at constant T, converts to a curve, of T_b vs. x_A at constant p. We could
draw this curve to show the mole fraction of a component in the solution or in the vapour
above the solution. What we usually do is put both curves on the same diagram and use it
to explain fractional distillation: see Fig. 5.20. The diagram shows that fractional distillation
is rather like a series of simple distillations. The distillate from the first distillation is rich in
the more volatile component. This is distilled to produce a second distillate even richer in
the more volatile component. The process is repeated many times until the final distillate
contains only a negligible amount of the less volatile component. In practice the whole
operation is carried out in a fractionating column on a laboratory scale or in a fractionating
tower on an industrial scale.

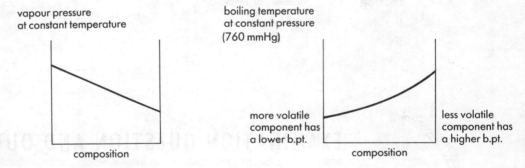

Fig. 5.19 Boiling point/
composition diagrams

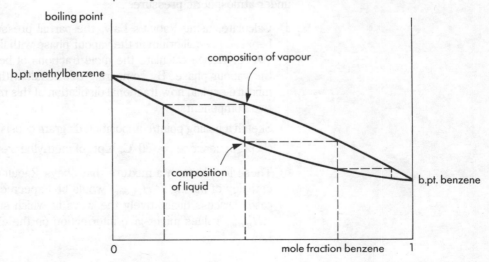

Fig. 5.20 Boiling point/liquid/
vapour/composition diagram

AZEOTROPES

Some solutions deviate so much from Raoult's law that their boiling point/composition diagrams show a maximum or a minimum: see Fig. 5.21. When you fractionally distil these solutions they give a distillate with a constant boiling point and a constant composition of both liquids. These constant boiling mixtures are called *azeotropes* (Greek – not changed on boiling). A solution of ethanol and water shows a positive deviation from Raoult's law and forms a minimum boiling azeotrope. A solution of nitric acid and water shows a negative deviation from Raoult's law and forms a maximum boiling azeotrope. You can interpret their boiling point/composition diagrams by treating the part on either side of the minimum or maximum as representing a simple boiling point/composition diagram. The temperature at the minimum or maximum corresponds to the boiling point of the azeotropic composition instead of only a single component.

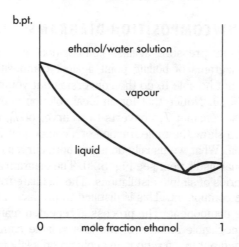

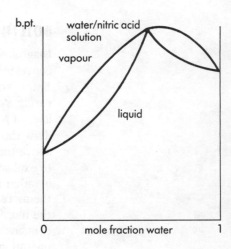

Fig. 5.21 Distillation of binary azeotropes

minimum b.pt. azeotrope
obtained on distillation

maximum b.pt. azeotrope
obtained on distillation

EXAMINATION QUESTION AND OUTLINE ANSWER

Q1 At 101 °C the saturated vapour pressures of benzene and of methylbenzene are 1521 mm Hg and 570 mm Hg respectively.
A mixture of 0.20 mole of benzene and 0.80 mole of methylbenzene boils at 101 °C under atmospheric pressure.

a) Calculate, using Raoult's Law, the partial pressures of benzene and of methylbenzene in equilibrium in the vapour phase with liquid of the above composition at 101 °C. Hence calculate the mole fractions of benzene and of methylbenzene in the vapour phase. By comparing this result with the original composition of the mixture explain how fractional distillation of this mixture allows a separation of the two components.

Sketch a boiling point/composition diagram consistent with the above information.

(B.pt. of benzene = 80 °C; b.pt. of methylbenzene = 111 °C.)   (*17 marks*)

b) These liquids form a mixture that obeys Raoult's Law closely. Explain why the enthalpy of mixing (ΔH_{mixing}) would be expected to be zero or very small in this case. Discuss qualitatively the ways in which substantially positive or negative ΔH_{mixing} values might give information on the extent of deviation from Raoult's Law.
(*8 marks*)
(ULSEB 1988)

Outline answer

Q1

(a) $P_{benzene}$ = 0.20 x 1521 mm Hg
                      = 304.2 mm Hg

$P_{methylbenzene}$ = 0.80 x 570 mm Hg
                      = 456 mm Hg

⟹ mole fraction benzene = $\dfrac{304}{304 + 456}$

= 0.4

Vapour is richer in benzene. If this vapour is condensed, the liquid could be distilled again to produce a vapour even richer in benzene.

(b) See text pages 58-59.

EXAMINATION QUESTION AND TUTOR'S ANSWER

Q2 a) Raoult discovered a simple relationship to exist between the relative lowering of vapour pressure and the concentration of an ideal solution expressed as the MOLE FRACTION of solute.

   i) State the meaning of the term MOLE FRACTION of solute.

$$\dfrac{\text{number of moles of solute}}{\text{total number moles solute and solvent}}$$

   ii) Name two liquids which form approximately ideal mixtures over the whole composition range.

   benzene and methylbenzene

   iii) On the axes below, sketch and label the boiling point-composition curves (at constant pressure) for mixtures of the two liquids given in ii). Your diagram should show both liquid and vapour compositions.

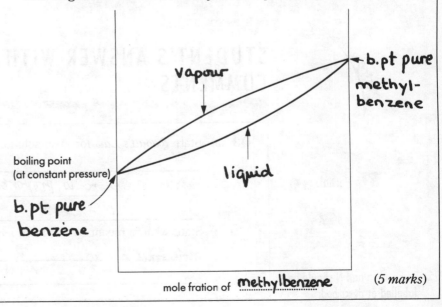

mole fration of **methylbenzene**   (5 marks)

b) Mixtures of cyclohexane (boiling point 81 °C) and ethanol (boiling point 78 °C) show positive deviations from Raoult's Law.

i) On the axes below sketch and label a vapour pressure-composition curve (at constant temperature) for the system.

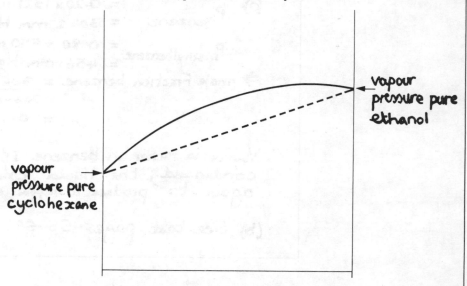

vapour pressure pure ethanol

vapour pressure pure cyclohexane

ii) Explain the deviations from Raoult's Law in terms of molecular interactions.

cyclohexane molecules have van der Waals forces only, so when they get between the ethanol molecules they reduce their hydrogen bonding. This makes the ethanol molecules escape more easily.

iii) Ethanol and cyclohexane, both at room temperature, are mixed. State whether the temperature of the mixture will rise, fall, or stay the same immediately after mixing. Explain your answer.

Fall — because some of the hydrogen bonds between the ethanol molecules break. This uses energy, so the temperature falls as heat is taken away.

(7)

(Total 12 marks)

(AEB 1988)

STUDENT'S ANSWER WITH EXAMINER'S COMMENTS

Q3 a) State *Raoult's Law* for ideal solutions.

"Partial."

Vapour pressure is proportional to mole fraction

$$p = x p^o$$

"Â"

b) State what is meant by the term *mole fraction*.

"Where p^o is the vapour pressure of the pure component A."

mole fraction is $x = \dfrac{n}{n+N}$

"n and N should be labelled or described."

c) Hexane and heptane form ideal liquid mixtures. The vapour pressures of the pure liquids at 50°C are $50\,kN\,m^{-2}$ for hexane, and $20\,kN\,m^{-2}$ for heptane. Calculate the mole fraction of heptane in the *liquid* when the mole fractions of hexane and heptane in the *vapour* are equal.

> Good calculation, but the answer should be clearly set out: mole fraction heptane is 5/7

If mole fractions in the vapour are equal, then their pressures are equal. So if x is the mole fraction of heptane in the liquid

$$x \times 20 = (1-x) \times 50$$
$$\Rightarrow x = 5/7$$

d) Fig. A shows the equilibrium vapour pressures of water and chlorobenzene as a function of temperature. Chlorobenzene and water are immiscible liquids.

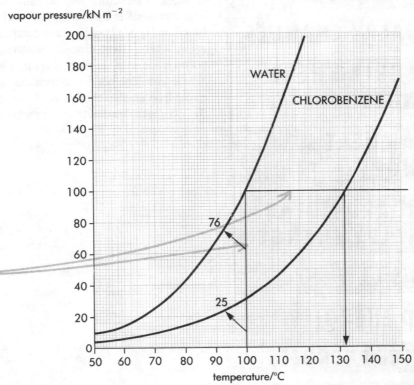

vapour pressure/kN m^{-2}

> These lines are not needed but it is a good idea to show your working on the graph.

Fig. A

i) From Fig. A, deduce the boiling point, at 1 atmosphere pressure ($101\,kN\,m^{-2}$) of

pure chlorobenzene, ____ 132·5 °C ____

a mixture of chlorobenzene and water. ____ 93°C ____

ii) Calculate the mole fraction of chlorobenzene in the distillate when a mixture of chlorobenzene and water is distilled at one atmosphere pressure.

number of moles chlorobenzene $\propto$ 25

" " " water $\propto$ 76

$\Rightarrow$ mole fraction is $\frac{25}{25+76}$ = 0.25 ✓

iii) Explain briefly the usefulness of the technique of steam distillation.

It can distil organic molecules that would decompose if heated to their normal boiling point. ✓

(WJEC 1988)

GETTING STARTED

The relating of energy changes of chemical and physical processes to changes in structure and bonding forms a central theme in chemistry. The emphasis is upon standard enthalpy changes and upon calculations based on Hess's law. Heats of combustion are measured in bomb or flame calorimeters. Heat changes involving solutions may be measured in a polystyrene cup. Energy diagrams and Born-Haber cycles are used to handle covalent bond energies and ionic lattice energies. The standard molar free energy change $\Delta G^{\ominus} = \Delta H^{\ominus} - T . \Delta S^{\ominus}$ is used to predict the energetic feasibility of a reaction.

ESSENTIAL PRINCIPLES

HEAT CHANGES

When you stir some concentrated sulphuric acid or some solid sodium hydroxide into water, the temperature rises. Likewise, if you add the aqueous acid to the aqueous alkali, you get a warm solution of sodium sulphate. But if you stir sodium hydrogencarbonate into aqueous citric acid, a large volume of carbon dioxide gas is evolved and the temperature of the solution falls.

EXOTHERMIC PROCESSES

In the neutralisation, heat is generated and taken up by the chemical system, so the temperature of the solution increases. However, if this solution of aqueous sodium sulphate is to return to the initial temperature of the aqueous acid and alkali before neutralisation, the heat energy liberated during the neutralisation must be dissipated to the surroundings. In other words, heat must be 'lost' or 'given out' by the system.

- A process where energy is transferred from the reacting system to the surroundings is called an 'exothermic' process (Greek: exo – outside; thermos – hot)

The formation of the aqueous solutions of sulphuric acid and sodium hydroxide, and their neutralisation reaction, are exothermic processes.

ENDOTHERMIC PROCESSES

In the reaction of citric acid with aqueous sodium hydrogencarbonate, energy is used by the reaction, so the temperature of the chemical system decreases. If the temperature before reaction is to be restored, the chemical system must gain heat from the surroundings. In other words, heat must be 'gained' or 'taken in' by the system.

- A process where energy is transferred from the surroundings to the reacting system is called an 'endothermic' process (Greek: endo – inside; thermos – hot)

ENTHALPY CHANGES

The reaction of magnesium with hydrochloric acid ($Mg(s) + 2HCl(aq) \rightarrow MgCl_2(aq) + H_2(g)$) is exothermic. When the reaction is carried out in a sealed, fixed volume, container, the heat 'given out' is about 473 kJ per mole of Mg used. In an open vessel, the heat 'given out' is about 470 kJ. The difference of 3 kJ arises because the one mole of hydrogen gas produced in the reaction uses about 3 kJ of energy to 'push aside' about 24 dm^3 of the surrounding air at constant pressure of 1 atm. This energy is released as 3 kJ of 'extra' heat if the gas is confined at constant volume in a sealed container.

- An enthalpy change (ΔH) is a heat change at constant pressure
- For exothermic processes ΔH is negative (heat 'taken away' from the system by the surroundings)
- For endothermic processes ΔH is positive (heat 'added' to the system from the surroundings)

MEASURING HEAT CHANGES

We determine the heat change (δh) of a process by measuring the temperature change (δT) it causes in a calorimeter and then multiplying the heat capacity (C) of the calorimeter and its contents by the temperature change: $\delta h = C \times \delta T$. Heat capacity means the amount of heat per degree temperature change. If we measure δh in kilojoules (kJ) and δT in Kelvin (K) then we need to know the heat capacity C in kilojoules per Kelvin (kJ K^{-1}):

$$\frac{\text{heat change}}{\text{kJ}} = \frac{\text{heat capacity}}{\text{kJ K}^{-1}} \times \frac{\text{temperature change}}{\text{K}}$$

THE BOMB CALORIMETER

The heat capacity of a bomb calorimeter may be found or calibrated by measuring the temperature rise produced by an accurately known amount of heat. This heat may be supplied either by burning, in the bomb, a measured amount of substance whose heat of combustion is well known e.g. benzoic acid or by heating the bomb with an accurately measured amount of electrical energy: see Fig. 6.1.

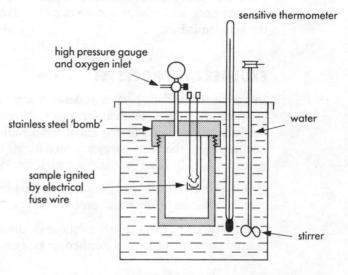

heat of complete combustion measured at constant volume

Fig. 6.1 Measuring heat of combustion at constant volume

- A bomb calorimeter measures the heat of combustion at constant volume
- $\Delta H = \Delta U + \Delta nRT$ where ΔH is the enthalpy change (heat change at constant pressure), ΔU is the internal energy change (heat change at constant volume), R is the gas constant, T is the absolute temperature and $\Delta n =$ no. of moles of gaseous products – no. of moles gaseous reactants

The bomb is filled with pure oxygen at high pressure to ensure that the combustion is rapid and complete.

THE FLAME CALORIMETER

The calorimeter in Fig. 6.2 is sometimes called a 'food' calorimeter because biologists use it to find the calorific values of edible organic substances. The heat capacity of the calorimeter may be determined using similar principles to those for the bomb calorimeter.

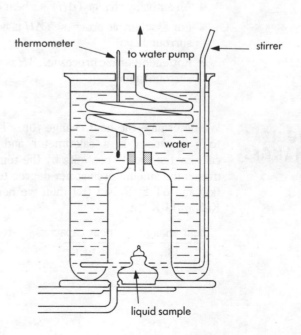

Fig. 6.2 Measuring heat of combustion at constant pressure

When my students use this apparatus to measure the enthalpy changes of combustion of some alcohols, they use propan-2-ol to calibrate the heat capacity.

■ A flame calorimeter measures heat of combustion at constant pressure

It is quite difficult to adjust the flow of air and the height of the burner's wick to give a steady, smokeless flame and prevent any incomplete combustion at all.

SIMPLE INSULATED CALORIMETERS

In your practical work you may use a glass vacuum flask or just a polystyrene drinking cup to measure heat changes involving solutions: see Fig. 6.3. Suppose you react hydrochloric acid with aqueous sodium hydroxide in the flask and measure the rise in temperature of the resulting aqueous sodium chloride. You could find the heat capacity of the vacuum flask and its contents by measuring the electrical energy needed to raise the temperature of the aqueous sodium chloride by the same number of degrees.

If you carry out the reaction in the polystyrene cup, you usually ignore its heat capacity because the polystyrene is such a good insulator and the mass of the cup is so small. Assume the aqueous sodium chloride has the same specific heat capacity as water $(4.2 J K^{-1} g^{-1})$ and calculate the heat capacity by multiplying 4.2 $(J K^{-1} g^{-1})$ by the mass (g) of the sodium chloride solution.

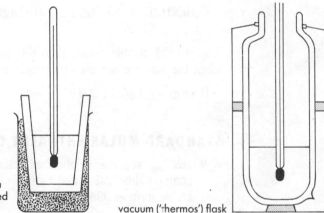

Fig. 6.3 Measuring enthalpy changes

polystyrene cup lagged with cotton wool and supported in a glass beaker

vacuum ('thermos') flask

MOLAR HEAT CHANGE

The amount of substance we burn in bomb and flame calorimeters or use in solutions in insulated calorimeters is usually quite small and much less than one mole. It should be obvious that the more of any one substance we burn, the more heat will be given out. Consequently we usually use the heat change we have found for a measured but small amount of reaction to calculate the molar heat change: see Fig. 6.4.

■ The molar heat of combustion is the heat evolved when one mole of a substance is completely burnt in oxygen

In some cases the term 'molar' refers to the 'amount of change' specified by the chemical equation representing the amounts of substances involved in the chemical change. Molar heat change is an imprecise term because it does not distinguish between heat change at constant pressure (ΔH) and at constant volume (ΔU). *For A-level chemistry the emphasis is upon molar enthalpy changes.*

When 50.0 cm³ of HCl(aq) and 50.0 cm³ of NaOH(aq), each of concentration 1.0 mol dm⁻³, are mixed in a polystyrene cup there is a temperature rise of 6.5 K. Calculate the molar heat of neutralisation.

The mass of the 100 cm³ of NaCl(aq) will be about 103 g
⇒ The heat capacity will be 4.2 × 103 ≈ 433 J K⁻¹
   50.0 cm³ of 1.0 mol dm⁻³ HCl(aq) contains 0.05 mol HCl
   0.05 mol HCl gives out 433 × 6.5 = 2.86 × 10³ J of heat
⇒ 1 mol HCl would give out 2.86 × 10³ × 1/0.05 = 57,200 J of heat
⇒ The molar heat of neutralisation is 57.2 kJ mol⁻¹ (cf HCl)

Fig. 6.4 Calculating a molar heat change

STANDARD MOLAR ENTHALPY CHANGE

Molar enthalpy change is more precise than molar heat change but it is still not precise enough because the ΔH value for a reaction may vary with the temperature, pressure, concentration and physical state of the reactants and products. Consequently we usually standardise these conditions and record in data books standard molar enthalpy changes ($\Delta H^{\ominus}$).

- $\Delta H_{298}^{\ominus}$ represents a standard molar enthalpy change for a process in which all substances are in their most stable forms at a pressure of 1 atm (101 kPa) and a temperature of 298 K (25 °C) and the concentration of any solution is 1 mol dm^{-3}

STANDARD MOLAR ENTHALPY CHANGE OF COMBUSTION $\Delta H_{c,298}^{\ominus}$

- $\Delta H_{c,298}^{\ominus}$ represents the heat change at constant pressure when one mole of substance (at 298 K and 1 atm) is completely burnt in oxygen to form products (at 298 K and 1 atm)

Values of standard enthalpies of combustion of many elements and compounds are listed in data books. They are particularly important for organic compounds where complete combustion of, say, $C_4H_9NH_2(l)$ means the formation of $CO_2(g)$, $H_2O(l)$ and $N_2(g)$. Although the burning will take place at a temperature well above 298 K, the hot products cool down to 298 K after the combustion. The heat they release on cooling is inlcuded in the 'heat of combustion':

$$C_4H_9NH_2(l) \; + \; 6\tfrac{3}{4}O_2(g) \; \rightarrow \; 4CO_2(g) \; + \; 5\tfrac{1}{2}H_2O(l) \; + \; \tfrac{1}{2}N_2(g);$$
$$\Delta H_{c,298}^{\ominus} \; = \; -3018 \, \text{kJ mol}^{-1}$$

The 3018 kJ of heat released to the surroundings includes $5\tfrac{1}{2} \times 41$ kJ of heat released when the water condenses to a liquid because:

$$H_2O(g) \; \rightarrow \; H_2O(l); \; \Delta H^{\ominus} \; = \; -41 \, \text{kJ mol}^{-1}$$

STANDARD MOLAR ENTHALPY CHANGE OF FORMATION $\Delta H_{f,298}^{\ominus}$

- $\Delta H_{f,298}^{\ominus}$ represents the heat change at constant pressure when one mole of a compound (at 298 K and 1 atm) is formed from its constituent elements in their most stable form at 298 K and 1 atm

These standard enthalpies of formation for both organic and inorganic compounds are also listed in data books. Notice the connection between the combustion of some elements and the formation of their oxides: e.g.

$$C(\text{graphite}) \; + \; O_2 \; \rightarrow \; CO_2(g); \; \Delta H_{298}^{\ominus} \; = \; -393.5 \, \text{kJ mol}^{-1}$$

The value of -393.5 kJ represents the standard enthalpy change of combustion of one mole of carbon in its most stable form of graphite and it is the standard enthalpy change of formation of one mole of carbon dioxide from its constituent elements in their standard states. From this you might guess that we may obtain the enthalpies of formation of oxides by direct measurement. But how do we measure the heat of formation of a compound (e.g. CH_4) whose elements do not undergo direct reaction? And why are $\Delta H_{f,298}^{\ominus}$ values so important that we list them in data books?

HESS'S LAW

Thermodynamics is the branch of science that deals with energy changes, especially heat changes, taking place in the course of physical and chemical processes. According to the first law of thermodynamics:

- Energy cannot be created and cannot be destroyed

66 Calculations using this law are very popular in examinations. 99

Hess's law of constant heat summation is a particular case of the more general first law of thermodynamics. It was first put forward by Germain Henri Hess in 1840. Since then it has been expressed in many different ways.

- The standard molar enthalpy change of a process is independent of the means or route by which that process takes place

We use Hess's law and, therefore, the first law of thermodynamics to calculate enthalpy changes that we could not determine by a direct measurement: see Fig. 6.5. You may use

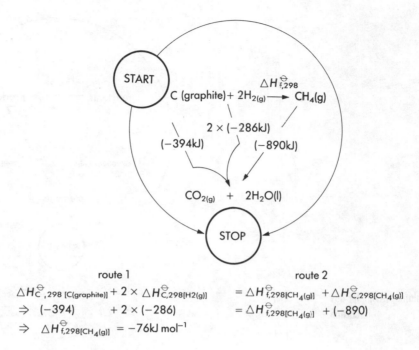

Fig. 6.5 Calculating $\Delta H_{f,298}^{\ominus}$ using Hess's law

route 1

$\Delta H_{C,298\ [C(graphite)]}^{\ominus} + 2 \times \Delta H_{C,298[H2(g)]}^{\ominus}$

$\Rightarrow \quad (-394) \qquad + 2 \times (-286)$

$\Rightarrow \quad \Delta H_{f,298[CH_4(g)]}^{\ominus} = -76\text{kJ mol}^{-1}$

route 2

$= \Delta H_{f,298[CH_4(g)]}^{\ominus} + \Delta H_{C,298[CH_4(g)]}^{\ominus}$

$= \Delta H_{f,298[CH_4(g)]}^{\ominus} + (-890)$

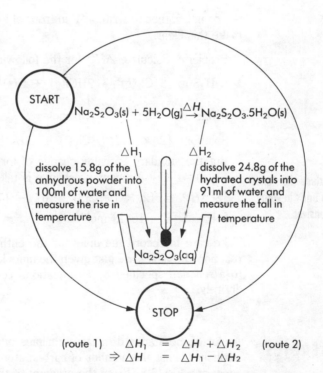

Fig. 6.6 Using Hess's law in a practical experiment

(route 1) $\quad \Delta H_1 = \Delta H + \Delta H_2 \quad$ (route 2)

$\Rightarrow \Delta H = \Delta H_1 - \Delta H_2$

Hess's law in your practical work to determine the enthalpy change of a reaction that cannot be measured directly: see Fig. 6.6. Notice that in each case I choose the start and finish of the routes so that they follow the same direction of the arrows for each step in the process.

■ Always put the sign with the numerical value of the enthalpy change; even for an endothermic process having a positive sign and value

STANDARD MOLAR ENTHALPY CHANGE OF REACTION $\Delta H_{r,298}^{\ominus}$

■ $\Delta H_{r,298}^{\ominus}$ represents the standard enthalpy change for a reaction represented by a stoichiometric chemical equation specifying the amounts of reactants and products involved in the reaction

We can use Hess's law and standard molar enthalpy changes of formation of compounds to calculate the standard molar enthalpy change of any reaction. And we can do this even for a purely hypothetical reaction that may never occur: see Fig. 6.7. Notice that the sign 'Σ' (Greek letter sigma) means the sum of all the molar enthalpy changes, each multiplied by the appropriate number of moles as shown by the equation. Remember that in equations

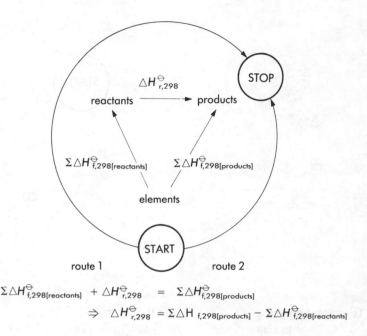

Fig. 6.7 Calculating $\Delta H_r^\ominus$ for any reaction

$$\Sigma \Delta H_{f,298[\text{reactants}]}^\ominus + \Delta H_{r,298}^\ominus = \Sigma \Delta H_{f,298[\text{products}]}^\ominus$$

$$\Rightarrow \Delta H_{r,298}^\ominus = \Sigma \Delta H_{f,298[\text{products}]}^\ominus - \Sigma \Delta H_{f,298[\text{reactants}]}^\ominus$$

we do not bother to write a '1' in front of a formula. The following simple example should make this clear.

Problem: Calculate $\Delta H_r^\ominus$ for the following reaction:

$$H_2S(g) + Cl_2(g) \rightarrow 2HCl(g) + S(s)$$
$$\uparrow \qquad\qquad\qquad\qquad \uparrow$$
one mole   two moles

Solution: $\{2 \times \Delta H_f^\ominus[HCl(s)] + \Delta H_f^\ominus[Cl_2(g)]\} - \{\Delta H_f^\ominus[H_2S(g)] + \Delta H_f^\ominus[S(s)]\}$

But the standard enthalpy change of formation of any element must be zero because $Cl_2(g) \rightarrow Cl_2(g)$ means no change is taking place.

> **Always include the sign and the units in your answer.**

$$\Rightarrow \Delta H_r^\ominus = \{2 \times (-92.3) + 0\} - \{1 \times (-20.6) + 0\}$$
$$= -164 \, kJ \, mol^{-1}$$

Take care to record the units for the enthalpy change as $kJ \, mol^{-1}$. In some textbooks, reaction enthalpies are just given the units kJ when the value of the enthalpy change refers to a reaction specified by an equation. You will certainly lose a mark if you omit units entirely.

BOND ENERGIES

A plot of the standard molar enthalpies of combustion of the lower alkanes and primary alcohols against the number of carbon atoms in their formulae gives two parallel straight lines: see Fig. 6.8. From the gradient of these lines you find that the difference in $\Delta H_c^\ominus$ from one homologue to the next is about $650 \, kJ \, mol^{-1}$ and corresponds to the combustion of a CH_2 group to form CO_2 and H_2O. This leads to the idea that 650 kJ represents the difference between the energy consumed to break the C—H bonds and those between the O-atoms in the O_2 molecules and the energy released when the C=O and H—O bonds form:

$$
\begin{array}{c}
H \\
| \\
-C- \\
| \\
H
\end{array}
+ 1\tfrac{1}{2}O{=}O(g) \rightarrow O{=}C{=}O(g) + H{-}O\diagdown_H
$$

This of course implies that there are definite 'bond energies' which must be provided to break bonds and which will be released when bonds form. Spectroscopic methods that are beyond A-level chemistry have been used to find the first and second standard bond dissociation enthalpies of water:

first   H—OH(g) $\rightarrow$ H(g) + OH(g); $\Delta H^\ominus = +498 \, kJ \, mol^{-1}$

second O—H(g) $\rightarrow$ O(g) + H(g); $\Delta H^\ominus = +430 \, kJ \, mol^{-1}$

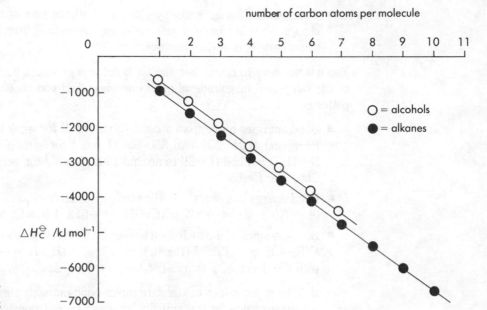

number of carbon atoms per molecule

○ = alcohols
● = alkanes

$\Delta H_c^{\ominus}$ /kJ mol^{-1}

Fig. 6.8 Variation in $\Delta H_c^{\ominus}$ for alkanes and alcohols

In each case energy must be provided. $\Delta H^{\ominus}$ is positive to break the bond between hydrogen and oxygen. But in the first case the bond is in a water molecule and in the second case it is in an hydroxyl free radical.

■ The precise strength of a bond (X—Y) between two atoms will depend upon the other atoms or groups of atoms attached to X and Y

If you look in a data book you will find the H—O bond energy listed as $+464$ kJ mol^{-1} which you may notice is $(498 + 430)/2$, the average value of the first and second standard bond dissociation enthalpies of water. Similarly, you will find the C—H bond energy for methane is 435 kJ mol^{-1} which is an average value of the standard dissociation enthalpies of the four bonds: namely, CH$_3$—H, CH$_2$—H, CH—H and C—H.

■ Bond energy is the *average* standard enthalpy change for the breaking of a mole of bonds in a gaseous molecule to form gaseous atoms

You will be expected to know that most bond energy values are average or mean values but you will not be expected to know anything about the experimental techniques to measure them. You should appreciate the use of Hess's law and standard molar enthalpy changes of atomisation of elements to calculate an average standard molar bond enthalpy: see Fig. 6.9.

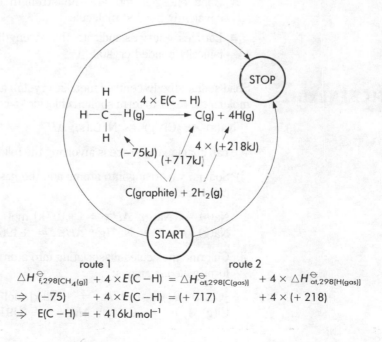

STOP

H
|
H—C—H(g) $\xrightarrow{4 \times E(C - H)}$ C(g) + 4H(g)
|
H

$4 \times (+218\text{kJ})$

(-75kJ) $(+717\text{kJ})$

C(graphite) + 2H$_2$(g)

START

route 1   route 2

Fig. 6.9 Calculating a mean bond enthalpy

$\Delta H^{\ominus}_{f,298[CH_4(g)]} + 4 \times E(C-H) = \Delta H^{\ominus}_{at,298[C(gas)]} + 4 \times \Delta H^{\ominus}_{at,298[H(gas)]}$

$\Rightarrow \quad (-75) \quad + 4 \times E(C-H) = (+717) \quad + 4 \times (+218)$

$\Rightarrow \quad E(C-H) = +416$ kJ mol^{-1}

- The standard molar enthalpy change of atomisation of an element is the enthalpy change when one mole of gaseous atoms are formed from the element under standard conditions. It is always positive.

You will not have to remember specific bond energy values but you should have some idea of the range and magnitude of bond energies. And you should remember certain simple patterns:

- Bond energies range from around $150\,\text{kJ mol}^{-1}$ for weak bonds (e.g. HO—OH, I—I, F—F and H_2N—NH_2) to 350–$550\,\text{kJ mol}^{-1}$ for strong bonds (e.g. C—C, C—H, N—H, O—H and H—Cl) to around $1000\,\text{kJ mol}^{-1}$ for very strong bonds (e.g. N≡N, C≡C and C≡O)

- Bond energy increases with the number of electron-pairs shared between two atoms (e.g. $E(\text{C—C}) = +347$, $E(\text{C=C}) = +612$, $E(\text{C≡C}) = +838\,\text{kJ mol}^{-1}$)

- Bond energies of the hydrogen halides decrease down the group ($E(\text{H—F}) = +568$, $E(\text{H—Cl}) = +432$, $E(\text{H—Br}) = +366$, $E(\text{H—I}) = +298\,\text{kJ mol}^{-1}$) in keeping with the decreasing thermal stability of the gases

You could be given values of standard molar bond enthalpy changes and asked to calculate an approximate value for the enthalpy change of a reaction: see Fig. 6.10.

Problem: Estimate the enthalpy change for the catalytic reforming of hexane to cyclohexane and hydrogen which is part of the process for the industrial production of petrol:
$C_6H_{14} \rightarrow C_6H_{12} + H_2$

Solution:

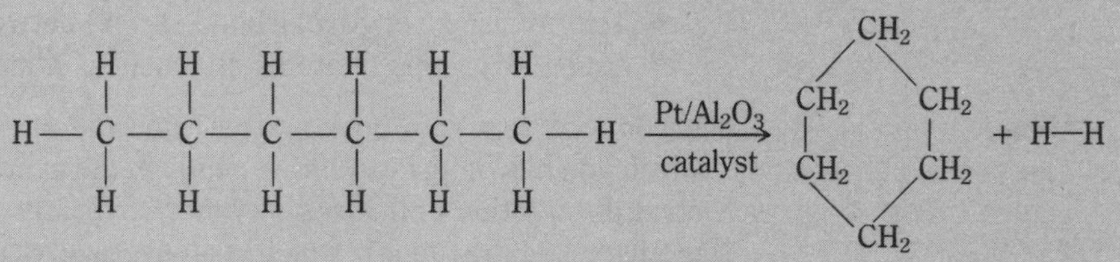

break 2 C—H bonds
⇒ 2 × 413 kJ used up
⇒ 826 kJ used up

form 2 C—C bond and 1 H—H bond
2 × 347 and 1 × 436 given out
1130 kJ given out

⇒ net release of 1130 − 826 = 304 kJ
⇒ $\Delta H_r^\ominus = -304\,\text{kJ mol}^{-1}$

Fig. 6.10 Using bond energies to estimate $\Delta H_r^\ominus$

- Bond energies indicate the strength of the forces holding together atoms in a covalently bonded molecule
- Lattices energies indicate the strength of the forces holding together ions in an ionically bonded crystal

LATTICE ENERGIES

Solid sodium (body-centred metallic crystal) and gaseous chlorine (simple diatomic molecules) react to form sodium chloride (face-centred ionic crystal):

$$Na(s) + \tfrac{1}{2}Cl_2(g) \rightarrow NaCl(s); \quad \Delta H_{f,298}^\ominus = -411\,\text{kJ mol}^{-1}$$

We can imagine the process involving the following steps:

i) Sodium vaporising into atoms and the gaseous atoms losing electrons to form gaseous cations:

$Na(s) \rightarrow Na(g); \quad \Delta H_{at}^\ominus = +107\,\text{kJ mol}^{-1}$ [atomisation energy]
$Na(g) \rightarrow e^- + Na^+(g); \quad \Delta H_i^\ominus = +496\,\text{kJ mol}^{-1}$ [ionisation energy]

ii) Chlorine molecules dissociating into atoms and the gaseous atoms gaining electrons to form gaseous anions:

$\tfrac{1}{2}Cl_2(g) \rightarrow Cl(g); \quad \Delta H_{at}^\ominus = +122\,\text{kJ mol}^{-1}$ [atomisation energy]
$Cl(g) + e^- \rightarrow Cl^-(g); \quad \Delta H_e^\ominus = -349\,\text{kJ mol}^{-1}$ [electron affinity]

iii) Gaseous sodium cations and gaseous chloride anions coming together to form a solid ionic crystal lattice:

$$Na^+(g) + Cl^-(g) \rightarrow NaCl(s); \quad \Delta H_l^\ominus = \text{the lattice energy}$$

This last imaginary or hypothetical step would be highly exothermic and the energy, if released as heat at constant pressure, would be called the molar lattice enthalpy change or, internal energy change, if the heat is released at constant volume. The reverse of this step would, of course, be highly endothermic.

> **"** Lattice energy depends upon the size and charge of the ions. **"**

■ Lattice energy may be defined as the heat given out when one mole of an ionic crystalline solid such as sodium chloride forms from its constituent ions in the gaseous state: $Na^+(g) + Cl^-(g) \rightarrow NaCl(s)$

We can draw an energy cycle and use Hess's law to calculate a value for the lattice energy: see Fig. 6.11. This cycle is usually called a 'Born-Haber' cycle. It is very instructive to present energy cycles like the Born-Haber cycle in such a way that the various energy changes are shown against a vertical scale with the arrows of endothermic processes pointing upwards and those for exothermic processes pointing downwards: see Fig. 6.12.

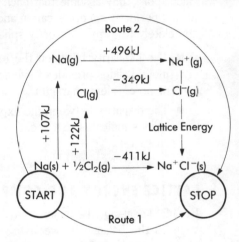

Route 1 = Route 2
(−411) = (+ 107) + (+ 496) + (−122) + (−349) + $\Delta H_l^\ominus$
⇒ $\Delta H_l^\ominus$ = (−411) − (+107) − (+496) − (+122) − (−349)
= −787 kJ mol⁻¹

Fig. 6.11 Calculating a lattice energy

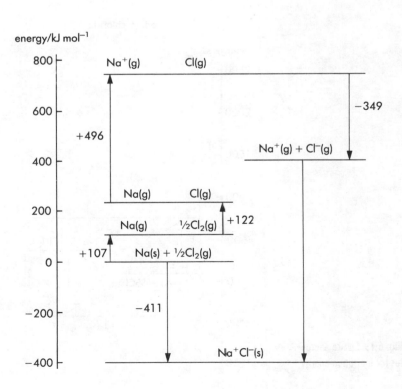

Fig. 6.12 Born-Haber cycle for sodium chloride

You will certainly not have to remember specific lattice energy values but you should have some idea of the range and magnitude of lattice energies in comparison with bond energies:

■ Lattice energies range from around 600 kJ mol⁻¹, beyond the strength of strong covalent bonds, to above 4000 kJ mol⁻¹, well beyond the strongest covalent bonds

You should remember and be able to suggest explanations for certain patterns in the values of lattice energies:

■ Lattice energies for doubly-charged ions are usually much more exothermic than those for singly-charged ions: $MgCl_2$ −2526; NaCl −787; Na_2O −2478 kJ mol⁻¹

■ Lattice energies become less exothermic as the size of the anion increases: LiF −1031; LiCl −848; LiBr −803; LiI −759 kJ mol⁻¹

■ Lattice energies become less exothermic as the size of the cation increases: LiF −1031; NaF −918; KF −817; RbF −783; CsF −747 kJ mol⁻¹

You should also appreciate that chemists have calculated completely theoretical values for the lattice energies of the alkali metal halides and silver halides. In these theoretical calculations they assume that each metal atom loses one or two electrons completely to form a perfectly spherical cation and that each non-metal atom gains one or two electrons completely to form a perfectly spherical anion.

■ The similarity between the experimental and theoretical lattice energies for alkali metal halides provides evidence that these compounds form ionic crystals with little or no covalent character

■ The disparity between the experimental and theoretical lattice energies for the silver halides indicates that these compounds have a considerable percentage of covalent character

LATTICE ENERGY AND COMPOUND FORMATION

Broadly speaking, the major contribution to the heat of formation of a compound may be seen as the balance between the energy consumed by the ionisation of the metal to gaseous cations and the energy released by the formation of the lattice. If you compare the Born-Haber cycles for NaCl and $MgCl_2$ you will see that the ionisation of Mg to Mg^+ and Mg^{2+} consumes far more energy than the ionisation of Na to Na^+ but that the $MgCl_2$ lattice energy is much more negative than the NaCl lattice energy. Consequently the standard molar enthalpy change of formation of $MgCl_2(s)$ is more negative than that of NaCl(s): see Fig. 6.13.

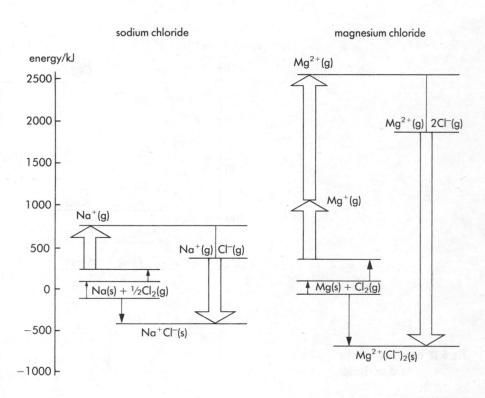

Fig. 6.13 Lattice energy compensates for ionisation energy

We can use this balance between ionisation energies and lattice energies to explain why MgCl(s) annd MgCl$_3$(s) do not exist. The heat of formation of MgCl would only be about -90 kJ mol^{-1} so we should expect this imaginary substance to disproportionate to form MgCl$_2$ (with its more exothermic lattice energy) and release more energy to achieve greater stability:

$$2\text{MgCl(s)} \rightarrow \text{Mg(s)} + \text{MgCl}_2\text{(s)}; \quad \Delta H = -470 \text{ kJ mol}^{-1}$$

The formation of MgCl$_3$ would be so endothermic ($\Delta H = +4000$ kJ mol^{-1}, because the ionisation of Mg^{2+} to Mg^{3+} involves a new inner electron shell and requires about 7700 kJ mol^{-1} of energy) that the compound would be too unstable to exist.

LATTICE ENERGY AND HEATS OF SOLUTION

When ionic substances dissolve in water the process may be exothermic or endothermic and the heat change is usually quite small because it is seen as the difference between two large energy values – the lattice energy and the hydration energy of the ions: see Fig. 6.14.

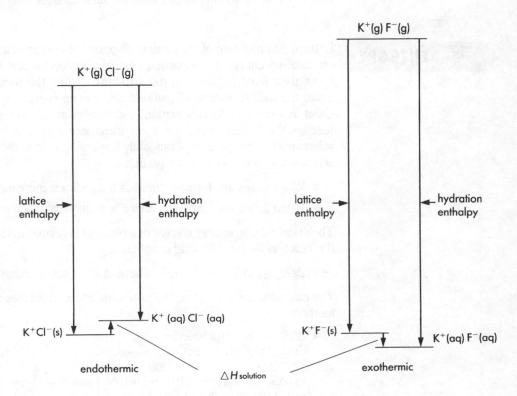

$\Delta G^{\ominus}$ AND ENERGETIC FEASIBILITY

Most chemical reactions are exothermic. The energetically unstable mixture of reactants forms an energetically stable mixture of products and energy is released to the surroundings. These reactions take place as soon as we put the reactants together unless a high activation energy barrier makes the mixture kinetically stable and the reaction very slow.

■ A spontaneous reaction is the reaction of an energetically unstable mixture even if the mixture is kinetically stable

The word 'spontaneous' means 'of its own accord'. It does NOT mean 'fast'. There are fast spontaneous reactions, slow spontaneous reactions and even spontaneous reactions that we expect will never occur. A spontaneous reaction is a reaction that could occur without having to be driven.

■ Most spontaneous chemical reactions are exothermic

Some chemical reactions are spontaneous but endothermic! For example:

$$\text{HCl(aq)} + \text{NaHCO}_3\text{(aq)} \rightarrow \text{NaCl(aq)} + \text{H}_2\text{O(l)} + \text{CO}_2\text{(g)}; \quad \Delta H^{\ominus} = +12 \text{ kJ mol}^{-1}$$

$\Delta H^{\ominus}$ is not always a reliable indication of the energetic instability of a reaction mixture.

However, another thermodynamic quantity is: this is the standard free energy change of a reaction, $\Delta G^{\ominus}$.

- A reaction will be spontaneous (energetically feasible) if its standard molar free energy change ($\Delta G^{\ominus}$) is negative

CALCULATING $\Delta G^{\ominus}$

You can calculate $\Delta G^{\ominus}$ in at least two ways.

- $\Delta G^{\ominus} = \Sigma\,\Delta G^{\ominus}(\text{products}) - \Sigma\,\Delta G^{\ominus}(\text{reactants})$

This is similar to the method you use for calculating $\Delta H^{\ominus}$: e.g.

$$H^+(aq)\ +\ HCO_3^-(aq)\ \rightarrow\ H_2O(l)\ +\ CO_2(g)$$

$\Delta H_f^{\ominus}/\text{kJ mol}^{-1}$	0	-692	-286	$-394 \Rightarrow \Delta H^{\ominus} = +12\,\text{kJ mol}^{-1}$
$\Delta G_f^{\ominus}/\text{kJ mol}^{-1}$	0	-587	-286	$-394 \Rightarrow \Delta G^{\ominus} = -93\,\text{kJ mol}^{-1}$

Notice that although this reaction actually takes in heat from the surroundings, it also releases a large volume of gas into the surroundings.

ENTROPY

Entropy is a measure of the random dispersal of energy of a system. When heated, a solid can take up energy as its particles (atoms, molecules and ions) vibrate more vigorously about their fixed positions in the structure. When the temperature reaches the melting point, the solid turns into a liquid that can take up energy by its particles vibrating, moving about in a limited way (translation) and turning around (rotation). When the temperature reaches the boiling point, the liquid turns into a gas. It is in the gaseous state that a substance is most capable of randomly dispersing heat in the form of vibrational, rotational and translational energy of its particles.

- When gases are formed, there is a significant increase in entropy
- When gases are 'used up', there is a significant decrease in entropy

The standard free energy change of a reaction is related to the standard enthalpy change of the reaction by the following equation:

- $\Delta G^{\ominus} = \Delta H^{\ominus} - T.\Delta S^{\ominus}$ where $\Delta S^{\ominus}$ is the standard molar energy change

You can calculate $\Delta S^{\ominus}$ from the entropies of the reactants and products by the following method:

- $\Delta S^{\ominus} = \Sigma\,S^{\ominus}(\text{products}) - \Sigma\,S^{\ominus}(\text{reactants})$
  $CoCl_2.6H_2O(s)\ +\ 6SOCl_2(l)\ \rightarrow\ CoCl_2(s)\ +\ 12HCl(g)\ +\ 6SO_2(g)$
  $S^{\ominus}$ 343 6×308 109 12×187 $6 \times 248\ \text{J mol}^{-1}\,\text{K}^{-1}$
  $\Rightarrow \Delta S^{\ominus}$ is $3841 - 2191 = +1650\,\text{J mol}^{-1}\,\text{K}^{-1}$ ($= 1.65\,\text{kJ mol}^{-1}\,\text{K}^{-1}$)
  and $\Delta H^{\ominus} = +392\,\text{kJ mol}^{-1}$ (calculated in the usual way)
  $\Rightarrow \Delta G^{\ominus}$ is $\{+392 - (298 \times 1.65)\} = -100\,\text{kJ mol}^{-1}$

Notice that here again is an endothermic reaction taking in heat whilst giving out a large volume of gas.

If we multiply the equation $\Delta G^{\ominus} = \Delta H^{\ominus} - T.\Delta S^{\ominus}$ by $-1/T$ it becomes:

$$-\frac{\Delta G^{\ominus}}{T} = -\frac{\Delta H^{\ominus}}{T} + \Delta S^{\ominus}_{\text{system}}$$

but $-\Delta H^{\ominus}/T$ is the standard molar entropy change of the surroundings ($\Delta S^{\ominus}_{\text{surroundings}}$), so $-\Delta G^{\ominus}/T$ is the total standard molar entropy change (of the chemical reaction system and its surroundings). If the value of $\Delta G^{\ominus}$ is negative, then $-\Delta G^{\ominus}/T$ is positive and $\Delta S^{\ominus}_{\text{total}}$ is positive:

- $\Delta S^{\ominus}_{\text{total}} = \Delta S^{\ominus}_{\text{surroundings}} + \Delta S^{\ominus}_{\text{system}}$

According to the second law of thermodynamics:

- the total entropy always increases if a spontaneous reaction occurs

It is from this second law of thermodynamics that we get the rule that $\Delta G^{\ominus}$ must be negative for a reaction to be spontaneous or energetically feasible.

EXAMINATION QUESTIONS AND OUTLINE ANSWERS

Q1 a) A gaseous oxide of chlorine, when heated, decomposes completely into chlorine gas and oxygen gas. It is found experimentally that two volumes of the oxide give two volumes of chlorine and one volume of oxygen, all volumes being measured under the same conditions of pressure and temperature. What law does this illustrate? What can you deduce about the formula of the oxide of chlorine? Indicate your reasoning, and name and state any further law you assume. *(7)*

b) i) Define *bond enthalpy*. *(2)*

ii) The standard enthalpy of formation of ammonia is $-46.0\,\text{kJ mol}^{-1}$ and that of hydrazine is $+42.0\,\text{kJ mol}^{-1}$. The enthalpy of formation of hydrogen from its atoms is $-436\,\text{kJ mol}^{-1}$ and that of nitrogen is $-712\,\text{kJ mol}^{-1}$.

Calculate the average bond enthalpies of the N—H and N—N single bonds and name the law you use. The structural formula of hydrazine is

(7)

c) Explain how the following data may be used to deduce information about the structure and stability of the benzene molecule.

C—H bond enthalpy: $410\,\text{kJ mol}^{-1}$
C—C bond enthalpy: $345\,\text{kJ mol}^{-1}$
C=C bond enthalpy: $610\,\text{kJ mol}^{-1}$
Enthalpy of formation of gaseous benzene from its gaseous isolated atoms: $-5530\,\text{kJ mol}^{-1}$. *(4)*

(ODLE 1988)

Q2 a) i) State *Hess's Law*. *(3)*

ii) Define the term *standard molar enthalpy change of formation*. *(3)*

b) Describe, giving brief details, a laboratory experiment to determine the enthalpy change of a named reaction. *(4)*

c) The following is a table of some standard molar enthalpy changes of formation.

Substance	$\Delta H_f^{\ominus}\,(298\,\text{K})/\text{kJ mol}^{-1}$
ClF(g)	-63.4
Cl(g) [i.e. $\frac{1}{2}Cl_2(g) \rightarrow Cl(g)$]	121.7
F(g) [i.e. $\frac{1}{2}F_2(g) \rightarrow F(g)$]	79.0

Use these values to calculate the molar bond dissociation energy of Cl—F. *(5)*

d) The standard molar enthalpy changes of combustion of red phosphorus and white phosphorus are $-735\,\text{kJ mol}^{-1}$ and $-753\,\text{kJ mol}^{-1}$ respectively for 1 mol of phosphorus atoms. In both cases phosphorus(V) oxide is the product. Calculate the standard molar enthalpy change of formation of white phosphorus from red phosphorus. All values are at 298 K.

State what effect an increase in temperature will have on the relative stabilities of red and white phosphorus. *(5)*

(WJEC 1988)

Outline answers

Q1

(a) Gay-Lussac's Law. (see text page 51)

$$Cl_xO_y(g) \longrightarrow \frac{x}{2}Cl_2(g) + \frac{y}{2}O_2(g)$$

2 vol   2 vol   1 vol

Avogadro's law

2 mol   2 mol   1 mol

$$\Rightarrow 1 : \frac{x}{2} = 2:2 \qquad \frac{x}{2} : \frac{y}{2} = 2:1$$

$\Rightarrow x = 2$ and $y = 1$

$\Rightarrow$ Formula is Cl_2O

(b)(i) See text page 71

(ii) $\frac{1}{2}N_2 + 1\frac{1}{2}H_2 \xrightarrow{-46.9} NH_3$

$$\left(\frac{-712}{2}\right) + \nwarrow \frac{3 \times (-436)}{2} \quad / \quad 3 \times E(N-H)$$

$$N + 3H$$

$$\Rightarrow E(N-H) = +352 \text{ kJ mol}^{-1}$$

$$N_2 + 2H_2 \xrightarrow{+42.0} N_2H_4$$

$$(-7/2) \nwarrow 2 \times (-436) \quad / \quad 4 \times (+352) + E(N-N)$$

$$2N + 4H$$

$$\Rightarrow E(N-N) = +218 \text{ kJ mol}^{-1}$$

(c) If benzene ring contained 3 single and 3 double bonds then enthalpy of formation from gaseous atoms would be $6 \times 410 + 3 \times 345 + 3 \times 610$

$$= -5325 \text{ kJ mol}^{-1}$$

But value is -5530 kJ mol^{-1}

So an extra 205 kJ of energy is released making molecule more stable.

Q2

(a) See text page 68

(b) " " " 68

(c)

$$Cl-F(g) \xrightarrow{\Delta H} Cl(g) + F(g)$$

$$(-63.4) \nwarrow \quad \nearrow \quad (+121.7) + (+79.0)$$

$$\frac{1}{2}Cl_2(g) + \frac{1}{2}F_2(g)$$

$$\Delta H = (+121.7) + (+79.0) - (-63.4)$$

$$= +264.1 \text{ kJ mol}^{-1}$$

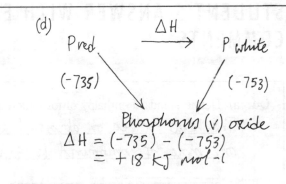

(d)

$$\Delta H$$

Pred $\xrightarrow{\hspace{2cm}}$ P white

(−735)   (−753)

Phosphorus (v) oxide

$$\Delta H = (-735) - (-753)$$
$$= +18 \, kJ \, mol^{-1}$$

white more stable at higher temperature
(Le chateliers principle − endothermic change
favoured at higher temperatures)

EXAMINATION QUESTION AND TUTOR'S ANSWER

Q3 The following sequence of reactions can be constructed for the formation of sodium chloride from its elements.

i)
$$Na(s) + \tfrac{1}{2}Cl_2(g) = Na(g) + \tfrac{1}{2}Cl_2(g) \qquad \Delta H_1 = +108 \, kJ \, mol^{-1}$$
$$Na(g) + \tfrac{1}{2}Cl_2(g) = Na^+(g) + Cl^-(g) \qquad \Delta H_2 = +257 \, kJ \, mol^{-1}$$
$$Na^+(g) + Cl^-(g) = NaCl(s) \qquad \Delta H_3 = -766 \, kJ \, mol^{-1}$$

a) Explain what you understand by Hess' Law

The total heat change for a reaction is the same and does not depend upon the route.

(2)

b) What is the significance of the negative and positive values of the enthalpy changes?

Negative means heat is lost to the surroundings. Positive means heat is gained from the surroundings.

(2)

c) Name the enthalpy change that has occurred in equation i).

Atomisation of sodium

(1)

d) Calculate the standard molar enthalpy of formation of sodium chloride.

$$\Delta H_1 + \Delta H_2 + \Delta H_3$$
$$(+108) + (+257) + (-766)$$
$$= -401 \, kJ \, mol^{-1}$$

(3)

e) Write an equation for the reaction which would give the ionisation energy of sodium.

$$Na \, (g) \rightarrow Na^+(g) + e^-$$

(2)

f) Write an equation for the reaction which would give energy of electron affinity of chlorine.

$$Cl \, (g) + e^- \rightarrow Cl^-(g)$$

(2)

(UCLES (AS-level specimen))

STUDENT'S ANSWER WITH EXAMINER'S COMMENTS

Q4 a) Define standard enthalpy of formation $\Delta H_f^{\ominus}$ of a compound.

66 **At 298K and 1 atm.** 99

The heat evolved or absorbed at constant pressure when one mole of compound is formed from its elements.

66 **In their most stable form.** 99

b) When ethanol burns in oxygen, carbon dioxide and water are formed.
   i) Write the equation which describes this reaction.

$$C_2H_5OH_{(l)} + 3O_2(g) \rightarrow 2CO_2(g) + 3H_2O(l)$$

   ii) Using the data
       $\Delta H_f^{\ominus}$ for ethanol (l)   = -277.0 kJ mol^{-1}
       $\Delta H_f^{\ominus}$ for carbon dioxide (g) = -393.7 kJ mol^{-1}
       $\Delta H_f^{\ominus}$ for water (l)   = -285.9 kJ mol^{-1}
       calculate the value of $\Delta H^{\ominus}$ for the combustion of ethanol.

66 **Take care with signs.** 99

$$\Delta H_c^{\ominus}(ethanol) = 2\times(-393.7) + 3\times(-285.9) - (-277.0)$$
$$= -1368.1 \; kJ \; mol^{-1}$$

(JMB (AS-level specimen))

66 **Units must be included for full marks.** 99

PRACTICE QUESTION

Q5 a) i) Define 'lattice energy' and write an equation for the change which represents the lattice energy of silver iodide. (State symbols must be shown.)   (3)

   ii) The theoretical value of the lattice energy of silver iodide, calculated from the attractive forces expected between silver ions and iodide ions in a lattice is -736 kJ mol^{-1}, while the value calculated from experimental data is -845 kJ mol^{-1}. Compare these values and give an explanation of the differences.   (2)

   iii) Predict an order for the values of the lattice energies of the compounds magnesium chloride, potassium iodide, sodium fluoride and sodium chloride, placing that with the largest negative value first. Briefly justify the order you give.   (3)

(WESSEX (specimen))

GETTING STARTED

The law of chemical equilibrium is of central importance to chemistry. It leads to the equilibrium constants, K_c and K_p and to an understanding of the optimum conditions for industrially important reversible reactions. For homogeneous equilibria in aqueous solution, the law leads to K_a, K_b, K_w and an understanding of acids, bases, salts, buffer solutions and indicators. It also leads to stability constants and an understanding of ligand exchange reactions for complex ions. For heterogeneous equilibria involving sparingly soluble salts, the law leads to the solubility product, K_{sp}, and to an understanding of ionic precipitation reactions.

CHAPTER 7

CHEMICAL EQUILIBRIA

COMPOSITION OF AN EQUILIBRIUM SYSTEM

LE CHATELIER'S PRINCIPLE

AQUEOUS ACID-BASE EQUILIBRIA

STRONG AND WEAK ACIDS AND BASES

ACID BASE STRENGTH AND PROTON TRANSFER

HYDROLYSIS OF SALTS

BUFFERS

INDICATORS

LIGAND EXCHANGE REACTIONS

IONIC PRECIPITATION REACTIONS

THE LAW OF
CHEMICAL
EQUILIBRIUM

ESSENTIAL PRINCIPLES

Equilibrium (Latin: aequus – equal, libra – balance) is the condition of constant properties that a system of one or more substances achieves when opposing kinetic molecular processes occur at exactly equal balancing rates. We usually call it a dynamic equilibrium and we measure intensive properties such as pressure, concentration and temperature to see if a system has reached equilibrium.

A physical or phase equilibrium involves changes of state of the same substances and is only heterogeneous. You will meet this if you have to deal with, for example, the distribution law and solvent extraction, Dalton's law of partial pressures and steam distillation, Raoult's law and fractional distillation, colligative properties and the determination of molar mass.

Chemical equilibria involve reversible reactions and may be homogeneous (one phase) or heterogeneous (two or more phases). Many chemical reactions are reversible but the favourite example in textbooks and examinations is the acid-catalysed hydrolysis of an ester:

$$CH_3CO_2C_2H_5(l) + H_2O(l) \rightleftharpoons CH_3CO_2H(l) + C_2H_5OH(l)$$

Many chemical reactions that you have met in your GCSE course may seem to be irreversible. But what about the industrially important synthesis of ammonia?

$$N_2(g) + 3H_2(g) \rightleftharpoons 2NH_3(g); \Delta H_{298}^{\ominus} = -92.4 \text{ kJ mol}^{-1}$$

At A-level you will probably encounter the historically important synthesis of hydrogen iodide:

$$H_2(g) + I_2(g) \rightleftharpoons 2HI(g); \Delta H_{298}^{\ominus} = -2.9 \text{ kJ mol}^{-1}$$

You will certainly have to deal with aqueous acid-base reactions (proton transfer), redox reactions (electron transfer) and complex ion formation (ligand transfer). They form an important part of your A-level chemistry course and require an understanding of the law of chemical equilibrium.

WHAT IS THE LAW OF CHEMICAL EQUILIBRIUM?

It is a statement about the composition of a chemical system. Here is one way we can put it using equations and words:

If a reversible reaction represented by the chemical equation

$$aA + bB \rightleftharpoons cC + dD \qquad \text{N.B products on right-hand side}$$

is at equilibrium at a constant temperature, T, then

$$\frac{[C]^c \times [D]^d}{[A]^a \times [B]^b} \text{ has a constant value, } K_c. \qquad \text{N.B. products in numerator}$$

> The law of chemical equilibrium is extremely important.
>
> Make sure you understand it.

The letters A, B, C and D stand for the chemical formulae of the reactants and products that are in the aqueous or liquid state. a, b, c and d are called the stoichiometric coefficients because they represent the numbers of moles of reactants and products in the balanced chemical equation. [A], [B], [C] and [D] represent the concentrations of the reactants and products. If the chemicals are all gaseous, then

$$\frac{p_C^c \times p_D^d}{p_A^a \times p_B^b} \text{ has a constant value, } K_p.$$

where p_A^a, p_B^b, p_C^c and p_D^d represent the partial pressures of the reactants and products.

You will, of course, realise that the letter T stands for the value of the temperature. In the same way you need to realise that the letter K stands for the value of the constant. We usually refer to K_c or K_p as the equilibrium constant value for the function of the concentrations or partial pressures or just the equilibrium constant. If we use a subscript (eq) to show that the concentration or partial pressure values are the values at equilibrium, then we can write

$$K_c = \frac{[C]_{eq}^c \times [D]_{eq}^d}{[A]_{eq}^a \times [B]_{eq}^b}$$

$$K_p = \frac{p_{C,eq}^c \times p_{D,eq}^d}{p_{A,eq}^a \times p_{B,eq}^b}$$

- The actual value and units of the equilibrium constant, K_c or K_p, depend upon the balanced chemical equation for the reversible reaction: see Fig. 7.1.

T/°C	BALANCED CHEMICAL EQUATION	K_p	UNITS	K_c	UNITS
80	$N_2O_4(g) \rightleftharpoons 2NO_2(g)$	4	atm	0.138	$mol\,dm^{-3}$
80	$\frac{1}{2}N_2O_4(g) \rightleftharpoons NO_2(g)$	2	$atm^{\frac{1}{2}}$	0.371	$mol^{\frac{1}{2}}dm^{-1\frac{1}{2}}$
80	$NO_2(g) \rightleftharpoons \frac{1}{2}N_2O_4(g)$	1/2	$atm^{-\frac{1}{2}}$	2.69	$mol^{-\frac{1}{2}}dm^{-1\frac{1}{2}}$
80	$2NO_2(g) \rightleftharpoons N_2O_4(g)$	1/4	atm^{-1}	7.25	$mol^{-1}dm^3$
830	$\frac{1}{2}H_2(g) + \frac{1}{2}I_2(g) \rightleftharpoons HI(g)$	5	–	5	–
830	$H_2(g) + I_2(g) \rightleftharpoons 2HI(g)$	25	–	25	–
830	$2HI(g) \rightleftharpoons H_2(g) + I_2(g)$	1/25	–	0.04	–

Fig. 7.1 K depends upon the chemical equation

Remember that we read a chemical equation from left (reactants) to right (products) and notice that in the expression for K_c or K_p the products are on the top in the numerator and the reactants on the bottom in the denominator. So if we write the equation for the reverse reaction by writing the equation the other way around, we turn the expression for K_c or K_p upside down.

- The value of K_c or K_p for a reaction is the reciprocal of the value of K_c or K_p for its reverse reaction and the units alter accordingly

Notice also that for the system of hydrogen, iodine and hydrogen iodide, the value of the equilibrium constant has no units and $K_c = K_p$, because the number of moles of reactants and products involved in the equilibrium constant expression balance out.

- If the number of terms raised to their appropriate powers in the top (numerator) and bottom (denominator) of the equilibrium constant expression balance out, then $K_c = K_p$ and the value of the equilibrium constant is non-dimensional, i.e. has no units

VARIATION OF K_c OR K_p WITH TEMPERATURE

- The value represented by K_c or K_p for a given chemical system depends upon the temperature at which the system is being kept: see Fig. 7.2

K_p/atm:	0.115	3.89	47.9	347	1700	6030
T/K:	298	350	400	450	500	550

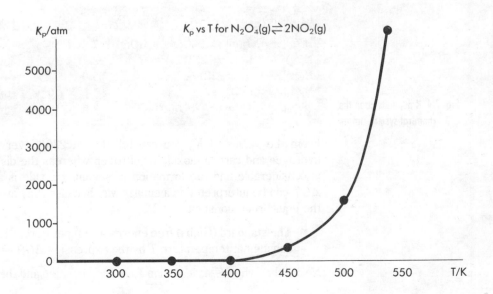

Fig. 7.2 K depends upon the temperature

If you warm some 'nitrogen dioxide' (prepared by decomposing lead(II) nitrate) you will see the gas become darker brown. If you cool the gas, you will see it become lighter brown in colour. The dissociation of colourless dinitrogen tetroxide into dark brown nitrogen dioxide is endothermic and reversible:

$$N_2O_4(g) \rightleftharpoons 2NO_2(g); \Delta H^\ominus = +58 \, kJ \, mol^{-1}$$

So when you raise the temperature, K_p increases in value. This means less colourless gas (N_2O_4), more dark gas (NO_2). Cooling makes the gas mixture lighter in colour because K_p for the dissociation decreases whilst K_p for the exothermic association, the reverse reaction increases:

$$2NO_2(g) \rightleftharpoons N_2O_4(g); \Delta H^\ominus = -58 \, kJ \, mol^{-1}$$

- K_c or K_p for the endothermic direction of a reaction will increase with increasing temperature

- K_c or K_p for the exothermic direction of a reaction will decrease with increasing temperature

Incidentally, a graph of $\log K_c$ or $\log K_p$ against the reciprocal of the absolute temperature is a straight line whose gradient gives a value for $-\Delta H^\ominus/R$ where R is the gas constant: see Fig. 7.3.

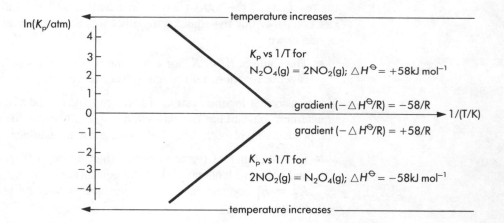

Fig. 7.3 log K related to temperature by $\Delta H^\ominus$

Notice that one graph is a reflection of the other because K_p for the forward reaction is the reciprocal of K_p for the reverse reaction and $\log (1/K_p)$ is the same as $-\log K_p$ so the gradients are equal but of opposite sign: i.e. $-58/R$ and $+58/R$.

VARIATION OF K_c OR K_p WITH THE CHEMICAL SYSTEM

- The value represented by K_c or K_p for a given temperature depends upon the chemical system being considered: see Fig. 7.4

CHEMICAL SYSTEM AT 500 KELVIN	K_p	UNITS	$\Delta H^\ominus_{500}/kJ$	$\Delta G^\ominus_{500}/kJ$
$H_2(g) + CO_2(g) \rightleftharpoons H_2O(g) + CO(g)$	7.8×10^{-3}	–	+41	+20
$N_2(g) + 3H_2(g) \rightleftharpoons 2NH_3(g)$	3.6×10^{-2}	atm^{-2}	−101	+14
$H_2(g) + I_2(g) \rightleftharpoons 2HI(g)$	25.0	–	−10	−13
$N_2O_4(g) \rightleftharpoons 2NO_2(g)$	1.7×10^3	atm	+57	−31
$2SO_2(g) + O_2(g) \rightleftharpoons 2SO_3(g)$	2.5×10^{10}	atm^{-1}	−200	−99

Fig. 7.4 K depends upon the chemical system chosen

From the values of K_p you can tell that at 230 °C very little reaction occurs between hydrogen and carbon dioxide or nitrogen whereas the dissociation of dinitrogen tetroxide is considerable and the formation of sulphur trioxide is almost complete. The values of $\Delta G^\ominus$ can be interpreted in a similar way because they are directly related to the values of the equilibrium constant.

- The standard (Gibbs) free energy at temperature T is related to the equilibrium constant at temperature T by the expression $\Delta G^\ominus = -RT \ln K$.

Notice that when K_p is less than 1, $\Delta G^\ominus$ is positive and the equilibrium favours reactants. And when K_p is greater than 1, $\Delta G^\ominus$ is negative and the equilibrium favours products:

- If $\Delta G^\ominus$ is negative, we say the reaction is energetically feasible
- If $\Delta G^\ominus$ is large and positive, say $> +60\,\text{kJ mol}^{-1}$, the equilibrium is strongly in favour of reactants
- If $\Delta G^\ominus$ is large and negative, say $< -60\,\text{kJ mol}^{-1}$, the equilibrium is strongly in favour of products

You can find out more about $\Delta G^\ominus$ and $\Delta H^\ominus$ in Chapter 6.

COMPOSITION OF AN EQUILIBRIUM SYSTEM

If we mix a fixed amount of hydrochloric acid, as catalyst, with ethanoic acid, ethanol, ethyl ethanoate and water in any known amounts, seal the mixture in a tube, leave it at a constant temperature to reach equilibrium and then analyse the contents by titrating with alkali, we find the value of the equilibrium constant, K_c is

$$\frac{[CH_3CO_2C_2H_5(l)]_{eq}[H_2O(l)]_{eq}}{[CH_3CO_2H(l)]_{eq}[C_2H_5OH(l)]_{eq}} \simeq 4 \text{ at about } 60\,°C$$

Problem: What amount of ester would be produced at equilibrium by mixing 1 mol of ethanol with a) 1 mol of ethanoic acid, and b) 2 mol of ethanoic acid?

Solution: The equation $CH_3CO_2H(l) + C_2H_5OH(l) \rightleftharpoons CH_3CO_2C_2H_5(l) + H_2O(l)$ tells us that one mole of ethanoic acid requires one mole of ethanol to produce one mole of ethyl ethanoate and one mole of water. If x mol of ester is produced at equilibrium, then x mol of water must also be produced and 1−x mol of unreacted ethanol must remain in the equilibrium mixture. In each case, x mol of ethanoic acid will have been used. Therefore the amount of ethanoic present at equilibrium in mixture a) must be 1−x mol and in mixture b) must be 2−x mol. If the volume of the system at equilibrium is V dm³, then the equilibrium concentrations of the components will be their equilibrium amounts divided by V: e.g. the equilibrium concentration of the ester will be x/V mol dm⁻³. We can substitute these values into the expression for the equilibrium constant:

mixture a):

$$\frac{\left(\frac{x}{V}\right) \times \left(\frac{x}{V}\right)}{\left(\frac{1-x}{V}\right) \times \left(\frac{1-x}{V}\right)} = 4$$

mixture b):

$$\frac{\left(\frac{x}{V}\right) \times \left(\frac{x}{V}\right)}{\left(\frac{2-x}{V}\right) \times \left(\frac{1-x}{V}\right)} = 4$$

These expressions simplify to

$\frac{x}{1-x} = 2$

or $3x = 2$

$\Rightarrow x = 2/3$

$\frac{x^2}{(2-x)(1-x)} = 4$

or $3x^2 - 12x + 8 = 0$

$\Rightarrow x = 0.85 \text{ (or 3.12)}$

Hence the amount of ethyl ethanoate at equilibrium would be:

mixture a) 0.67 mol   mixture b) 0.85 mol

The above problem illustrates the important practical point that the yield of ester from an alcohol may be increased by using an excess of the organic acid. Now compare the amounts of organic acid and alcohol at equilibrium in the two cases:

mixture a) 0.33 mol ethanoic acid and 0.33 mol ethanol
mixture b) 1.15 mol ethanoic acid and 0.15 mol ethanol

These values are simply related to the equilibrium concentrations because the volume term is constant and illustrate the following general point:

- If you increase the equilibrium concentration or partial pressure of one reactant the equilibrium values of the other reactants decrease

CHANGE OF COMPOSITION WITH CONCENTRATION

Suppose we have a mixture at 60 °C in which the amount of ethyl ethanoate = the amount of water = 0.67 mol and the amount of ethanoic acid = the amount of ethanol = 0.33 mol. The mixture will be at equilibrium because

$$\frac{[CH_3CO_2C_2H_5(l)][H_2O(l)]}{[CH_3CO_2H(l)][C_2H_5OH(l)]} = \frac{\left(\dfrac{0.67}{V}\right) \times \left(\dfrac{0.67}{V}\right)}{\left(\dfrac{0.33}{V}\right) \times \left(\dfrac{0.33}{V}\right)} = 4$$

This is the mixture obtained by mixing 1 mol of ethanoic acid and 1 mol of ethanol and allowing them to reach equilibrium at 60 °C. If we now add a further 1 mol ethanoic acid to this equilibrium mixture to make the amount of $CH_3CO_2H(l)$ = 1.33 mol, the mixture will no longer be at equilibrium because the value of $\left(\dfrac{0.67}{V}\right) \times \left(\dfrac{0.67}{V}\right) \div \left(\dfrac{1.33}{V}\right) \times \left(\dfrac{0.33}{V}\right)$ is about 1 and *not* equal to 4, the value of K_c. The concentrations of ester and water are too low and the concentrations of the acid and alcohol are too high for equilibrium. We say that the equilibrium has been disturbed.

When the mixture eventually attains the 'new' equilibrium, it will contain more ester (0.85 mol), more water (0.85 mol) and less ethanol (0.15 mol) than it did in the 'old' equilibrium mixture. Because some of the ethanol and added ethanoic acid has been converted to ester and water by the forward (left to right) reaction to restore the equilibrium, we often say that the equilibrium position has shifted from left to right and opposed the increase in acid concentration.

- If the conditions of a reversible reaction are altered to change the concentration of a component (reactant or product) and the equilibrium is disturbed, the composition may alter to restore equilibrium

CHANGE OF COMPOSITION WITH PRESSURE

When the reversible reaction involves gases we can usually alter the partial pressures of the components and the composition of the system by altering the volume. For the dissociation $N_2O_4(g) = 2NO_2(g)$ at about 80 °C the value of K_p = 4 atm. If a syringe at 80 °C contains nitrogen dioxide with a partial pressure of 1.0 atm and dinitrogen tetroxide with a partial pressure of 0.25 atm, the gas mixture will have a total pressure of 1.25 atm and be at equilibrium because

$$\frac{p_{NO_2}^2}{p_{N_2O_4}} = \frac{1.0 \times 1.0}{0.25} = 4.0\,atm$$

If we decrease the volume of the mixture from 100 to 50 cm^3, keeping the temperature constant, we should expect, according to Boyle's law, to double the total pressure and the partial pressures. If this were to happen, the 50 cm^3 of gas mixture would have a total pressure of 2.5 atm but not be at equilibrium. The reason is that $\dfrac{p_{NO_2}^2}{p_{N_2O_4}}$ would be 8.0 atm, not 4.0 atm (because p_{NO_2} = 2.0 atm and $p_{N_2O_4}$ = 0.5). Molecules of $NO_2(g)$ would have to associate to lower the partial pressure of nitrogen dioxide to 1.65 atm, and raise the partial pressure of the dinitrogen tetroxide to 0.68 atm in the 50 cm^3 of mixture in order to restore the equilibrium:

$$\frac{p_{NO_2}^2}{p_{N_2O_4}} = \frac{1.65 \times 1.65}{0.68} = 4.0\,atm$$

So when we compress the 100 cm^3 of equilibrium gas mixture at constant temperature to 50 cm^3 the total pressure increases from 1.25 atm to 2.33, and not 2.50 atm, because the equilibrium position of $N_2O_4(g) = 2NO_2(g)$ shifts from right to left.

- If the total number of moles of gaseous reactants is greater than that of the gaseous products, an increase in the total pressure is accompanied by a shift in the composition in favour of products

This is why the industrial synthesis of ammonia by the Haber process is carried out at 200 atm:

$$N_2(g) + 3H_2(g) \rightleftharpoons 2NH_3(g)$$

4 mol gaseous reactants   2 mol gaseous products

If we compress to half its volume an equilibrium mixture of hydrogen, iodine and hydrogen iodide gases at constant temperature:

> 66 K_c and K_p change with temperature. 99

$$H_2(g) + I_2(g) \rightleftharpoons 2HI(g),$$

the total pressure doubles since the ratio of the partial pressure stays the same.

> 66 They do NOT change with concentration or pressure. 99

- If the total number of moles of gaseous reactants is the same as that of the gaseous products, the equilibrium composition does not change when the total pressure changes

LE CHATELIER'S PRINCIPLE

The law of chemical equilibrium and the expression for the equilibrium constant deals *quantitatively* with the way the composition of an equilibrium system may change with temperature, concentration and pressure. Le Chatelier's principle deals *qualitatively* with these effects.

- If you alter the conditions of a reversible reaction and disturb the equilibrium, the composition of the mixture may change to restore the equilibrium and to minimise the effect of altering the conditions

Examiners frequently ask you to predict how a reaction at equilibrium would change with temperature, concentration or pressure and to give your reasoning. You should refer to Le Chatelier's principle and explain how it applies to the given reaction. Remember that the value of K_c or K_p for a given reaction does NOT change with concentration or pressure. It changes only with the temperature.

- For a given chemical system kept at a given constant temperature, the value represented by K_c or K_p is a constant

CALCULATIONS

You could be asked to do simple calculations involving K_c or K_p and the equilibrium composition. The problems fall into two types:

1) Calculation of a value for K_c or K_p from data on the equilibrium composition: see Fig. 7.5

Question: At 60 °C and a total pressure of 1 atmosphere dinitrogen tetroxide is 50.0% dissociated into nitrogen dioxide:

$$N_2O_4(g) \rightleftharpoons 2NO_2(g)$$

Calculate the equilibrium constant, K_p, at this temperature.

(UCLES 1988)

Answer: $N_2O_4(g) \rightarrow 2NO_2(g)$

           1 mol $\rightarrow$ 2 mol

so   0.5 mol $\rightarrow$ 1 mol

$\Rightarrow$ if 50% of 1 mol N_2O_4 dissociates the resulting gas mixture would contain 0.5 mol (undissociated) N_2O_4 and 1 mol NO_2, so the amounts of the two gases would be in the ratio 1 : 2

$\Rightarrow$ partial pressure of $N_2O_4 = \frac{1}{3}$ atm and of $NO_2 = \frac{2}{3}$ atm

$$K_p = \frac{p^2_{NO_2}}{p_{N_2O_4}} = \frac{(\frac{2}{3})^2}{(\frac{1}{3})} \quad \text{Hence } K_p = 4 \text{ atm}$$

Fig. 7.5 Calculating K_p

2) Derivation of information about the equilibrium composition by calculations using a given value of K_c or K_p: see Fig. 7.6

> Question: For the esterification reaction
> $$C_2H_5CO_2H(l) + C_2H_5OH(l) \rightleftharpoons C_2H_5CO_2C_2H_5(l) + H_2O(l)$$
> $K_c = 8.0$ at $90\,°C$
> Calculate the mass of ester obtained by heating $37.0\,g$ of propanoic acid with $46.0\,g$ of ethanol to equilibrium at $90\,°C$.
> $[A_r(H) = 1.0; A_r(C) = 12.0; A_r(O) = 16.0]$
> Answer: Molar masses/$g\,mol^{-1}$: $C_2H_5CO_2H = 74.0$ $C_2H_5OH = 46.0$
> Initial amounts/mol: $37.0 \div 74.0 = 0.5$ $46.0 \div 46.0 = 1.0$
> If x mol of ester is formed at equilibrium then x mol of water will also be formed and there will be $(0.5 - x)$ mol acid and $(1 - x)$ mol alcohol left at equilibrium.
>
> $$\frac{[C_2H_5CO_2C_2H_5(l)]_{eq}[H_2O(l)]_{eq}}{[C_2H_5CO_2H(l)]_{eq}[C_2H_5OH(l)]_{eq}} = 8.0 = \frac{\left(\frac{x}{V}\right)^2}{[(0.5-x)/V][(1.0-x)/V]}$$
> $\Rightarrow x^2 = 8.0(0.5-x)(1.0-x)$ and this gives: $7x^2 - 12x + 4 = 0$
> using $x = \left\{\frac{-b \pm \sqrt{b^2 - 4ac}}{2a}\right\}$ to solve $ax^2 + bx + c = 0$ gives
> $x = 0.45$ (or 1.26 but x cannot be greater than 1.0)
> Molar mass of $C_2H_5CO_2C_2H_5$ is $102\,g\,mol^{-1}$
> Hence $(0.45 \times 102) = 45.9\,g$ of ethyl propanoate formed.

Fig. 7.6 Calculating composition from K_c

AQUEOUS ACID-BASE EQUILIBRIA

The following equilibrium exists in water and all aqueous solutions:
$$H_2O(l) + H_2O(l) \rightleftharpoons H_3O^+(aq) + OH^-(aq)$$

According to the law of chemical equilibrium:

$$\frac{[H_3O^+(aq)]_{eq}[OH^-(aq)]_{eq}}{[H_2O(l)]_{eq}[H_2O(l)]_{eq}} = K_c \text{ (at constant temperature)}$$

The ionisation is so slight that the concentration of the water molecules is regarded as constant. And aqueous ionic equilibria are established so rapidly that all concentrations are regarded as equilibrium values. So we write:

- $[H_3O^+(aq)][OH^-(aq)] = K_w$ (called the ionic product for water).
- $K_w = 1 \times 10^{-14}$ (or $0.000\,000\,000\,000\,01$) $mol^2\,dm^{-6}$ at $25\,°C$.

Even expressed in standard form as a negative power of ten, the value of K_w is an inconvenient number. So we define pK_w as follows:

- $pK_w = -\log_{10}(K_w/mol^2\,dm^{-6}) = 14$ at $25\,°C$.

Notice that p (from the German word potenz – power) is always in lower case, even when it is the first letter in a sentence.

In pure water the concentration of $H_3O^+(aq)$ must be equal to the concentration of $OH^-(aq)$. So it follows that:

- $[H_3O^+(aq)] = [OH^-(aq)] = 1 \times 10^{-7}$ mol dm^{-3} in pure water at $25\,°C$.
- If $[H_3O^+(aq)] = [OH^-(aq)]$ then the water or solution is neutral.

You should realise that the ionic product of water varies with temperature (e.g. $K_w = 1 \times 10^{-13}$ $mol^2\,cm^{-6}$ at $60\,°C$), but unless you are told otherwise assume that the K_w-value for $25\,°C$ is to be used, i.e. 1×10^{-14} $mol^2\,dm^{-6}$.

STRONG ACIDS AND BASES

Hydrogen chloride gas dissolves rapidly in water and ionises completely into hydrogen ions and chloride ions:
$$HCl(g) + H_2O(l) \rightarrow H_3O^+(aq) + Cl^-(aq)$$

So in a 0.1 mol dm^{-3} solution of hydrochloric acid the concentration of $H_3O^+(aq)$ is 0.1 $(= 1 \times 10^{-1})$ mol dm^{-3}. But $[H_3O^+(aq)] \times [OH^-(aq)]$ is always 1×10^{-14} $mol^2\,dm^{-6}$. So in a 0.1 mol dm^{-3} solution of hydrochloric acid, the concentration of $OH^-(aq)$ is 1×10^{-13} mol dm^{-3}.

- If $[H_3O^+(aq)] > [OH^-(aq)]$ then the aqueous solution is acidic

In acidic solutions the $H_3O^+(aq)$ concentration usually varies from about 1×10^{-1} mol dm^{-3}

down to 1×10^{-7} mol dm^{-3}. A Danish biochemist called Sørensen proposed a more convenient scale of acidity by defining pH as follows:

- pH $= -\log_{10}([H_3O^+(aq)]/\text{mol dm}^{-3})$

The pH of acidic solutions usually varies from about 1 (where $[H_3O^+(aq)] = 1\,\text{mol dm}^{-3}$) to 7 (where $[H_3O^+(aq)] = 1 \times 10^{-7}\,\text{mol dm}^{-3}$).

Sodium hydroxide is an ionic solid that dissolves readily in water to give aqueous sodium ions and hydroxide ions:

$$\text{NaOH(s)} \; \{+\; H_2O(l)\} \rightarrow Na^+(aq) + OH^-(aq)$$

So in a 0.1 mol dm^{-3} solution of sodium hydroxide the concentration of $OH^-(aq)$ is $0.1 \; (= 1 \times 10^{-1})$ mol dm^{-3}. But $[H_3O^+(aq)] \times [OH^-(aq)]$ is always 1×10^{-14} mol^2 dm^{-6}. So in a 0.1 mol dm^{-3} solution of sodium hydroxide the concentration of $H_3O^+(aq)$ is 1×10^{-13} mol dm^{-3}.

- If $[H_3O^+(aq)] < [OH^-(aq)]$ then the aqueous solution is alkaline

In alkaline solutions the $OH^-(aq)$ concentration usually varies from about 1×10^{-1} mol dm^{-3} down to 1×10^{-7} mol dm^{-3}. We can obtain a more convenient scale of alkalinity by defining pOH as follows:

- pOH $= -\log_{10}([OH^-(aq)]/\text{mol dm}^{-3})$

The pOH of alkaline solutions usually varies from about 1
(where $[OH^-(aq)] = 1\,\text{mol dm}^{-3}$) to 7 (where $[OH^-(aq)] = 1 \times 10^{-7}\,\text{mol dm}^{-3}$).

When you multiply two numbers together you add their logarithms. But you have seen that $K_w = [H_3O^+(aq)][OH^-(aq)] = 1 \times 10^{-14}$ mol^2 dm^{-6} and that 'p' means $-\log_{10}$. This leads to the following important relationship:

- $pK_w = pH + pOH = 14$

You should now see that in an alkaline solution the pH will usually vary from 7 (where $[H_3O^+(aq)] = 1 \times 10^{-7}$ mol dm^{-3}) to 13 (where $[H_3O^+(aq)] = 1 \times 10^{-13}$ mol dm^{-3}).

pH SCALE OF ACIDITY

In principle we could have acidic solutions with a pH less than 0 or alkaline solutions with a pH greater than 14. In practice we work mostly with solutions whose pH values range from 1 to 13: see Fig. 7.7.

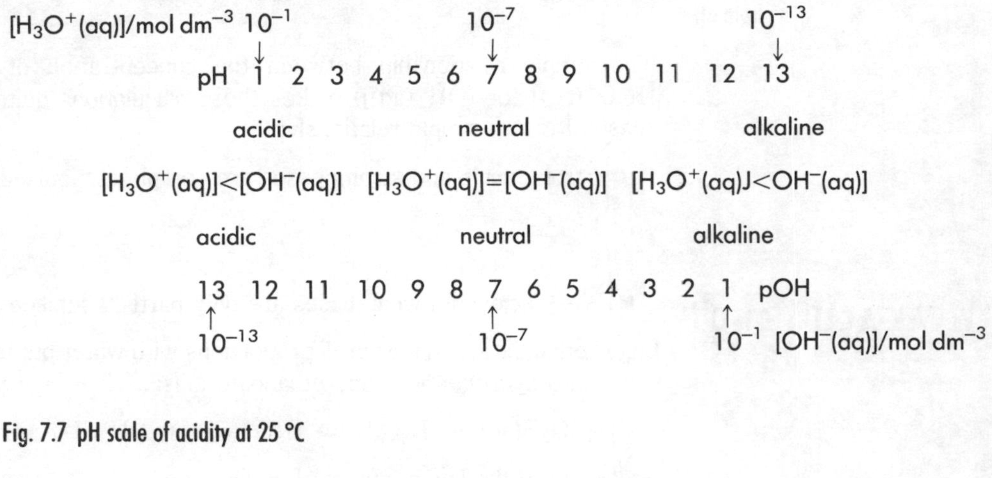

$[H_3O^+(aq)]/\text{mol dm}^{-3}$ 10^{-1} 10^{-7} 10^{-13}

pH   1   2   3   4   5   6   7   8   9   10   11   12   13

   acidic   neutral   alkaline

$[H_3O^+(aq)] < [OH^-(aq)]$ $[H_3O^+(aq)] = [OH^-(aq)]$ $[H_3O^+(aq)] < OH^-(aq)]$

   acidic   neutral   alkaline

13   12   11   10   9   8   7   6   5   4   3   2   1   pOH

10^{-13} 10^{-7} 10^{-1} $[OH^-(aq)]/\text{mol dm}^{-3}$

Fig. 7.7 pH scale of acidity at 25 °C

CALCULATIONS

You could be asked to do simple calculations involving pH and the composition of a solution. The problems fall into two types:

1) Calculation of a value for pH from the concentration of a strong aqueous acid or alkali: see Fig. 7.8.

> pH always has a small 'p' (PH is wrong).

Question: Calculate the pH of an aqueous solution of hydrochloric acid of concentration $0.1 \, \text{mol} \, \text{dm}^{-3}$

(AEB)

Answer: Hydrochloric acid is a strong acid that is completely ionised in water giving $1 \, \text{mol} \, H_3O^+(aq)$ for every $1 \, \text{mol} \, HCl$

$\Rightarrow \quad [H_3O^+(aq)] = 0.1 \, \text{mol} \, \text{dm}^{-3}$
$\Rightarrow \quad pH = -\log_{10}([H_3O^+(aq)]/\text{mol} \, \text{dm}^{-3})$
$\quad = 1.0$

Question: Calculate the pH of aqueous sodium hydroxide of concentration $0.02 \, \text{mol} \, \text{dm}^{-3}$

Answer: Sodium hydroxide is a strong base that is completely ionised in water giving $1 \, \text{mol} \, OH^-(aq)$ for every $1 \, \text{mol} \, NaOH$

$\Rightarrow \quad [OH^-(aq)] = 0.02 \, \text{mol} \, \text{dm}^{-3}$
$\Rightarrow \quad pOH = -\log_{10}([OH^-(aq)]/\text{mol} \, \text{dm}^{-3})$
$\quad = 1.70$
But $pH + pOH = pK_w = 14$
$\Rightarrow \quad pH = 12.30$

Fig. 7.8 Calculating pH from concentration

2) Calculation of the concentration of a strong aqueous acid or alkali from the pH of the solution: see Fig. 7.9.

Question: Calculate the concentration of an aqueous solution of nitric acid whose pH = 1.7

Answer: $pH = -\log_{10}([H_3O^+(aq)]/\text{mol} \, \text{dm}^{-3})$
$\Rightarrow \quad \log_{10}([H_3O^+(aq)]/\text{mol} \, \text{dm}^{-3}) = -1.70$
$\Rightarrow \quad [H_3O^+(aq)] = 0.02 \, \text{mol} \, \text{dm}^{-3}$
But nitric acid is a strong acid that is completely ionised in water giving $1 \, \text{mol} \, H_3O^+(aq)$ for every $1 \, \text{mol} \, HNO_3$
$\Rightarrow \quad$ concentration of $HNO_3(aq)$ is $0.02 \, \text{mol} \, \text{dm}^{-3}$

Question: Calculate the concentration of an aqueous solution of barium hydroxide whose pH is 12

Answer: $pH + pOH = pK_w = 14$
$\Rightarrow \quad pOH = 2$
$pOH = -\log_{10}([OH^-(aq)]/\text{mol} \, \text{dm}^{-3})$
$\Rightarrow \quad \log_{10}([OH^-(aq)]/\text{mol} \, \text{dm}^{-3}) = -2$
$\Rightarrow \quad [OH^-(aq)] = 0.01 \, \text{mol} \, \text{dm}^{-3}$
But barium hydroxide is a strong base that is completely ionised in water giving $2 \, \text{mol} \, OH^-(aq)$ for every $1 \, \text{mol} \, Ba(OH)_2$
$\Rightarrow \quad$ concentration of $Ba(OH)_2(aq)$ is $0.005 \, \text{mol} \, \text{dm}^{-3}$

Fig. 7.9 Calculating concentration from pH

The simple relationship between the concentration of the acid (or alkali) and the $[H_3O^+(aq)]$ (or $[OH^-(aq)]$) makes these calculations quite straightforward. Here is the reason for that simple relationship:

■ Strong acids and strong bases are completely ionised in water

WEAK ACIDS AND BASES

■ Weak acids and weak bases are only partially ionised in water

Liquid ethanoic acid mixes in all proportions with water but fewer than 10% of its molecules ionise into hydrogen ions and ethanoate ions:

$$CH_3CO_2H(aq) + H_2O(l) \rightleftharpoons H_3O^+(aq) + CH_3CO_2^-(aq)$$

According to the law of chemical equilibrium,

$$\frac{[H_3O^+(aq)][CH_3CO_2^-(aq)]}{[CH_3CO_2H(aq)][H_2O(l)]} = K_c \text{ (at constant temperature)}$$

The concentration of the water is approximately constant, so we write:

$$\blacksquare \quad \frac{[H_3O^+(aq)][CH_3CO_2^-(aq)]}{[CH_3CO_2H(aq)]} = K_a \text{ (the acid dissociation constant)}$$

At 25 °C the value of K_a for ethanoic acid is 1.7×10^{-5} mol dm^{-3}. We can obtain a more convenient number by defining pK_a as follows:

- p$K_a = -\log_{10}(K_a/\text{mol dm}^{-3})$
- pK_a for ethanoic acid is 4.8 and is typical for carboxylic acids

Ammonia gas dissolves rapidly in water but less than 10% ionises into ammonium ions and hydroxide ions:

$$NH_3(aq) + H_2O(l) \rightleftharpoons NH_4^+(aq) + OH^-(aq)$$

According to the law of chemical equilibrium

$$\frac{[NH_4^+(aq)][OH^-(aq)]}{[NH_3(aq)][H_2O(l)]} = K_c \text{ (at constant temperature)}$$

The concentration of the water is approximately constant, so we write:

- $\dfrac{[NH_4^+(aq)][OH^-(aq)]}{[NH_3(aq)]} = K_b$ (base dissociation constant)

At 25 °C the value of K_b for aqueous ammonia is 1.8×10^{-5} mol dm^{-3}. We can obtain a more convenient number by defining pK_b as follows:

- p$K_b = -\log_{10}(K_b/\text{mol dm}^{-3})$
- pK_b for aqueous ammonia is 4.8 and similar for aliphatic amines

CALCULATIONS

You could be asked to do calculations involving pH, K_a or K_b and the composition of a solution. For certain values of K_a or K_b the mathematics could become quite tricky and involve solving a quadratic equation. You will probably not encounter these difficulties at A-level because the values of K_a and K_b will be about 1×10^{-5} mol dm^{-3} and will allow you to make the following simplifying assumptions:

> 66 Calculations on pH and K_a are very popular with examiners. 99

1) The weak acid (or alkali) supplies all the $H_3O^+(aq)$ ions (or $OH^-(aq)$ ions) for the solution; the amount coming from the water being negligible
2) The weak acid (or alkali) ionises so little that it can be regarded as being unionised when calculating the concentration of its molecules

The problems could fall into three types. Calculation of (1) the pH from the concentration and K_a (or K_b) of a weak aqueous acid (or alkali), (2) the value of K_a (or K_b) for a weak acid (or alkali) from the concentration and pH of its solution, (3) the concentration of a weak aqueous acid (or alkali) from its K_a (or K_b) and pH. The first type of calculation is the most common: see Fig. 7.10. The second type occurs in connection with titration curves and the third with buffer solutions: see later.

Question: Calculate the pH of an aqueous solution of ethanoic acid of concentration 0.1 mol dm^{-3}. ($K_a = 1.7 \times 10^{-5}$ mol dm^{-3})   (AEB)

Answer: $CH_3CO_2H(aq) + H_2O(l) = H_3O^+(aq) + CH_3CO_2^-(aq)$
$$\frac{[H_3O^+(aq)][CH_3CO_2^-(aq)]}{[CH_3CO_2H(aq)]} = 1.7 \times 10^{-5} \text{ mol dm}^{-3} \dotfill \text{(A)}$$
If acid provides all the $H_3O^+(aq)$ ions then
$$[H_3O^+(aq)] = [CH_3CO_2^-(aq)]$$
If acid dissociation into ions is negligible then
$$[CH_3CO_2H(aq)] = 0.1 \text{ mol dm}^{-3}$$
Hence equation (A) becomes
$$\frac{[H_3O^+(aq)][H_3O^+(aq)]}{0.1} = 1.7 \times 10^{-5} \text{ mol dm}^{-3}$$
$\Rightarrow ([H_3O^+(aq)])^2 = 1.7 \times 10^{-6} \text{ mol}^2 \text{ dm}^{-6}$
$\Rightarrow [H_3O^+(aq)] = \sqrt{1.7} \times 10^{-3} \text{ mol dm}^{-3}$
$\Rightarrow \text{pH} = -\log_{10}([H_3O^+(aq)]/\text{mol dm}^{-3})$
$= 2.9$

Fig. 7.10 Calculating pH from concentration and K_a or K_b

Here are the Brønsted-Lowry definitions of an acid and a base:

- An acid is a molecule or ion that can donate a proton
- A base is a molecule or ion that can accept a proton
- A reaction of an acid with a base is called a neutralisation

Notice that the product(s) of a 'neutralisation' reaction may not be neutral and may not have a pH $= 7$.

When an acid loses a proton it forms what we call its conjugate base: e.g.

$$NH_4^+(aq) \rightarrow H^+ + :NH_3(aq)$$
acid   conjugate base

If an acid is strong (because it loses a proton easily), then the conjugate base must be weak (because it does not accept a proton easily).

When a base accepts a proton it forms what we call its conjugate acid: e.g.

$$:NH_3(aq) + H^+ \rightarrow NH_4^+(aq)$$
base   conjugate acid

If a base is strong (because it accepts a proton easily) then the conjugate acid must be weak (because it loses a proton easily).

K_a and pK_a are measures of the strength of a weak acid on a scale from $K_a = 1 \times 10^0$ (p$K_a = 0$ for the $H_3O^+(aq)$ ion) to 1×10^{-14} mol dm^{-3} (p$K_a = 14$ for the $H_2O(l)$ molecule): see Fig. 7.11.

ACID	=	H⁺	+	CONJUGATE BASE		K_a/mol dm^{-3}	pK_a
$H_3O^+(aq)$				$H_2O(l)$		1.0	0
$H_2SO_3(aq)$				$HSO_3^-(aq)$		1.5×10^{-2}	1.8
$HSO_4^-(aq)$				$SO_4^{2-}(aq)$		1.0×10^{-2}	2.0
$H_3PO_4(aq)$				$H_2PO_4^-(aq)$		7.9×10^{-3}	2.1
$HF(aq)$				$F^-(aq)$		5.6×10^{-4}	3.3
$HNO_2(aq)$				$NO_2^-(aq)$		4.7×10^{-4}	3.3
$CH_3CO_2H(aq)$				$CH_3CO_2^-(aq)$		1.7×10^{-5}	4.8
$H_2CO_3(aq)$				$HCO_3^-(aq)$		4.3×10^{-7}	6.4
$H_2PO_4^-(aq)$				$HPO_4^{2-}(aq)$		6.2×10^{-8}	7.2
$NH_4^+(aq)$				$NH_3(aq)$		5.6×10^{-10}	9.3
$C_6H_5OH(aq)$				$C_6H_5O^-(aq)$		1.3×10^{-10}	9.9
$HCO_3^-(aq)$				$CO_3^{2-}(aq)$		4.8×10^{-11}	10.3
$HPO_4^{2-}(aq)$				$PO_4^{3-}(aq)$		4.4×10^{-13}	12.4
$H_2O(l)$				$OH^-(aq)$		1.0×10^{-14}	14.0

(Left vertical arrow labelled INCREASING ACID STRENGTH; right vertical arrow labelled INCREASING BASE STRENGTH)

Fig. 7.11 Table of acid strengths

We could use K_b and pK_b for a scale of base strength but a simple relationship between K_a and K_b and, therefore, between pK_a and pK_b makes this unnecessary:

- For a weak acid and its conjugate base $K_a \times K_b = K_w = 1 \times 10^{-14}$ mol^2 dm^{-6} and therefore pK_a + pK_b = pK_w = 14 (at 25 °C)

For example, pK_b for ammonia would be $(14 - 9.3) = 4.7$.

DISPLACEMENT REACTIONS

- A strong acid with a low pK_a can transfer its protons to the conjugate base of a weaker acid with a higher pK_a

Acids with p$K_a < 10.3$ are stronger than the hydrogencarbonate ion (acting as a proton donor) so they will react with its conjugate base (carbonate ion): e.g.

$$CH_3CO_2H(aq) + CO_3^{2-}(aq) \rightarrow CH_3CO_2^-(aq) + HCO_3^-(aq)$$

And acids with a p$K_a < 6.4$ are stronger than carbonic acid so they will react with its conjugate base (hydrogencarbonate ion acting as a proton acceptor): e.g.

$$CH_3CO_2H(aq) + HCO_3^-(aq) \rightarrow CH_3CO_2^-(aq) + H_2CO_3(aq)$$

The displaced carbonic acid is unstable and readily decomposes into water and carbon dioxide (detected by limewater).

The relative strengths of carboxylic acids, carbonic acid and phenol would explain why a) you use the action on a carbonate to distinguish between carboxylic acids and phenol, and b) you can 'dissolve' phenol in aqueous sodium hydroxide and liberate it by passing carbon dioxide into the solution.

■ A strong base with a low pK_b can accept protons from the conjugate acid of a weaker base with a higher pK_b

You should be able to produce reasoning similar to that above for the displacement of acids to explain why acids and acidic gases (such as CO_2 and SO_2) 'dissolve' so readily in aqueous sodium hydroxide. You should also be able similarly to explain why you test for ammonium compounds by warming them with aqueous sodium hydroxide:

$$\{Na^+(aq)\} \; OH^-(aq) \; + \; NH_4^+(aq) \; \rightarrow \; H_2O(l) \; + \; NH_3(aq) \; \{Na^+(aq)\}$$

The sodium ion is a 'spectator' ion and plays no part in the reaction.

■ Acid displacement reactions are sometimes described as the reaction of a strong acid with the salt of a weaker acid

■ Base displacement reactions are sometimes described as the reaction of a strong base with the salt of a weaker base

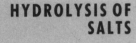

HYDROLYSIS OF SALTS

■ If a salt gives either an acidic (pH $<$ 7) or alkaline (pH $>$ 7) aqueous solution, the effect is sometimes explained as hydrolysis of the salt

When the pK_a is less than 7 we tend to refer to the proton donor and conjugate base as the acid and its salts. When the pK_a value is greater than 7 we tend to speak of the proton acceptor and conjugate acid as the base and its salts. For example, aqueous ammonia is an alkali and ammonium chloride a salt. But you should expect the pH of aqueous ammonium chloride to be less than 7 because the chloride ion is a very weak base and the ammonium cation is a moderately good proton donor:

$$NH_4^+(aq) \; + \; H_2O(l) \; = \; H_3O^+(aq) \; + \; NH_3(aq)$$

Similarly, aqueous ethanoic acid is an acid and sodium ethanoate a salt. But you should expect the pH of aqueous sodium ethanoate to be more than 7 because the sodium ion is not an acid and the ethanoate anion is a moderately good proton acceptor:

$$CH_3CO_2^-(aq) \; + \; H_2O(l) \; = \; CH_3CO_2H(aq) \; + \; OH^-(aq)$$

You could guess the pH of aqueous ammonium ethanoate to be equal to 7 because the donation of protons to the water by the ammonium ions would be balanced by reception of protons from the water by the ethanoate ions.

CHANGE IN pH WITH COMPOSITION

You will know that if you add enough aqueous sodium hydroxide to hydrochloric acid you can make a neutral (pH $=$ 7) solution of sodium chloride:

$$NaOH(aq) \; + \; HCl(aq) \; \rightarrow \; NaCl(aq) \; + \; H_2O(l)$$

If you don't add quite enough alkali the salt solution will have some hydrochloric acid in it and the pH will be less than 7. If you add too much alkali the salt solution will have some sodium hydroxide in it and the pH will be more than 7. You could titrate 0.1 mol dm^{-3} NaOH(aq) from a burette into 20.0 cm^3 of 0.1 mol dm^{-3} HCl(aq) in a conical flask: see Fig. 7.12.

If you use 20.0 cm^3 of the alkali the solution would be neutral (pH $=$ 7). If you use 19.9 cm^3 of the alkali (two drops too few) the solution would still be acidic. If you use 20.1 cm^3 of the alkali (two drops too many) the solution would now be alkaline. You could estimate the pH of these two solutions: see Fig. 7.13.

You could also use a pH meter to monitor the acidity of the hydrochloric acid as you add the sodium hydroxide. A graph of pH against the total volume of alkali (added from the burette) is usually known as a pH-titration curve: see Fig. 7.14.

■ The shape of a pH-titration curve depends upon the concentration of acid and alkali

When we do acid-base titrations we usually use solutions of concentration 0.1 mol dm^{-3} to make the end-point as sharp as possible. For example, two drops either side of the end

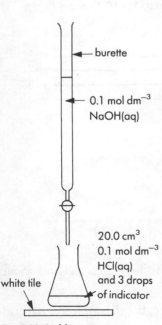

burette

0.1 mol dm^{-3}
NaOH(aq)

20.0 cm^3
0.1 mol dm^{-3}
HCl(aq)
and 3 drops
of indicator

white tile

Fig. 7.12 Acid-base titration

Question: Calculate the pH of a solution obtained by adding $19.9\,cm^3\,0.1\,mol\,dm^{-3}\,NaOH(aq)$ to $20.0\,cm^3\,0.1\,mol\,dm^{-3}\,HCl(aq)$

Answer: Volume of solution is $19.9\,+\,20.0\,\simeq\,40.0\,cm^3$
Solution contains $0.1\,cm^3$ of $0.1\,mol\,dm^{-3}\,HCl(aq)$ unreacted
$\Rightarrow [HCl(aq)] \simeq 0.1\,mol\,dm^{-3}$ diluted by a factor of $0.1:40.0$
$\Rightarrow [H_3O^+(aq)]$ is $0.1 \times 0.1/40.0 = 2.5 \times 10^{-4}\,mol\,dm^{-3}$
$pH = -\log_{10}([H_3O^+(aq)]/mol\,dm^{-3})$
$\qquad = 3.6$

Question: Calculate the pH of a solution obtained by adding $20.1\,cm^3\,0.1\,mol\,dm^{-3}\,NaOH(aq)$ to $20.0\,cm^3\,0.1\,mol\,dm^{-3}\,HCl(aq)$

Answer: Volume of solution is $20.1\,+\,20.0\,\simeq\,40.0\,cm^3$
Solution contains $0.1\,cm^3$ of $0.1\,mol\,dm^{-3}\,NaOH(aq)$ unreacted
$\Rightarrow [NaOH(aq)] \simeq 0.1\,mol\,dm^{-3}$ diluted by a factor of $0.1:40.0$
$\Rightarrow [OH^-(aq)]$ is $0.1 \times 0.1/40.0 = 2.5 \times 10^{-4}\,mol\,dm^{-3}$
$pOH = -\log_{10}([OH^-(aq)]/mol\,dm^{-3})$
$\qquad = 3.6$
$pH + pOH = pK_w = 14$
$\Rightarrow \quad pH = 14 - 3.6$
$\qquad\qquad = 10.4$

Fig. 7.13 Calculating pH from composition

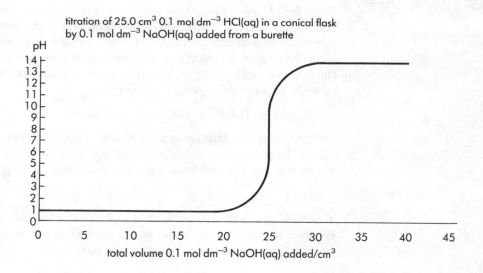

titration of $25.0\,cm^3\,0.1\,mol\,dm^{-3}\,HCl(aq)$ in a conical flask by $0.1\,mol\,dm^{-3}\,NaOH(aq)$ added from a burette

Fig. 7.14 pH-titration curve

total volume $0.1\,mol\,dm^{-3}\,NaOH(aq)$ added/cm^3

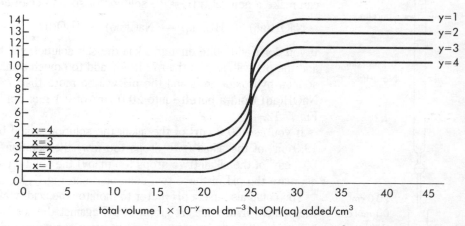

titration of $25.0\,cm^3\,1 \times 10^{-x}\,mol\,dm^{-3}\,Hcl(aq)$ in a conical flask by $1 \times 10^{-y}\,mol\,dm^{-3}$ NaOH(aq) added from a burette

Fig. 7.15 Effect of concentration upon pH-titration curve

total volume $1 \times 10^{-y}\,mol\,dm^{-3}\,NaOH(aq)$ added/cm^3

point produces a change of about 7 pH units with $0.1\,mol\,dm^{-3}$ solutions and only about 3 pH units with $0.001\,mol\,dm^{-3}$ solutions.

■ The shape of a pH-titration curve depends upon the strength (pK_a or pK_b) of the acid or the alkali used

You cannot titrate a weak acid (e.g. aqueous ethanoic acid) with a weak base (e.g. aqueous ammonia) because the pH does not change sharply at the end-point.

INTERPRETING pH-TITRATION CURVES

From a titration curve you should be able to deduce whether alkali is being added to acid (or acid to alkali) and whether the acid (or alkali) is strong or weak: see Fig. 7.16. In analytical laboratories, the common practice is to avoid putting the alkali in the burette. Can you think why? In examinations and textbooks you will probably find the curves will be for the titration of an acid in the conical flask by an alkali from the burette. This simplifies the interpretation.

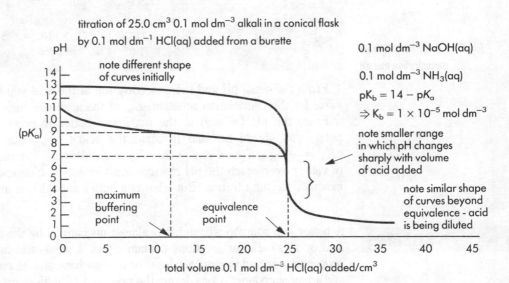

Fig. 7.16 pH-titration curve of strong and weak alkali by strong acid

You should be able to read from the graph the initial pH of the acid being titrated and the total volume of alkali required to reach the end-point or equivalence point. From the reacting volumes of acid and alkali, and the concentration of the alkali, you should be able to calculate the concentration of the acid titrated: see Fig. 7.17 and Fig. 7.18.

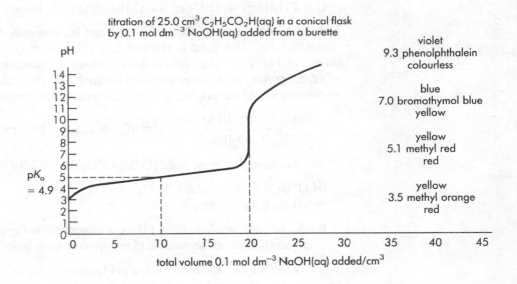

Fig. 7.17 pH-titration curve of aqueous propanoic acid

The concentration of the NaOH(aq) is given as $0.10 \, \text{mol dm}^{-3}$. And from the pH-titration curve, $20.0 \, \text{cm}^3$ of the alkali reacts with $25.0 \, \text{cm}^3$ of the propanoic acid.

Since the volume of acid is greater than the volume of alkali, the acid must be less concentrated than the alkali.

So the concentration of the monoprotic acid must be $0.10 \times \dfrac{20.0}{25.0}$

Fig. 7.18 Calculating concentration from titration data

$= 0.080 \, \text{mol dm}^{-3}$

The propanoic acid concentration is $0.080 \, \text{mol dm}^{-3}$ and its pH is 3.0.

$$C_2H_5CO_2(aq) + H_2O(l) \rightleftharpoons H_3O^+(aq) + CH_3CO_2^-(aq)$$

$$\frac{[H_3O^+(aq)][C_2H_5CO_2^-(aq)]}{[C_2H_5CO_2H(aq)]} = K_a \quad \dots\dots\dots\dots\dots\dots\dots\dots\dots\dots\dots\dots\dots\dots\dots\dots\dots\dots\dots \quad (A)$$

Assume acid provides all the $H_3O^+(aq)$ ions so that

$$[H_3O^+(aq)] = [C_2H_5CO_2^-(aq)]$$

Assume acid dissociation into ions is negligible so that

$$[C_2H_5CO_2H(aq)] = 0.080 \, \text{mol dm}^{-3}$$

Hence equation (A) becomes $\dfrac{[H_3O^+(aq)][H_3O^+(aq)]}{0.080} = K_a$

$$pH = -\log_{10}([H_3O^+(aq)]/\text{mol dm}^{-3})$$
$$= 3.0$$
$$\Rightarrow [H_3O^+(aq)] = 1 \times 10^{-3} \, \text{mol dm}^{-3}$$
$$\Rightarrow K_a = (1 \times 10^{-3})(1 \times 10^{-3})/0.08$$
$$= 1.25 \times 10^{-5} \, \text{mol dm}^{-3}$$

Fig. 7.19 Calculating K_a from concentration and pH

From the initial pH and the concentration of the acid you should be able to calculate a value for the dissociation constant, K_a, of the acid: see Fig. 7.19.

From the pH (= 4.9) at the maximum buffering point, half-way towards the end-point, you should be able to state the acid's pK_a value (= 4.9) and deduce from $pK_a = -\log_{10}(K_a/\text{mol dm}^{-3})$ that $K_a = 1.26 \times 10^{-5} \, \text{mol dm}^{-3}$. And from the range of values over which the pH changes most rapidly you should be able to select a suitable indicator for the titration. But what is a buffer and what is an indicator?

BUFFERS

A buffer is a solution whose pH is almost unchanged by the addition of small amounts of acid or alkali. Buffer solutions contain either a weak acid and its conjugate base (e.g., propanoic acid and propanoate ions) or a weak base and its conjugate acid (e.g., ammonia and ammonium ions). Considering the first type of buffffer system, if you add any acid to it, the conjugate base of the buffer combines with the hydrogen ions from the added acid:

$$C_2H_5CO_2^-(aq) + H^+(aq) \rightarrow C_2H_5CO_2H(aq)$$

If you add any alkali, the acid of the buffer reacts with the hydroxide ions from the added alkali:

$$C_2H_5CO_2H(aq) + OH^-(aq) \rightarrow C_2H_5CO_2^-(aq) + H_2O(l)$$

For maximum buffering effect, the acid and its conjugate base should have the same concentrations. This point is reached in a pH-titration curve halfway towards the end-point: see Fig. 7.17. The curve is flattest where the concentration of propanoic acid must equal the propanoate ion concentration because enough sodium hydroxide has been added to convert half of the original amount acid into sodium propanoate. For propanoic acid:

$$\frac{[C_2H_5CO_2^-(aq)][H_3O^+(aq)]}{[C_2H_5CO_2H(aq)]} = K_a = 1.25 \times 10^{-5} \, \text{mol dm}^{-3}$$

At the maximum buffering point $[C_2H_5CO_2^-(aq)] = [C_2H_5CO_2H(aq)]$, so

$$[H_3O^+(aq)] = K_a = 1.25 \times 10^{-5} \, \text{mol dm}^{-3}$$
$$\Rightarrow pH_{(\text{max buffer})} = pK_a = 4.9$$

- pK_a for any weak acid = pH of a maximum buffering solution containing equal concentrations of the weak acid and its conjugate base.

You can use similar arguments for a pH-titration curve of a weak base (e.g. ammonia) titrated by a strong acid (e.g. hydrochloric acid): see Fig. 7.16. For aqueous ammonia:

$$\frac{[NH_4^+(aq)][OH^-(aq)]}{[NH_3(aq)]} = K_b = 1.8 \times 10^{-5} \, \text{mol dm}^{-3}$$

At the maximum buffering point $[NH_4^+(aq)] = [NH_3(aq)]$, so

$$[OH^-(aq)] = K_b = 1.8 \times 10^{-5} \, \text{mol dm}^{-3}$$
$$\Rightarrow pOH_{(\text{max buffer})} = pK_b = 4.7$$
$$\Rightarrow pH_{(\text{max buffer})} \text{ is } 14 - 4.7 = 9.3$$

■ pK_b for any weak base = pOH of a maximum buffering solution containing equal concentrations of the weak base and its conjugate acid

Notice that 9.3 is the pK_a value of the ammonium ion, $NH_4^+(aq)$, the conjugate acid of aqueous ammonia. So the solution of $NH_4^+(aq)$ and $NH_3(aq)$ could be considered as a buffer of the acid, $NH_4^+(aq)$, and its conjugate base, $NH_3(aq)$:

$$\frac{[H_3O^+(aq)][NH_3(aq)]}{[NH_4^+(aq)]} = K_a = 5.6 \times 10^{-10} \text{ mol dm}^{-3}$$

At the maximum buffering point $[NH_4^+(aq)] = [NH_3(aq)]$, so

$$[H_3O^+(aq)] = K_a = 5.6 \times 10^{-10} \text{ mol dm}^{-3}$$
$$\Rightarrow pH_{(max \ buffer)} = 9.3$$

If the concentrations of acid and conjugate base (or base and conjugate acid) are not quite equal, the buffer solution will not be at its maximum buffering point and the pH of the buffer will not equal pK_a (or pK_b). If you know what the concentrations are, you can easily calculate the pH of the buffer: see Fig. 7.20.

Question: Calculate the pH of a buffer solution obtained by adding 19.0 cm³ 0.1 mol dm⁻³ NaOH(aq) to 50.0 cm³ 0.1 mol dm⁻³ $KH_2PO_4(aq)$. For aqueous dihydrogenphosphate(V) ion $K_a = 6.2 \times 10^{-8}$ mol dm⁻³.

Answer: Volume of solution is 19.0 + 50.0 = 69 cm³
$KH_2PO_4(aq) + NaOH(aq) \rightarrow KNaHPO_4(aq) + H_2O(l)$
or $H_2PO_4^-(aq) + OH^-(aq) \rightarrow HPO_4^{2-}(aq) + H_2O(l)$
The sodium hydroxide converts 19 cm³ (of the 50 cm³) of the aqueous acid (the dihydrogenphosphate(V) ion) into the aqueous conjugate base (the hydrogenphosphate(V) ion). This leaves 31 cm³ of the acid unreacted.
$\Rightarrow$ ratio $[H_2PO_4^-(aq)]$ to $[HPO_4^{2-}(aq)]$ is 31:19
But $\dfrac{[H_3O^+(aq)][HPO_4^{2-}(aq)]}{[H_2PO_4^-(aq)]} = 6.2 \times 10^{-8}$ mol dm⁻³
$\Rightarrow [H_3O^+(aq)]19/31 = 6.2 \times 10^{-8}$ mol dm⁻³
$\Rightarrow [H_3O^+(aq)] = 1.01 \times 10^{-7}$ mol dm⁻³
But pH $= -\log_{10}([H_3O^+(aq)]/\text{mol dm}^{-3})$
$\Rightarrow$ pH = 7.0

Fig. 7.20 Calculating the pH of a buffer solution

■ For a buffer solution at or near its maximum point the calculation can be carried out using the following Henderson–Hasselbalch equations:

$$pH_{(buffer)} = pK_a - \log \frac{[\text{weak acid}]}{[\text{conjugate base}]} \quad \dots\dots\dots\dots (1)$$

$$\text{or} \quad pH_{(buffer)} = pK_a + \log \frac{[\text{conjugate base}]}{[\text{weak acid}]} \quad \dots\dots\dots\dots (2)$$

which are alternative versions of the following general equation:

$$\frac{[H_3O^+(aq)][\text{conjugate base}]}{[\text{weak acid}]} = K_a$$

The Henderson–Hasselbalch equations (1) and (2) can be derived by taking logarithms to the base 10:

$$\log [H_3O^+(aq)] + \log \frac{[\text{conjugate base}]}{[\text{weak acid}]} = \log K_a$$

substituting pH $= -\log ([H_3O^+(aq)]/\text{mol dm}^{-3})$ and $pK_a = -\log (K_a/\text{mol dm}^{-3})$:

$$\Rightarrow -pH_{(buffer)} + \log \frac{[\text{conjugate base}]}{[\text{weak acid}]} = -pK_a$$

then rearranging. Equation (2) is the simplest version to remember because there must be more conjugate base than weak acid to make the value of log [conjugate base]/[weak acid] positive and adding more conjugate base will add a positive number to the pH to increase its value.

(A) 0.1 mol dm^{-3} aqueous ethanoic acid
(B) 0.1 mol dm^{-3} aqueous ethanoic acid/sodium ethanoate buffer
(W) distilled water

volume of alkali added/cm^3	0.0	1.0	10.0	15.0
pH of (A) + alkali	2.89	3.05	3.82	4.02
percentage change in pH	0	6	32	39
pH of (B) + alkali	4.77	4.78	4.82	5.04
percentage change in pH	0	0.2	1	6
pH of (W) + alkali	7.00	11.0	12.0	12.1
percentage change in pH	0	57	71	73

Fig. 7.21 Adding 0.1 mol dm^{-3} NaOH(aq) to 100 cm^3 of different liquids

If you add only small amounts of acid or alkali, the ratio of the concentrations of the acid and its conjugate base does not change very much. Consequently the logarithm of the ratio changes even less and so the pH of the buffer is hardly changed: see Fig. 7.21.

■ The pH of a buffer solution is governed by the pK_a and the ratio of the concentrations of weak acid to conjugate base

If you add large amounts of acid or alkali in relation to the concentration of weak acid and conjugate base in the buffer solution, the pH will change. A tiny drop of buffer solution will not cope with a bucket of concentrated acid!

■ The capacity of a buffer is governed by the concentrations of the weak acid and its conjugate base

In principle, a concentrated aqueous solution of a strong acid (or alkali) should be regarded as a buffer with a very low (or high) pH value. In practice, we talk of buffer solutions as being concentrated aqueous solutions of weak acids (or bases) and their conjugate bases (or acids) with pH values near to 7. You should appreciate the importance of phosphate buffers and be able to use a data book to show that the dihydrogenphosphate(V) ion ($H_2PO_4^-$(aq) as a weak acid) and the hydrogenphosphate(V) ion (HPO_4^{2-}(aq) as its conjugate base) gives a pH of about 7.2 at the maximum buffering point.

INDICATORS

■ An acid-base indicator is a substance whose colour changes with pH

Acid-base indicators are usually coloured organic compounds whose molecules can lose or gain protons. The colour change accompanying the proton loss or gain is connected with a change in molecular structure. You may think of an acid-base indicator as a weak acid. The aqueous weak acid has one colour and its aqueous conjugate base has a different colour: see Fig. 7.22.

NAME	WEAK ACID FORM	pK_a	CONJUGATE BASE FORM
methyl orange	red	3.5	yellow
methyl red	red	5.1	yellow
bromothymol blue	yellow	7.0	blue
phenolphthalein	colourless	9.3	violet

Fig. 7.22 Acid-base indicators

Indicators are usually not very water-soluble so their brightly coloured solutions are only very dilute. And we dilute them further by putting only two or three drops of indicator solution into our conical flask. Because of these low concentrations, indicators do not act as buffers. In other words, their buffer capacity is negligible. Even in an extremely dilute aqueous solution of the indicator, the weak acid (HIn) and conjugate base (In$^-$) will be in equilibrium:

$$\frac{[H_3O^+(aq)][In^-(aq)]}{[HIn(aq)]} = K_a$$

Note that the dissociation constant of an indicator may be represented by the symbol K_{in}

MID-POINT COLOUR AND pH RANGE

If I shine a blue light and a yellow light onto a white wall, it looks green to my eyes. But if the blue light is at least ten times more intense than the yellow light the white wall just looks blue. As a rule our eyes will not detect one of the colours if it contributes less than 10%, and the other colour contributes more than 90%, to the overall colour.

- The colour of an aqueous acid-base indicator is governed by the ratio of the concentrations of its weak acid and conjugate base

Suppose HIn(aq) makes the solution yellow and In$^-$(aq) makes it blue. How will the colour of the solution change as the ratio [HIn(aq)] to [In$^-$(aq)] changes?

- pH $=$ pK_{in} when an acid-base indicator shows its mid-point colour
- An acid-base indicator changes colour over a range of about 2 pH units, one unit either side of the pH at its mid-point colour: see Fig. 7.23.

yellow	green	blue
[HIn] $\geqslant$ [In$^-$] $\times$ 10	[HIn] $=$ [In$^-$]	[HIn] $\leqslant$ [In$^-$] $\div$ 10
[H$_3$O$^+$(aq)] $\geqslant K_a \times 10$	[H$_3$O$^+$(aq)] $= K_a$	[H$_3$O$^+$(aq)] $\leqslant K_a \div 10$
pH$_{(yellow)}$ $\leqslant$ pK_a $-$ 1	pH$_{(green)}$ $=$ pK_a	pH$_{(blue)}$ $\geqslant$ pK_a $+$ 1

Fig. 7.23 pH range of an indicator

SELECTING AN INDICATOR

- For an acid-base indicator to be suitable for a titration, its pK_{in} should be equal to the pH at the end-point where pH changes most sharply

If you run 0.1 mol dm^{-3} NaOH(aq) from a burette into 0.1 mol dm^{-3} aqueous propanoic acid containing methyl orange, the pink solution gradually turns colour to orange and then yellow before you even use half the required alkali. If, however, the acid contains bromothymol blue, the yellow solution suddenly changes to blue almost at the end-point. And if it contains phenolphthalein, the colourless solution suddenly turns violet just beyond the end-point: see Fig 7.17.

- You cannot titrate a weak acid with a weak alkali, or vice versa, because the pH does not change suddenly at the end-point so no indicator will give a sharp change in colour

LIGAND EXCHANGE REACTIONS

The law of chemical equilibrium can be applied to reactions in which a strong (Lewis base) ligand replaces a weaker ligand to form a more stable complex ion. For example, when colourless aqueous ammonia is added to pale blue aqueous copper(II) sulphate, a dark blue solution is formed:

$$4NH_3(aq) + [Cu(H_2O)_6]^{2+}(aq) \rightarrow [Cu(NH_3)_4(H_2O)_2]^{2+}(aq) + 4H_2O(l)$$

You may simplify this equation and write an expression for its equilibrium constant as follows:

$$4NH_3(aq) + Cu^{2+}(aq) \rightleftharpoons Cu(NH_3)_4^{2+}(aq)$$

$$\frac{[Cu(NH_3)_4^{2+}(aq)]}{[NH_3(aq)]^4[Cu^{2+}(aq)]} = K_c = 1.3 \times 10^{13} \text{ mol}^{-4} \text{ dm}^{12}$$

The tetraaminecopper(II) cation is more stable than the copper(II) aqua ion: see Chapter 13. The equilibrium constant K for this displacement of water molecules from an aqua complex by a stronger ligand is called the stability constant of the new complex. Because the stability constants are rather large we usually use the logarithms of their values: see Fig. 7.24.

If you add aqueous 'EDTA' (EthyleneDiamineTetraAcetic acid) to aqueous copper(II) sulphate the solution stays pale blue but the water molecules are replaced by the hexadentate ligand:

$$EDTA^{4-}(aq) + Cu^{2+}(aq) \rightarrow Cu(EDTA)^{2-}(aq)$$

$$\frac{[Cu(EDTA)^{2-}(aq)]}{[EDTA^{4-}(aq)][Cu^{2+}(aq)]} = K_c = 6.3 \times 10^{18} \text{ mol}^{-1} \text{ dm}^3$$

Since the stability constant for Cu(EDTA)$^{2-}$(aq) is larger than that for Cu(NH$_3$)$_4^{2+}$(aq)

LIGAND	AQUEOUS COMPLEX ION	STABILITY CONSTANT	LOG K
Cl^-	$CuCl_4^{2-}$	4.2×10^5	5.6
NH_3	$Cu(NH_3)_4^{2+}$	1.3×10^{13}	13.1
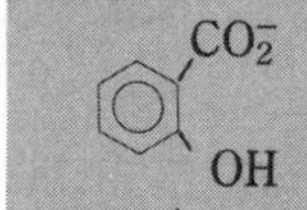 [structure]	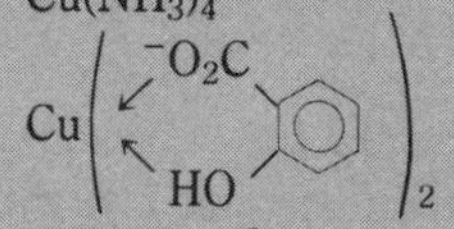 [Cu structure]	7.9×10^{16}	16.9
$EDTA^{4-}$	$Cu(EDTA)^{2-}$	6.3×10^{18}	18.8
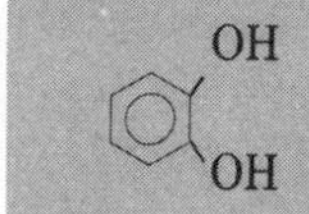 [structure]	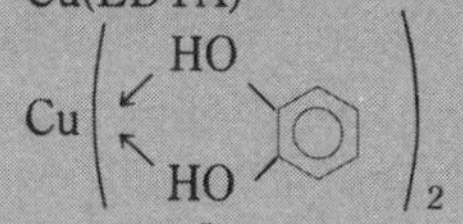 [Cu structure]	1.0×10^{25}	25.0
NH_3	$Co(NH_3)_6^{2+}$	2.5×10^4	4.4
NH_3	$Ni(NH_3)_6^{2+}$	1.0×10^8	8.0
NH_3	$Ag(NH_3)_2^+$	1.7×10^7	7.2

Fig. 7.24 Stability constants of complex ions

you should predict that the hexadentate EDTA ligand would replace the monodentate NH_3 ligand:

$$EDTA^{4-}(aq) + Cu(NH_3)_4^{2+}(aq) \rightarrow Cu(EDTA)^{2-}(aq) + 4NH_3(aq)$$

And you could even predict the value for the following equilibrium constant:

$$\frac{[Cu(EDTA)^{2-}(aq)][NH_3(aq)]^4}{[EDTA^{4-}(aq)][Cu(NH_3)_4^{2+}(aq)]} = K_x = 4.8 \times 10^5 \text{ mol}^3 \text{ dm}^{-9}$$

You can obtain this expression for K_x by dividing the expression for the $Cu(EDTA)^{2-}(aq)$ stability constant by that for the $Cu(NH_3)_4^{2+}(aq)$ stability constant.

So K_x will be $6.3 \times 10^{18} \div 1.3 \times 10^{13}$.
Hence $\log K_x = \log 6.3 \times 10^{18} - \log 1.3 \times 10^{13}$
$\Rightarrow \log K_x$ will be $18.8 - 13.1 = 5.7$

IONIC PRECIPITATION REACTIONS

In your practical work you can often detect and identify metal ions present in an aqueous solution by adding drops of a suitable reagent (e.g. aqueous silver nitrate) and observing the formation and appearance of a precipitate.

$$Ag^+(aq) + Cl^-(aq) \rightarrow AgCl(s)$$

And sometimes you can confirm the identity by the disappearance of the precipitate upon formation of a complex (e.g. adding aqueous ammonia):

$$AgCl(s) + 2NH_3(aq) \rightarrow Ag(NH_3)_2^+(aq) + Cl^-(aq)$$

We refer to the precipitates as sparingly soluble substances. We apply the law of chemical equilibrium to the equation written to show the slight dissolving of the substance: e.g.

$$AgCl(s) \rightarrow Ag^+(aq) + Cl^-(aq)$$

$$\frac{[Ag^+(aq)][Cl^-(aq)]}{[AgCl(s)]} = \text{constant}$$

But since the 'concentration' of a substance in a solid phase corresponds to the density and is constant (at constant temperature), we write:

$$[Ag^+(aq)][Cl^-(aq)] = K_{sp} \text{ (the solubility product)}$$

The units of K_{sp} depend upon the formula of the sparingly soluble substance. In the case of the silver chloride, they are $\text{mol}^2 \text{ dm}^{-6}$. The values of K_{sp} range from small to very small: see Fig. 7.25.

Values for K_{sp} are usually obtained by determining the very low ion concentrations using EMF measurements.

■ For substances of similar formula, K_{sp} has the same units and the smaller the value of K_{sp} is, the less soluble is the substance

CONDITIONS FOR IONIC PRECIPITATION

A white precipitate of sparingly soluble silver chloride usually forms the moment you add a

AgCl	$2 \times 10^{-10} \, mol^2 \, dm^{-6}$	CaCO$_3$	$5 \times 10^{-9} \, mol^2 \, dm^{-6}$
AgBr	$5 \times 10^{-13} \, mol^2 \, dm^{-6}$	BaSO$_4$	$1 \times 10^{-10} \, mol^2 \, dm^{-6}$
AgI	$8 \times 10^{-17} \, mol^2 \, dm^{-6}$	CaSO$_4$	$2 \times 10^{-5} \, mol^2 \, dm^{-6}$
Mg(OH)$_2$	$2 \times 10^{-11} \, mol^3 \, dm^{-9}$	Al(OH)$_3$	$1 \times 10^{-32} \, mol^4 \, dm^{-12}$
Ca(OH)$_2$	$1 \times 10^{-5} \, mol^3 \, dm^{-9}$	Fe(OH)$_3$	$8 \times 10^{-40} \, mol^4 \, dm^{-12}$
MgCO$_3$	$1 \times 10^{-5} \, mol^2 \, dm^{-6}$		

Fig. 7.25 Solubility products

drop of aqueous silver nitrate to an aqueous chloride. You expect a fast reaction between oppositely charged aqueous ions. But not all precipitates form instantly. Some (e.g. hydroxides and carbonates of the alkaline earth metals) may be very slow to form. And of course you should not expect a precipitate to form if the solution is unsaturated.

■ The expression and value for K_{sp} specifies the conditions needed for the precipitation of a sparingly soluble substance to be feasible: see Fig. 7.26

$[Ca^{2+}(aq)][SO_4^{2-}(aq)] < 2 \times 10^{-5}$	unsaturated solution of CaSO$_4$
$[Ca^{2+}(aq)][SO_4^{2-}(aq)] = 2 \times 10^{-5}$	saturated solution of CaSO$_4$
$[Ca^{2+}(aq)][SO_4^{2-}(aq)] > 2 \times 10^{-5}$	supersaturated solution of CaSO$_4$

Fig. 7.26 Precipitation of calcium sulphate

TESTING FOR HALIDES

When you mix aqueous silver nitrate and sodium chloride the $Ag^+(aq)$ and $Cl^-(aq)$ ion concentrations will be so high that $[Ag^+(aq)][Cl^-(aq)]$ reaches $1 \times 10^{-10} \, mol^2 \, dm^{-6}$ (the K_{sp} value). Solid silver chloride precipitates to establish equilibrium with $[Ag^+(aq)][Cl^-(aq)] = K_{sp}$. If the $[Cl^-(aq)]$ at equilibrium is, say, $0.001 \, mol \, dm^{-3}$, then $[Ag^+(aq)]$ would be $K_{sp}/0.001 \, mol \, dm^{-3}$: i.e. $2 \times 10^{-7} \, mol \, dm^{-3}$. If you apply the same reasoning for silver iodide, then $[Ag^+(aq)]$ would be $8 \times 10^{-13} \, mol \, dm^{-3}$. These values allow us to explain the use of ammonia to confirm the identity of the halide.

If you add ammonia to the silver iodide suspension, $8 \times 10^{-13} \, mol \, dm^{-3}$ would be a much lower value of $[Ag^+(aq)]$ than that required for the $Ag^+(aq)$ in equilibrium with the $Ag(NH_3)^+(aq)$. So the complex ion does not form and the silver iodide precipitate does not redissolve. If you add ammonia to the silver chloride suspension, $2 \times 10^{-7} \, mol \, dm^{-3}$ would be a much higher value of $[Ag^+(aq)]$ than that required for the $Ag^+(aq)$ in equilibrium with the $Ag(NH_3)^+(aq)$. So the complex ion forms and lowers the $[Ag^+(aq)]$. This shifts the equilibrium and causes the silver chloride precipitate to redissolve.

EXAMINATION QUESTION AND OUTLINE ANSWER

Q1 a) Define pH and state how you would measure the pH of a solution. (3)

b) Give an approximate pH value (or a pH range) for each of the following solutions. In each case explain your answer.
   i) aqueous 0.1 M FeCl$_3$
   ii) aqueous 0.1 M Na$_2$SO$_4$
   iii) aqueous 0.1 M Na$_2$CO$_3$ (11)

c) The following data show how the pH changes during a titration when aqueous 0.100 M NaOH is added to 10 cm^3 of aqueous ethanoic (acetic) acid.

volume 0.100 M NaOH/cm^3	0.0	1.0	2.0	4.0	6.0	7.0	8.0	8.5	10.0	14.0
pH	2.9	4.0	4.3	4.7	5.2	5.5	6.4	11.2	12.0	12.4

   i) Use the data to plot a titration curve on the graph paper provided.
   ii) State the pH range for the end point and explain why the pH changes rapidly in the region of the end point.
   iii) Calculate the initial concentration of the ethanoic acid.
   iv) Using any suitable pH value, calculate K_a for ethanoic acid. Show your working. (16)

(JMB 1988)

Outline answer

Q1

(a) see text (p 89)

(b) (i) 2 to 4 because $Fe(H_2O)_6^{3+}$ is a weak acid
   (ii) 7 because sodium sulphate is the salt of a strong acid and strong base
   (iii) 9 to 11 because CO_3^{2-} is a weak base

(c) (i) Plot the points carefully using 1 cm on the y-axis for 0.5 pH unit and on the x-axis for 1 cm³ alkali added. Label both axes
   (ii) 6.4 to 11.2
   (iii) end-point at 8.25 cm³ of 0.100M NaOH for 10 cm³ of acid
   $\Rightarrow$ concentration of acid is $\frac{8.25}{10} \times 0.100$
   $= 0.0825$ M
   (iv) $Ka = \dfrac{10^{-2.9} \times 10^{-2.9}}{0.0825}$
   $= 3.05 \times 10^{-5}$ mol dm⁻³

EXAMINATION QUESTION AND TUTOR'S ANSWER

Q2 a) i) Define *acid* and *base* in terms of electron-pair transfer. (2)

   Acid is an electron-pair acceptor

   Base is an electron-pair donor

   ii) Illustrate your answer by considering the reaction of NH_3 and BF_3. (2)

   $H_3N : \searrow BF_3 \longrightarrow H_3N \rightarrow BF_3$
   base   acid   dative bond

b) i) For an aqueous solution of benzoic acid, C_6H_5COOH, of concentration 0.010 mol dm⁻³, calculate to two significant figures the hydrogen ion concentration and the pH, given that $K_a = 6.3 \times 10^{-5}$ mol dm⁻³. (4)

   $\dfrac{[H^+]^2}{0.01} = 6.3 \times 10^{-5} \Rightarrow [H^+] = \sqrt{6.3 \times 10^{-7}}$

   $\Rightarrow [H^+] = 7.9 \times 10^{-4}$ mol dm⁻³ and pH = 3.1

   ii) Would the pH of this solution increase or decrease if sodium benzoate were added? Explain the reason for your answer. (2)

   Increase because the benzoate ion is the conjugate base of the benzoic acid molecule

c) i) Explain the meaning of the statement
$K_w = 10^{-14} \, mol^2 \, dm^{-6}$ at 25 °C. (2)

$[H^+(aq)][OH^-(aq)] = K_w = 10^{-14} \, mol^2 \, dm^{-6}$

because at 25°C $[H^+] = [OH^-] = 10^{-7} \, mol \, dm^{-3}$ for pure

water at equilibrium

ii) Comment on the observation that the enthalpies of neutralisation of aqueous solutions of strong acids and bases are approximately constant. (2)

$H_3O^+(aq) + OH^-(aq) \rightarrow 2H_2O(l)$

Same reaction for all strong acid and strong base

neutralizations

d) When phosphorus pentachloride is heated it dissociates and at a given temperature the following equilibrium is established

$PCl_5(g) \rightleftharpoons PCl_3(g) + Cl_2(g).$

At constant pressure p, the fraction of PCl_5 dissociated is α.

i) Obtain expressions for the partial pressures of the three species in terms of p and α. (2)

$PCl_5(g) \rightleftharpoons PCl_3(g) + Cl_2(g)$

total $\quad 1-\alpha \qquad \alpha \qquad \alpha = 1+\alpha \, mol$

$\Rightarrow PP \left(\frac{1-\alpha}{1+\alpha}\right) P \quad \left(\frac{\alpha}{1+\alpha}\right) P \quad \left(\frac{\alpha}{1+\alpha}\right) P$

ii) Obtain an expression for K_p in terms of α and p. (1)

$K_p = \frac{P_{PCl_3} \times P_{Cl_2}}{P_{PCl_5}} = \frac{\left(\frac{\alpha}{1+\alpha}\right)^2 P^2}{\left(\frac{1-\alpha}{1+\alpha}\right)P} \Rightarrow \frac{\alpha^2 P}{(1-\alpha)(1+\alpha)} = K_p$

iii) What would be the effect upon α and upon K_p of decreasing p? (2)

α would increase but K_p does NOT change (because it is constant at constant T).

iv) The dissociation of PCl_5 is endothermic. Explain the effect upon K_p of decreasing the temperature. Name and state the principle you use in your explanation. (4)

Le Chatelier's principle that equilibrium alters to oppose any constraint. Value of K_p will decrease with a decrease in T ie. move in exothermic direction.

(ODLE 1988)

STUDENT'S ANSWER WITH EXAMINER'S COMMENTS

Q3 The dissociation of hydrogen iodide and the combination of iodine with hydrogen were studied by Bodenstein from 1893 to 1899.
This is a homogeneous equilibrium system.

$H_2(g) + I_2(g) \rightleftharpoons 2HI(g)$

Known amounts of hydrogen and iodine were sealed in a glass bulb. The bulb was kept at a temperature of 445°C until equilibrium was reached. The glass bulb was then rapidly cooled and broken open under alkali.

a) i) Suggest how it would be possible to know that the mixture has reached equilibrium.

> **"Purple"**

No further change in colour of the iodine ✓

ii) Why is the glass bulb rapidly cooled after equilibrium has been reached?

To freeze the reaction at equilibrium so it does not change. ✓

iii) Why is the glass bulb broken open under alkali?

> **"It also absorbs the iodine."**

To neutralize the acid and stop it escaping ✓

(5)

b) i) What is meant by a *homogeneous* equilibrium?

Reactants and products are all in the same state - e.g. all gases ✓

ii) Write balanced equations which illustrate TWO other homogeneous equilibria not involving halogens.

> **"Good."**

$CH_3CO_2H(l) + C_2H_5OH(l) \rightleftharpoons CH_3CO_2C_2H_5(l) + H_2O(l)$ ✓

$H_2O(l) \rightleftharpoons H^+(aq) + OH^-(aq)$ ✓

(3)

c) i) Write an expression for the equilibrium constant K_c, for the reaction of hydrogen with iodine.

> **"The states should be shown: eg. [HI(g)]"**

$$K_c = \frac{[HI]^2_{eq}}{[H_2]_{eq}[I_2]_{eq}}$$

ii) Calculate the value of K_c, using the following equilibrium concentrations which were obtained by Bodenstein for the reaction of hydrogen with iodine at 445°C.

Equilibrium concentrations/mol dm^{-3}		
$H_2(g)$	$I_2(g)$	$HI(g)$
2.06	13.4	36.98

> **"Give the answer to an appropriate number of significant figures."**

$$\frac{(36.98)^2}{2.06 \times 13.4}$$

$K_c = 49.5407$

(3)

d) The values for the equilibrium constant K_c for the following reactions of the halogens with hydrogen at 1000°C are shown in the table below.

Reaction	K_c
$H_2(g) + F_2(g) \rightleftharpoons 2HF(g)$	1×10^{24}
$H_2(g) + Cl_2(g) \rightleftharpoons 2HCl(g)$	2×10^8
$H_2(g) + I_2(g) \rightleftharpoons 2HI(g)$	13

i) Why are no units given for K_c in the above table?

K_c is a ratio because the units cancel ✓

ii) By comparing the values of K_c at 1000 °C and 445 °C deduce whether the reaction between hydrogen and iodine is endothermic or exothermic.

Kc at 445°C is 49·5 and at 1000°C is 13 so reaction is exothermic ✓

iii) The effect of pressure on each of these equilibrium systems is the same. Explain why this is so.

They all have same volume of gases as reactants and products. The volume does not change, so no pressure effect. ✓

iv) Using the above values of K_c comment on the strength of hydrogen-halogen bonds.

H-F bond is strongest

H-Cl bond is weaker

H-I bond is weakest ✓

(7)

(NISEC 1988)

PRACTICE QUESTIONS

Q4 a) Write down an expression for the acid dissociation constant (K_a) for the equilibrium

$$CH_3COOH(aq) \rightleftharpoons H^+(aq) + CH_3COO^-(aq)$$

and calculate pK_a, given that K_a is 1.78×10^{-5} mol l^{-1} at 25 °C.

Expression for K_a _____

Calculation of pK_a _____

b) i) Explain briefly why an aqueous solution containing ethanoic (*acetic*) acid and its sodium salt constitutes a buffer system which is able to minimise the effect of added hydrogen ions.

ii) Why are buffers important in biological systems?   (4)

(JMB 1988)

Q5 a) What do you understand by the term *equilibrium constant*?   (2)

At 60 °C and a total pressure of 1 atmosphere dinitrogen tetraoxide is 50.0% dissociated into nitrogen dioxide:

$$N_2O_4(g) \rightleftharpoons 2NO_2(g)$$

Calculate the equilibrium constant K_p, at this temperature.   (6)

b) Define pH.   (2)

Calculate the pH of
i) 0.100 mol dm^{-3} aqueous ethanoic acid,   (3)
ii) a mixture of equal volumes of 0.100 mol dm^{-3} aqueous ethanoic acid and 0.100 mol dm^{-3} aqueous sodium ethanoate.   (4)

What are solutions of the type used in ii) called? Give one example of their practical application.   (3)
[Dissociation constant, K_a, of ethanoic acid = 1.75×10^{-5} mol dm^{-3}]

(UCLES)

Q6 a) Explain the terms *strong* and *weak* as applied to acids, using aqueous hydrochloric acid and aqueous ethanoic acid as examples.   (5)

b) Calculate the pH of an aqueous solution of

   i) hydrochloric acid of concentration 0.1 mol dm^{-3};   *(3)*

   ii) ethanoic acid of concentration 0.1 mol dm^{-3}.

      (K_a (ethanoic acid) $= 1.7 \times 10^{-5}$ mol dm^{-3})   *(5)*

c) Draw titration curves to show the changes in pH as an excess of aqueous sodium hydroxide of concentration 0.1 mol dm^{-3} is added to

   i) 25 cm^3 of 0.1 mol dm^{-3} hydrochloric acid;

   ii) 25 cm^3 of 0.1 mol dm^{-3} ethanoic acid.

For **each** titration name an indicator which would be suitable to detect the end-point. Briefly explain your choice.   *(12)*

          (AEB)

CHAPTER 8

GETTING STARTED

Redox reactions are regarded as electrochemical processes involving electron-transfer. The emphasis is upon electrochemical cells, cell diagrams and their interpretation in terms of electrode processes and overall cell redox reactions. E^θ (standard electrode potential) values may be used to calculate e.m.f.s (electromotive force) of electrochemical cells, to interpret redox reactions, to predict their energetic feasibility and to determine values for their equilibrium constants and free energy changes. The dependence of e.m.f upon concentration as embodied in the Nernst equation provides a means of measuring pH.

ELECTRO-CHEMISTRY

REDOX REACTIONS

METAL DISPLACEMENT REACTIONS

E.M.F. OF AN ELECTROCHEMICAL CELL

NON-METAL DISPLACEMENT REACTIONS

METAL AND NON-METAL COMBINATION REACTIONS

OTHER REDOX REACTIONS

MEASURING pH

ESSENTIAL PRINCIPLES

REDOX REACTIONS

Oxidation could be seen as the gain of oxygen, loss of hydrogen or of electrons, and reduction could be seen as the converse. In electrochemistry we regard all redox reactions as chemical processes involving electron transfer:

- Reduction is the gain of electrons by an oxidant
- Oxidation is the loss of electrons by a reductant
- Redox is the transfer of electrons from a reductant to an oxidant

METAL DISPLACEMENT REACTIONS

If you stand an iron nail in aqueous copper(II) sulphate, you see a red coating of copper appear on the nail. If you stir iron powder into aqueous copper(II) sulphate, the blue solution becomes green and warm. You should be able to write the ordinary equation for this metal displacement reaction:

$$Fe(s) + CuSO_4(aq) \rightarrow FeSO_4(aq) + Cu(s); \; \Delta H^\ominus = -154 \, kJ \, mol^{-1}$$

The sulphate ions are 'spectator' ions, so you should be able to write the ionic equation for the displacement reaction:

$$Fe(s) + Cu^{2+}(aq) \rightarrow Fe^{2+}(aq) + Cu(s)$$

You should now be able to see that this is a redox reaction in which two moles of electrons are transferred from one mole of iron atoms to one mole of copper(II) ions. The iron acts as reductant by losing electrons and the aqueous copper(II) ions act as oxidant by gaining those electrons:

$$Fe(s) \rightarrow Fe^{2+}(aq) + 2e^- \cdots > 2e^- + Cu^{2+}(aq) \rightarrow Cu(s)$$

If we carry out this reaction in an electrochemical cell so that the iron does not come into direct contact with the aqueous copper(II) ions, then a maximum of about 144 kJ (out of the 154 kJ) could be released as electrical energy by the electrons travelling through the wires of an electrical circuit as they transfer from the iron atoms to the aqueous copper(II) ions: see Fig. 8.1.

This experiment could be demonstrated: see Fig. 8.1(a). The porous plug acts as a bridge to allow electrolytic contact between the two solutions but to stop them mixing. You could do this experiment: see Fig. 8.1(b). The strip of filter paper soaked in saturated aqueous potassium nitrate acts as a 'salt bridge' to make electrolytic contact between the two solutions.

If we repeat the experiment but dip an iron nail into aqueous zinc sulphate, we see no sign of any displacement reaction. However, if we set up the electrochemical cell using zinc

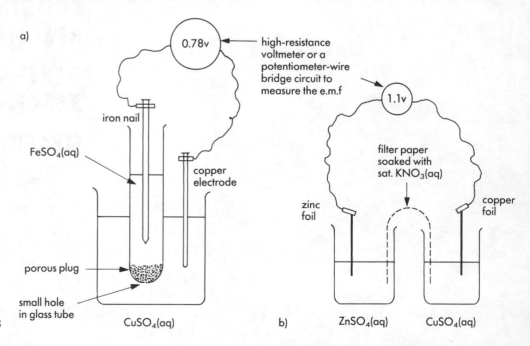

Fig. 8.1 Electrochemical cells

foil, instead of copper foil, dipping into aqueous zinc sulphate, instead of aqueous copper(II) sulphate, we detect an electrical current BUT the electrons now travel from the zinc atoms to the aqueous iron(II) ions:

$$Zn(s) \rightarrow Zn^{2+}(aq) + 2e^- \cdots \cdots> 2e^- + Fe^{2+}(aq) \rightarrow Fe(s)$$

From this we may predict that the following displacement reaction should occur if we dip zinc foil into aqueous iron(II) sulphate:

$$Zn(s) + Fe^{2+}(aq) \rightarrow Zn^{2+}(aq) + Fe(s)$$

If you stir zinc powder into aqueous iron(II) sulphate, the mixture gets warm and the grey, non-magnetic powder turns to a black magnetic powder:

$$Zn(s) + FeSO_4(aq) \rightarrow ZnSO_4(aq) + Fe(s)$$

In this reaction, zinc acts as reductant by losing electrons and the aqueous iron(II) ions act as oxidant by gaining electrons.

If we were to contemplate a third experiment to dip an iron nail into aqueous iron(II) sulphate, we certainly should not expect to see any displacement. There is no reason why iron should displace itself from aqueous iron(II) ions. And if we were to contemplate setting up an electrochemical cell in which both halves of the cell are identical, we certainly should not expect to detect any current or measure a voltage.

REACTIVITY SERIES OF METALS

Because zinc displaces iron from aqueous iron(II) sulphate and because iron displaces copper from aqueous copper(II) sulphate, we may put these metals into an order of reactivity and their aqueous cations into an order of stability. And because zinc and iron react with $H_3O^+(aq)$ ions to displace hydrogen gas but copper does not, we may include $H_2(g)$ in the order of reactivity and $H_3O^+(aq)$ in the order of stability: see Fig. 8.2.

from most reactive	$Zn(s)$	$Fe(s)$	$H_2(g)$	$Cu(s)$	to least reactive
from most stable	$Zn^{2+}(aq)$	$Fe^{2+}(aq)$	$H_3O^+(aq)$	$Cu^{2+}(aq)$	to least stable

Fig. 8.2 Reactivity of metals

You may remember using a reactivity series of metals in your GCSE chemistry, for example, in order to help you understand why zinc can protect iron from rusting.

From the reactivity series in Figure 8.2 we may predict that zinc metal could displace copper from aqueous copper(II) ions but copper metal could not displace zinc or iron from their aqueous cations:

$$Zn(s) + Cu^{2+}(aq) \rightarrow Zn^{2+}(aq) + Cu(s)$$

We may also predict the reaction between zinc and aqueous copper(II) sulphate to be more exothermic than that between iron and aqueous copper(II) sulphate:

$$Zn(s) + CuSO_4(aq) \rightarrow ZnSO_4(aq) + Cu(s); \Delta H^{\ominus} = -219 \, kJ \, mol^{-1}$$

What, if anything, may we predict about an electrochemical cell whose overall cell reaction is this metal displacement reaction? The only thing we can say so far is that electrons will travel in the wires of the electrical circuit from the zinc atoms of the electrode in one half of the cell to the aqueous copper(II) ions in the solution in the other half of the cell.

E.M.F. OF AN ELECTROCHEMICAL CELL

If we decrease the current we draw from an electrochemical cell we find that the voltage of the cell increases and reaches a maximum when the current finally reaches zero.

■ The e.m.f. (electromotive force) of an electrochemical cell is the maximum potential difference (voltage) between the electrodes

We can measure the e.m.f.'s of our cells, using a very high resistance voltmeter or a potentiometer circuit with a very sensitive galvanometer to detect very tiny currents: see Fig. 8.3.

When we do make these measurements we find that the e.m.f. of a cell depends upon the concentration and temperature of the solutions in the cell.

Electrochemists usually standardise the concentration of electrolytes at $1 \, mol \, kg^{-1}$ of solvent, the temperature at $298.15 \, K$ ($25°C$) and, if the cell reaction involves a gas, the

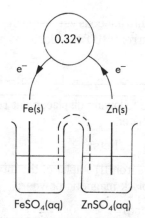
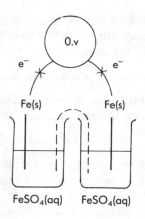
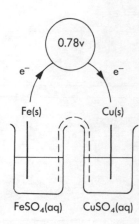

Fig. 8.3 Direction of flow of electrons in electrochemical cells

pressure at 101.325 kPa (1 atm). They call the e.m.f., measured in volts under these standard conditions, the standard e.m.f. of the cell. We simplify this for the purposes of A-level chemistry.

■ The *standard* e.m.f. of an electrochemical cell is its maximum voltage at a temperature of 25 °C, a pressure of 1 atm and with solutions of concentration equal to 1 mol dm^{-3}.

ELECTROCHEMICAL CELL DIAGRAMS

The standard e.m.f. of a Daniell cell is 1.10 V. In it, the copper electrode is positive and the zinc electrode is negative. If we were to connect the metal electrodes together by a wire, elelctrons lost by the atoms of the zinc electrode would flow through the wire to the other half-cell to be gained by the aqueous copper(II) ions. In an examination you could be asked to draw an apparatus diagram of the actual electrochemical cell and to indicate the direction of flow of electrons you would expect upon wiring the electrodes together. However, you are much more likely to be asked to 'draw a cell diagram' and to nominate the positive or negative electrode: see Fig. 8.4.

■ The sign given to the e.m.f. associated with the cell diagram of an electrochemical cell refers to the diagram's right-hand half-cell

apparatus diagrams

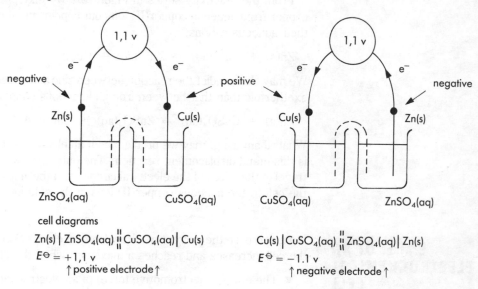

cell diagrams

Zn(s) | ZnSO₄(aq) ‖ CuSO₄(aq) | Cu(s)

$E^{\ominus} = +1,1 \text{ v}$

↑ positive electrode ↑

Cu(s) | CuSO₄(aq) ‖ ZnSO₄(aq) | Zn(s)

$E^{\ominus} = -1.1 \text{ v}$

↑ negative electrode ↑

Fig. 8.4 Apparatus and cell diagrams

THE STANDARD HYDROGEN ELECTRODE

We obviously cannot dip a rod of hydrogen gas into aqueous hydrogen ions and attach a wire to it! So how do we make a hydrogen electrode? see Fig. 8.5.

Platinum is used as the electrical contact (with the hydrogen gas and aqueous hydrogen ions simultaneously) because it is too inert to act as a reductant. The platinum surface is covered with finely divided platinum deposited electrolytically as a black coating. At this

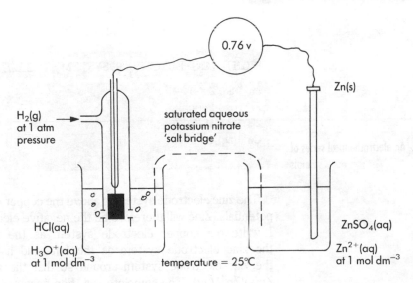

Fig. 8.5 Hydrogen electrode

$$Pt[H_2(g)] \mid 2H_3O^+(aq) \overset{\shortparallel}{\shortparallel} Zn^{2+}(aq) \mid Zn(s) \quad E^{\ominus} = -0.76 \text{ v}$$

electrode, the half-cell electron-transfer reaction occurs between the hydrogen gas and the aqueous hydrogen ions:

$$H_2(g) + 2H_2O(l) \rightleftharpoons 2H_3O^+(aq) + 2e^-$$
or $\quad H_2(g) \rightleftharpoons 2H^+(aq) + 2e^-$

■ The standard hydrogen electrode is the half-cell consisting of pure hydrogen gas, at 25 °C and 1 atm pressure, bubbling past a platinised platinum electrode dipping into a 1 mol dm^{-3} solution of $H_3O^+(aq)$

STANDARD ELECTRODE POTENTIALS

An electrochemical cell is made of two half-cells. If one half-cell is a hydrogen electrode then the e.m.f. of the electrochemical cell is called the *electrode potential* of the other half-cell. For the electrochemical cell represented by the cell diagram

$$Pt[H_2(g)]\mid 2H^+(aq)\overset{..}{}Cu^{2+}(aq)\mid Cu(s)$$

the e.m.f. is called the *electrode potential* of the $Cu^{2+}(aq)\mid Cu(s)$ electrode system. If both half-cells are under standard conditions of temperature, pressure and concentration, then the e.m.f. of the electrochemical cell is called the *standard electrode potential* of the $Cu^{2+}(aq)\mid Cu(s)$ system, it's value is 0.34 V and copper is the positive electrode.

■ The standard electrode potential ($E^{\ominus}$) is defined as the e.m.f. of an electrochemical cell represented by a cell diagram in which a standard hydrogen electrode is shown as the left-hand half-cell

The standard electrode potential of the system $Zn^{2+}(aq)\mid Zn(s)$ is -0.76 V because the standard e.m.f. of the following electrochemical cell is 0.76 V, the zinc electrode is negative and the sign of the e.m.f. refers to the right-hand side of the cell diagram:

$$Pt[H_2(g)]\mid 2H^+(aq)\overset{..}{}Zn^{2+}(aq)\mid Zn(s) \qquad E^{\ominus} = -0.76 \text{ V}$$

You will find the right-hand side of these electrochemical cells, together with the sign and values of their standard e.m.f.'s listed in data books under standard electrode (reduction) potentials.

■ The standard electrode potential for the hydrogen electrode must be zero because the e.m.f. of $Pt[H_2(g)]\mid 2H^+(aq)\overset{..}{}2H^+(aq)\mid[H_2(g)]Pt$ will be zero under standard conditions

COMBINING STANDARD ELECTRODE POTENTIALS

Using $E^{\ominus}$ values to predict reactions is very important.

If we list our three metals and hydrogen, together with their aqueous cations, according to their standard electrode potentials we will have an electrochemical series: see Fig. 8.6. Remember we predicted that the displacement reaction between zinc and aqueous copper(II) ions would be more exothermic than that between iron and aqueous copper(II) ions. We may now predict that in a standard electrochemical cell, the $Zn(s) + Cu^{2+}(aq)$ reaction would give a higher e.m.f. than the $Fe(s) + Cu^{2+}(aq)$ reaction. And we may also predict the value of each e.m.f. as follows:

RIGHT-HAND ELECTRODE SYSTEM	$E^{\ominus}$/V
$Zn^{2+}(aq)|Zn(s)$	−0.76
$Fe^{2+}(aq)|Fe(s)$	−0.44
$2H^{+}(aq)|[H_2(g)]Pt$	0.00
$Cu^{2+}(aq)|Cu(s)$	+0.34

Fig. 8.6 An electrochemical series of metals

The zinc electrode system is above the copper electrode system in the table of electrode potentials. Zinc will therefore be the negative electrode and copper will be the positive. So I write the copper electrode system as the right-hand half-cell $Cu^{2+}(aq)|Cu(s)$ and the zinc electrode system as the left-hand half-cell $Zn^{2+}(aq)|Zn(s)$. But I must turn the zinc electrode system around so that the zinc metal is on the extreme left: thus $Zn(s)|Zn^{2+}(aq)$. The complete cell then becomes:

$$Zn(s)|Zn^{2+}(aq)\vdots Cu^{2+}(aq)|Cu(s) \quad \text{and} \quad E^{\ominus}_{cell} = +1.10\,V$$

You could think of the negative (−0.76 V) zinc system repelling electrons and the positive (+0.34 V) copper system attracting electrons. With one half-cell 'pushing' and the other 'pulling' electrons, the two half-cells are working together to move the electrons in the same direction, so you find the sum of 0.76 and 0.34 to get 1.10 V. In the same way you should be able to write:

$$Fe(s)|Fe^{2+}(aq)\vdots Cu^{2+}(aq)|Cu(s) \quad \text{and} \quad E^{\ominus}_{cell} = +0.78\,V$$

If you write the cell diagram so that the positive electrode is on the right-hand side then you can read the cell reaction from left to right from the diagram. For example, here is the cell diagram of an electrochemical cell made by combining the zinc and iron half-cells:

$$Zn(s) \quad | \quad Zn^{2+}(aq) \quad \vdots \quad Fe^{2+}(aq) \quad | \quad Fe(s) \quad \text{and} \quad E^{\ominus}_{cell} = +0.32\,V$$
$$Zn(s) \rightarrow Zn^{2+}(aq) \qquad Fe^{2+}(aq) \rightarrow Fe(s)$$

so $Zn(s) + Fe^{2+}(aq) \rightarrow Zn^{2+}(aq) + Fe(s)$

Notice that this time both half-cells are 'pushing' the electrons in opposite directions and not working together, so you find the difference between 0.76 and 0.44 to get 0.32 V.

The standard electrode potential for $Ag^{+}(aq)|Ag(s)$ is +0.80 V so if we combine this half-cell with the copper half-cell we would obtain the following electrochemical cell:

$$Cu(s) \quad | \quad Cu^{2+}(aq) \quad \vdots \quad Ag^{+}(aq) \quad | \quad Ag(s) \quad \text{and} \quad E^{\ominus}_{cell} = +0.46\,V$$
$$Cu(s) \rightarrow Cu^{2+}(aq) \qquad Ag^{+}(aq) \rightarrow Ag(s)$$

so $Cu(s) + 2Ag^{+}(aq) \rightarrow Cu^{2+}(aq) + 2Ag(s)$

Notice that the 'equation' for the reduction of the silver cations to silver atoms must be doubled before it can be combined with the 'equation' for the oxidation of copper atoms to copper(II) cations. Can you see why?

■ The number of electrons lost in the ion-electron half-equation for one electrode reaction must match the number of electrons gained in the ion-electron half-equation for the other electrode reaction before the two half-equations can be combined to obtain the ionic equation for the overall cell reaction

Notice also that this time both half-cells are 'pulling' the electrons in opposite directions and not working together, so again you find the difference between 0.80 and 0.34 to get 0.46 V.

Here is a simple rule for combining two electrode potentials to obtain the numerical value (without the sign) of the e.m.f. of an electrochemical cell:

■ If the two electrode potentials have the same sign (both + or both −) find the difference between the numbers

■ If the two electrode potentials have opposite signs (one + and the other −) ignore the signs and find the sum of the numbers

The rule to obtain the value and sign of the standard e.m.f. of an electrochemical cell from the standard electrode potentials of its two half-cells is as follows:

■ $E_{cell} = E^{\ominus}_{\text{right-hand half-cell}} - E^{\ominus}_{\text{left-hand half-cell}}$

NON-METAL DISPLACEMENT REACTIONS

If you add aqueous chlorine to aqueous potassium iodide, aqueous potassium chloride and iodine are formed:

$$Cl_2(aq) + 2KI(aq) \rightarrow 2KCl(aq) + I_2(aq)$$
$$\text{or}\quad Cl_2(aq) + 2I^-(aq) \rightarrow 2Cl^-(aq) + I_2(aq)$$

The aqueous iodide anions act as reductant by losing electrons and the aqueous chlorine molecules act as oxidant by gaining those electrons:

$$2I^-(aq) \rightarrow I_2(aq) + 2e^- \cdots > 2e^- + Cl_2(aq) \rightarrow 2Cl^-(aq)$$

These displacement reactions can occur in electrochemical cells:

$$Pt|2I^-(aq),I_2(aq)\vdots Cl_2(aq),2Cl^-(aq)|Pt \qquad (E^{\ominus}_{cell} = +0.82\,V)$$

Note that the platinum is neither consumed in nor generated by the cell reaction; the metal is providing chemically inert electrical contacts with the solutions. (Actually chlorine will attack platinum and the water but you may ignore any problems this causes with electrochemical measurements.)

In your practical work you may do test-tube reactions with halogens and aqueous halides (but not fluorine and fluorides) that enable you to put the non-metal molecules and their aqueous anions into an order of reactivity and stability: see Fig. 8.7.

Fig. 8.7 Reactivity of non-metals

| from most reactive | $F_2(aq)$ | $Cl_2(aq)$ | $Br_2(aq)$ | $I_2(aq)$ | to least reactive |
| from most stable | $2F^-(aq)$ | $2Cl^-(aq)$ | $2Br^-(aq)$ | $2I^-(aq)$ | to least stable |

From this reactivity series we may predict that aqueous bromine will displace iodine from aqueous iodide anions but will not displace chlorine or fluorine from their aqueous anions:

$$Br_2(aq) + 2I^-(aq) \rightarrow 2Br^-(aq) + I_2(aq)$$

We usually judge the reactivity of halogen molecules as oxidants gaining electrons and the reactivity of metal atoms as reductants losing electrons. So if we want to judge the reactivity of the metals and non-metals on the same basis, we should rewrite the reactivity table for non-metal anions acting as reductants: see Fig. 8.8.

Fig. 8.8 Reactivity of aqueous non-metal anions

| from most reactive | $2I^-(aq)$ | $2Br^-(aq)$ | $2Cl^-(aq)$ | $2F^-(aq)$ | to least reactive |
| from most stable | $I_2(aq)$ | $Br_2(aq)$ | $Cl_2(aq)$ | $F_2(aq)$ | to least stable |

Now we can list the aqueous anions, together with their aqueous molecules, according to their standard electrode potentials and we obtain an electrochemical series for non-metals: see Fig. 8.9.

Fig. 8.9 An electrochemical series of non-metals

RIGHT-HAND ELECTRODE SYSTEM	$E^{\ominus}$/V
$2H^+(aq)|[H_2(g)]Pt$	0.00
$I_2(aq),2I^-(aq)|Pt$	+0.54
$Br_2(aq),2Br^-(aq)|Pt$	+1.09
$Cl_2(aq),2Cl^-(aq)|Pt$	+1.36
$F_2(aq),2F^-(aq)|Pt$	+2.87

You should be able to use these electrode potentials to predict the standard e.m.f. for an electrochemical cell based on our predicted reaction between aqueous bromine molecules and aqueous iodide anions:

$$Pt|2I^-(aq),I_2(aq)\vdots Br_2(aq),2Br^-(aq)|Pt \qquad E^{\ominus}_{cell} = +0.55\,V$$

REACTIVITY SERIES FOR METALS AND NON-METALS

If we judge the reactivity of metals and aqueous non-metal anions as reductants losing electrons, we may put the two reactivity series together into one table: see Fig. 8.10. And we may do the same with the two electrochemical series: see Fig. 8.11.

from most reactive reductant						to least reactive reductant	
$Zn(s)$	$Fe(s)$	$H_2(g)$	$Cu(s)$	$2I^-(aq)$	$2Br^-(aq)$	$2Cl^-(aq)$	$2F^-(aq)$
$Zn^{2+}(aq)$	$Fe^{2+}(aq)$	$H_3O^+(aq)$	$Cu^{2+}(aq)$	$I_2(aq)$	$Br_2(aq)$	$Cl_2(aq)$	$F_2(aq)$
from most stable oxidant						to least stable oxidant	

Fig. 8.10 Reactivity of metals and non-metals

RIGHT-HAND ELECTRODE SYSTEM	$E^{\ominus}/V$
$Zn^{2+}(aq)\|Zn(s)$	-0.76
$Fe^{2+}(aq)\|Fe(s)$	-0.44
$2H^+(aq)\|[H_2(g)]Pt$	0.00
$Cu^{2+}(aq)\|Cu(s)$	$+0.34$
$I_2(aq),2I^-(aq)\|Pt$	$+0.54$
$Br_2(aq),2Br^-(aq)\|Pt$	$+1.09$
$Cl_2(aq),2Cl^-(aq)\|Pt$	$+1.36$
$F_2(aq),2F^-(aq)\|Pt$	$+2.87$

Fig. 8.11 An electrochemical series of metals and non-metals

METAL AND NON-METAL COMBINATION REACTIONS

You should be able to predict from the standard electrode potentials (see Fig. 8.11) that zinc atoms could react with aqueous iodine molecules to form aqueous zinc cations and iodide anions. You should also be able to predict the standard e.m.f. of the electrochemical cell based on that reaction and write down the cell diagram:

$$:Zn^{2+}(aq)|Zn(s) \quad E^{\ominus} = -0.76\,V \qquad :I_2(aq),2I^-(aq)|Pt \qquad E^{\ominus} = +0.54\,V$$
$$\Rightarrow -0.76\,V \quad Zn(s)|Zn^{2+}(aq): \qquad :I_2(aq),2I^-(aq)|Pt \qquad +0.54\,V$$
$$\Rightarrow Zn(s)|Zn^{2+}(aq)\vdots I_2(aq),2I^-(aq)|Pt \qquad \text{and } E^{\ominus}_{cell} = +1.30\,V$$

OTHER REDOX REACTIONS

In the Revised Nuffield Advanced Science BOOK OF DATA there are 111 right-hand electrode systems and standard electrode potentials listed. In the 65th Edition of the CRC Handbook of Chemistry and Physics (the 'rubber book') there are over 350 different electrode systems listed. Some of these may look more complicated than those above but you combine them in just the same way to write a cell diagram, predict the standard e.m.f. and predict the direction of the cell reaction: e.g.

Study this example carefully.

(A) $[4H^+(aq) + SO_4^{2-}(aq)],[H_2SO_3(aq) + H_2O(l)]|Pt \qquad E^{\ominus} = +0.17\,V$
(B) $[MnO_4^-(aq) + 8H^+(aq)],[Mn^{2+}(aq) + 4H_2O(l)]|Pt \qquad E^{\ominus} = +1.51\,V$
$\Rightarrow$ half-cell (B) will be the positive electrode put on the right:

$$:[MnO_4^-(aq) + 8H^+(aq)],[Mn^{2+}(aq) + 4H_2O(l)]|Pt$$
$$[+7] \qquad\qquad\qquad [+2] \qquad [\text{ox.nos.}]$$
$$\downarrow \qquad\qquad\qquad\qquad \downarrow$$
$$MnO_4^-(aq) + 8H^+(aq) \xrightarrow{+5e^-} Mn^{2+}(aq) + 4H_2O(l)$$

and half-cell (A) will be 'turned around' and be the negative electrode put on the left:

$$Pt|[H_2SO_3(aq) + H_2O(l)],[4H^+(aq) + SO_4^{2-}(aq)]:$$
$$[+4] \qquad\qquad\qquad\qquad [+6] \quad [\text{ox.nos.}]$$
$$\downarrow \qquad\qquad\qquad\qquad\qquad \downarrow$$
$$H_2SO_3(aq) + H_2O(l) \xrightarrow{2e^-} 4H^+(aq) + SO_4^{2-}(aq)$$

before the two half-cells are put together to make the complete electrochemical cell with a standard e.m.f. of $1.51 - 0.17 = +1.34\,V$:

$$Pt|[H_2SO_3(aq) + H_2O(l)],[4H^+(aq) + SO_4^{2-}(aq)]\vdots[MnO_4^-(aq) + 8H^+(aq)], [Mn^{2+}(aq) + 4H_2O(l)]|Pt$$
$$\uparrow \qquad\qquad\qquad\qquad \uparrow \qquad\quad \uparrow \qquad\qquad\qquad\qquad \uparrow$$
$$[+4] \qquad\qquad\qquad\qquad [+6] \qquad [+7] \qquad\qquad\qquad [+2] \quad [\text{ox.nos.}]$$

(When you have written the cell diagram you should check that the oxidation numbers of the molecules or ions nearest to the centre of the diagram are higher than the oxidation numbers of the ions, molecules or atoms nearest to the terminals. This is a simple check to avoid a *common error by candidates* in exams: $Zn^{2+}(aq)|Zn(s)::Cu^{2+}(aq)|Cu(s)$ scores **no marks**!)

And before you combine the two ion-electron half-equations you must make sure that the number of electrons gained (by the manganate(VII) anions being reduced to manganese(II) cations) is balanced by the number of electrons lost (by the sulphurous acid molecules being oxidised to sulphate anions):

$$5 \times \{H_2SO_3(aq) + H_2O(l) \xrightarrow{5 \times 2e^-} 4H^+(aq) + SO_4^{2-}(aq)\}$$
$$\Rightarrow 5H_2SO_3(aq) + 5H_2O(l) \rightarrow 2OH^+(aq) + 5SO_4^{2-}(aq) \ \ldots\ldots (A)$$

$$2 \times \{MnO_4^-(aq) + 8H^+(aq) \xrightarrow{2 \times 5e^-} Mn^{2+}(aq) + 4H_2O(l)\}$$
$$\Rightarrow 2MnO_4^-(aq) + 16H^+(aq) \rightarrow 2Mn^{2+}(aq) + 8H_2O(l) \ \ldots\ldots (B)$$

Adding and cancelling equations (A) and (B) then gives

$$5H_2SO_3(aq) + 2MnO_4^-(aq) \rightarrow 2Mn^{2+}(aq) + 3H_2O(l) + 4H^+(aq) + 5SO_4^{2-}(aq)$$

SPONTANEOUS REDOX REACTIONS

If the e.m.f. of an electrochemical cell is not zero then the cell reaction will take place when you short-circuit the cell by wiring its terminals together. Electrons will travel through the wire from the negative half-cell losing them to the positive half-cell gaining them. For example, if you short-circuit the following cell

$$Zn(s)|Zn^{2+}(aq)::Pb^{2+}(aq)|Pb(s) \qquad E^{\ominus}_{cell} = +0.63\,V$$

the spontaneous cell reaction that will occur is:

$$Zn(s) + Pb^{2+}(aq) \rightarrow Zn^{2+}(aq) + Pb(s)$$

If you put a zinc rod into aqueous lead(II) nitrate, the zinc is oxidised to aqueous zinc cations as it reduces the lead(II) cations to metallic lead. The lead often forms a 'tree' of shiny metallic lead crystals. On the other hand, nothing happens if you put a piece of lead into aqueous zinc nitrate. If you write the above cell diagram the other way around, you must change the sign of the e.m.f.:

$$Pb(s)|Pb^{2+}(aq)::Zn^{2+}(aq)|Zn(s) \qquad E^{\ominus}_{cell} = -0.63\,V$$

The negative sign tells you to read the cell diagram from right to left to find the spontaneous cell reaction. Put another way, the negative sign tells you that, reading the cell diagram from left to right, lead will NOT spontaneously react with aqueous zinc cations under standard conditions.

STANDARD GIBBS FREE ENERGY CHANGE

■ For the spontaneous electrochemical cell reaction, $\Delta G^{\ominus} = -zFE^{\ominus}_{cell}$ where z is the number of moles of electrons transferred per mole of redox reaction specified by the equation for the cell reaction and F is the Faraday constant

In the above displacement reaction, 2 electrons are transferred from a zinc atom to an aqueous lead(II) cation, so $z = 2$. The approximate value of F is $96,500\,C\,mol^{-1}$. Hence $\Delta G^{\ominus}$ is $-2 \times 96,500 \times 0.63\,J\,mol^{-1}$. So we may write:

$$Zn(s) + Pb^{2+}(aq) \rightarrow Zn^{2+}(aq) + Pb(s); \ \Delta G^{\ominus} = -122\,kJ\,mol^{-1}$$

This large negative value for $\Delta G^{\ominus}$ means that the reaction is energetically very feasible. By contrast, the reverse reaction has an equally large positive value for $\Delta G^{\ominus}$ and is not energetically feasible:

$$Pb(s) + Zn^{2+}(aq) \rightarrow Pb^{2+}(aq) + Zn(s); \ \Delta G^{\ominus} = +122\,kJ\,mol^{-1}$$

EQUILIBRIUM CONSTANT, K_c

$E^{\ominus}_{cell}$, $\Delta G^{\ominus}$ and the equilibrium constant, K_c, for spontaneous cell reactions are inter-related:

$$\Delta G^{\ominus} = -zFE^{\ominus}_{cell} \quad and \quad \Delta G^{\ominus} = -RT lnK_c \quad so \quad E^{\ominus}_{cell} = (RT/zF)lnK_c$$

where F is the Faraday constant (96,500 coulombs per mole)

R is the Gas constant (8.31 joules per Kelvin per mole)

and T is the standard temperature (298 Kelvin)

Consequently, you could use the standard e.m.f. of an electrochemical cell to calculate a value for the equilibrium constant of the cell reaction. For the above displacement, $E_{cell}^{\ominus} = +0.63\,V$ and $z = 2$.

So,

$$So\ \ 0.63 = \frac{8.31 \times 298}{2 \times 96,500} \times \ln K_c \quad \text{hence} \quad \ln K_c = \frac{0.63 \times 2 \times 96,500}{8.31 \times 298} = 49.1$$

This makes the value of K_c approximately 2×10^{21}. According to the law of chemical equilibrium,

$$K_c = \frac{[Zn^{2+}(aq)]}{[Pb^{2+}(aq)]} = 2 \times 10^{21}$$

This means the displacement goes to completion and there will be effectively no detectable amount of unreacted aqueous lead(II) cations in contact with metallic zinc.

MEASURING pH

You know that the standard e.m.f. of the following cell will be zero:

$$Pt[H2(g)] \mid 2H^+(aq) \vdots 2H^+(aq) \mid [H_2(g)]Pt \qquad E_{cell}^{\ominus} = 0$$
$$H_2(g) \rightleftharpoons 2H^+(aq) \qquad 2H^+(aq) \rightleftharpoons H_2(g)$$

But what would we expect to happen to the e.m.f. if we alter the $H^+(aq)$ ion concentration in the right-hand half-cell from its value of 1 mol dm^{-3}?

If we increase in $[H^+(aq)]$ above 1 mol dm^{-3} then the right-hand half-cell should be more able to gain electrons. This means that the e.m.f. of the cell (E_{cell}) should increase to above its standard value ($E_{cell}^{\ominus}$). The connection between E_{cell}, $E_{cell}^{\ominus}$ and $[H^+(aq)]$ is given by the Nernst equation:

$$E_{cell} = E_{cell}^{\ominus} + (2.3RT/F)\log[H^+(aq)]$$

This equation tells us that a graph of E_{cell} plotted against $\log[H^+(aq)]$ should be a straight line. But pH = $-\log[H^+(aq)]$. So a graph of E_{cell} plotted against pH should also be a straight line: see Fig. 8.12.

Consequently we may use the measurement of the e.m.f. of an electrochemical cell to measure the pH of a solution. We do not actually need to use a hydrogen electrode. Nowadays we use a combination glass/Ag-AgCl reference electrode. Your teacher or lecturer may show you one or you can look at a photograph in a catalogue of scientific equipment. The 'electrode' is really an electrochemical cell with the solution missing from one of its half-cells. You simply connect it to a high resistance voltmeter (often wrongly called the pH meter), dip the 'electrode' into the test solution to complete the half-cell then read the pH from the e.m.f. scale; calibrated in pH units.

Fig. 8.12 Measuring pH with an electrochemical cell

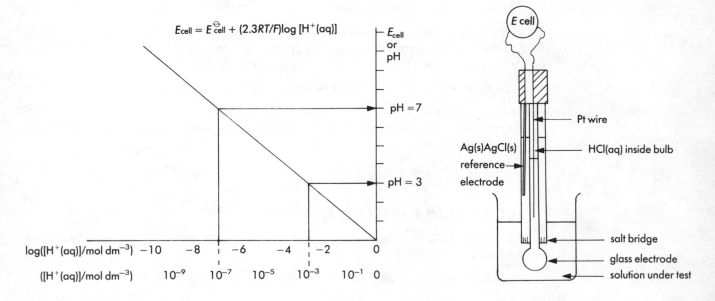

EXAMINATION QUESTIONS AND OUTLINE ANSWERS

Q1

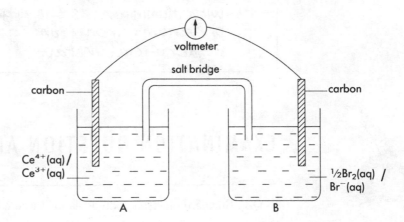

The half-reaction $Ce^{4+}(aq) + e^- \rightarrow Ce^{3+}(aq)$ has a Standard Reduction Potential of $+1.45\,V$.

a) In which direction will electrons flow in the external circuit?
(Data Booklet, page 6) *(1)*

b) What will gradually happen to the colour intensity of the solution in beaker B as the cell is operating? *(1)*

c) Write a balanced ionic equation for the overall reaction. *(1)*

(3)

(SEB 1988)

Q2 A cell consists of aluminium and iron electrodes in contact with aqueous solutions of an aluminium and of an iron salt respectively. Make a sketch of the cell, and give the direction of flow of the electrons externally and the processes occurring at each electrode. Given:

$$Fe^{2+} + 2e^- \rightarrow Fe \qquad E^\theta = +0.44\,V$$
$$Al^{3+} + 3e^- \rightarrow Al \qquad E^\theta = -1.66\,V$$

calculate the initial e.m.f. of the cell under standard conditions. What practical application is made of this cell system? *(7)*

(ODLE 1988)

Outline answers

Q1

(a) Data booklet gives the standard reduction potential of $\frac{1}{2}Br_2(aq) + e^- \rightarrow Br^-(aq)$ as $+1.07\,V$
$\Rightarrow$ electrons will flow from the Br^- (at the $\frac{1}{2}Br_2/Br^-$ electrode) to the Ce^{4+} (at the Ce^{4+}/Ce^{3+} electrode)

(b) It will get a darker red-brown as more bromine molecules form

(c) $Ce^{4+}(aq) + Br^-(aq) \rightarrow Ce^{3+}(aq) + \frac{1}{2}Br_2(aq)$

Q2 See Fig. 8.3
e.m.f. is $(-1.66) - (-0.44)$
$= -1.22$ V
With aluminium as the negative electrode.
Preventing iron from rusting — aluminium is
a sacrificial metal.

EXAMINATION QUESTION AND TUTOR'S ANSWER

Q3 Standard electrode potentials for three electrode systems are given below.

$$Na^+ + e^- = Na; \qquad E^\ominus = -2.71\,V$$
$$Fe^{3+} + e^- = Fe^{2+}; \qquad E^\ominus = +0.76\,V$$
$$Ce^{4+} + e^- = Ce^{3+}; \qquad E^\ominus = +1.61\,V$$

The best oxidising agent present above is the cerium(IV) ion, Ce^{4+}. Use the above data to decide which of the *species* present in the above equations is the best reducing agent. State how you arrive at your choice. *(3)*

best reducing agent Na

reasons Na has the greatest tendency to lose
electrons because Na$^+$ (compared to Fe^{3+} and
Ce^{4+}) has the least tendency to gain electrons.

(ODLE 1988)

STUDENT'S ANSWER WITH EXAMINER'S COMMENTS

Q4 This question is about electrode potentials.
  a) The incompletely labelled diagram below shows an experimental arrangement by which the standard electrode potential of the Fe^{3+}/Fe^{2+} couple may be determined.

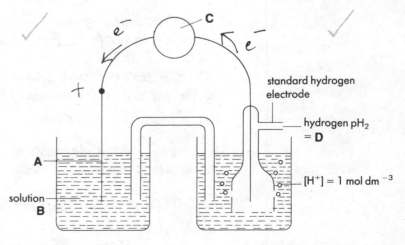

What is represented by A–C?

A Platinum wire ✓

Good. Do not confuse this with an iron wire.

B 1m solution $Fe^{3+}(aq)$ and $Fe^{2+}(aq)$ ✓

C high-resistance voltmeter ✓

What is the value of D?

D 1 atm ✓   (4)

b) Given the standard electrode potentials:

Fe^{3+}/Fe^{2+} $E^{\ominus} = 0.77\,V$
$H^+/\frac{1}{2}H_2$ $E^{\ominus} = 0.00\,V$

Mark on the diagram
 i) the positive pole of the cell
 ii) the direction of electron flow in the external circuit.   (2)

c) State, and explain, two changes that would _increase_ the electrode potential of the hydrogen electrode above its standard value of 0.00 V.

Increase the concentration of $H^+(aq)$ ions ✓

Increase the pressure of the $H_2(g)$ gas ✗

This would make the electrode generate more electrons – $H_2(g) \rightarrow 2H^+(aq) + 2e^-$

                                                                (4)

d) i) Calculate the e.m.f. of the cell
      $Pt|H_2(g)|H^+(aq) \vdots Fe^{3+}(aq), Fe^{2+}(aq)|Pt$

      0·00   0.77

      $\Rightarrow$ E cell is + 0·77 v

Good. Always show the sign and units.

  ii) How and why would this e.m.f. change if a low resistance voltmeter were used so that current was drawn from the cell?

The emf would decrease because according to Ohm's law the voltage (v) related to the current (i) by the resistance (R) : $V = R \times i$ ✓ (4)

e) The electrode potentials given in b) suggest that a certain chemical reaction is possible. What is this reaction? Give two reasons why this reaction might not occur in practice.

$Fe^{3+}(aq) + \frac{1}{2}H_2(g) \rightarrow Fe^{2+}(aq) + H^+(aq)$

Reaction may be very slow. ✓

Hydrogen gas is not soluble in water. ✓

                                                                (3)
                                                (_Total 17 marks_)
                                                   (ULSEB 1988)

CHAPTER

9

KINETICS

GETTING STARTED

Chemical kinetics is the study of reaction rates and the influence on reaction rates of such factors as concentration, pressure, temperature and catalysts. Rate equations for zero, first and second order reactions are determined experimentally and may provide evidence for reaction mechanisms. The Arrhenius equation connects the rate constant with temperature and provides a means of determining the activation energy of a reaction. Homogeneous and heterogeneous catalysts provide alternative reaction pathways with lower activation energies.

ESSENTIAL PRINCIPLES

RATE OF REACTION

During any chemical reaction the concentrations of reactants decrease and the concentrations of products increase with time. We usually express the rate of a reaction as a change of concentration (d[ ]) in a given interval of time (dt) so that the rate (d[ ]/dt) is related to the slope of a concentration/time graph and its units are mol dm^{-3} s^{-1}: see Fig. 9.1.

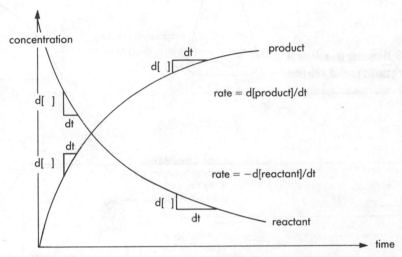

Fig. 9.1 Rate of reaction from slopes

Notice that the gradients of a graph of reactant concentration against time have negative values (because the d[ ] values are negative). So if we follow the change in concentration of a reactant we must use −d[ ]/dt to change the sign and have positive rate values. Can you see why? By definition, reactants turn into products as the reaction goes forward (not backwards). And as time goes by, reactions slow down and eventually stop, i.e. positive rates get smaller and eventually become zero.

MEASURING RATES OF REACTION

We determine the change in concentration of a reactant or product by observing and measuring a convenient property of the chemical mixture. As part of your practical work you should do at least one kinetics experiment. The chemical reactions you study will not be extremely fast or slow; they will proceed at a rate that you can conveniently measure: see Fig. 9.2.

PROPERTY MEASURED	EXAMPLE OF REACTION
gas volume	$Mg(s) + 2HCl(aq) \rightarrow MgCl_2(aq) + H_2(g)$
mass	$CaCO_3(s) + 2HCl(aq) \rightarrow CaCl_2(aq) + H_2O(l) + CO_2(g)$
concentration	$CH_3CO_2C_2H_5(l) + H_2O(l) \rightarrow CH_3CO_2H(aq) + C_2H_5OH(aq)$
colour intensity	$CH_3COCH_3(aq) + I_2(aq) \rightarrow CH_3COCH_2I(aq) + HI(aq)$
intervals of time	$2HI(aq) + H_2O_2(aq) \rightarrow I_2(aq) + 2H_2O(l)$

Fig. 9.2 Methods of measuring rates of reaction

Your experiments will not require highly sophisticated equipment but can be done with standard laboratory apparatus: see Fig. 9.3 and Fig. 9.4.

Some reactions can be followed by sampling. For example, you can measure the concentration of ethanoic acid produced during the hydrolysis of ethyl ethanoate by taking samples of the reaction mixture with a pipette, chilling them in ice-cold distilled water (this slows the reaction down) and titrating with aqueous sodium hydroxide of known concentration.

If you have a colorimeter you could determine the concentration of iodine (reacting with propanone in acid solution) by measuring the intensity of the red-brown colour. And if you have a conductivity apparatus you could determine the concentration of the hydroiodic acid produced during the reaction. Your textbook may also refer to a polarimeter (to measure changes in the angle of rotation of the plane of polarized light) and a dilatometer (to measure very small changes in volume of the liquid reaction mixture). You do not need

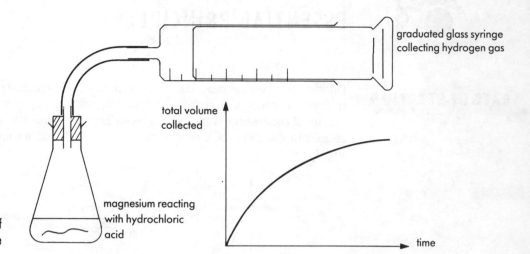

Fig. 9.3 Measuring the volume of
gaseous product with time

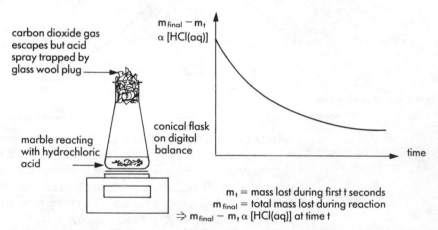

Fig. 9.4 Methods of measuring
rates of reaction

m_t = mass lost during first t seconds
m_{final} = total mass lost during reaction
$\Rightarrow m_{final} - m_t \propto [HCl(aq)]$ at time t

actually to have used these instruments in order to understand how they may be used to follow the progress of a chemical reaction. Notice that an instrumental method based on colorimetry, conductivity, dilatometry or polarimetry is usually better than sampling and titrating the reaction mixture even though you might need a calibration graph to convert your instrument readings into values for the concentration of a reactant or product. When you have measured the concentrations at definite times you usually have to plot a graph of concentration on the y-axis (ordinate) against time on the x-axis (abscissa).

■ Rates of reaction may be determined from the gradients of a graph of concentration of a reactant or product against time

Drawing tangents to a curve can be very inaccurate. More accurate values for the gradients may be obtained by calculation (using a program in an electronic calculator or computer) but the procedure is more complicated. For some chemical reactions, a value proportional to the rate may be obtained by measuring the time interval required for a definite amount of reaction to take place. Reactions studied in this way are sometimes called 'clock' reactions.

As aqueous hydrogen peroxide oxidises hydroiodic acid to iodine the initially colourless solution becomes coloured (pale yellow at first, then orange and finally red-brown):

$$2HI(aq) + H_2O_2(aq) \rightarrow 2H_2O(l) + I_2(aq) \text{ [red-brown]}$$

However, if you add some aqueous sodium thiosulphate, it reduces the iodine back to an iodide and the solution becomes colourless again:

$$2S_2O_3^{2-}(aq) + I_2(aq) \rightarrow S_4O_6^{2-}(aq) + 2I^-(aq) \text{ [colourless]}$$

If you put some aqueous sodium thiosulphate in with the hydroiodic acid before you add the hydrogen peroxide, the reaction mixture will not become coloured until all the sodium thiosulphate is used up. And if you include some starch the solution will turn blue as soon as a trace of free iodine is present.

With this knowledge you can devise an experiment to study the rates of oxidation of HI by H_2O_2. You add the same very small measured amount of aqueous sodium thiosulphate

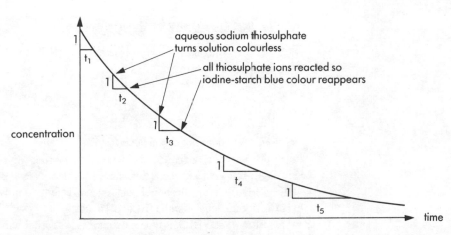

Fig. 9.5 Determining rates from time intervals

to the (HI(aq), H_2O_2(aq) and starch) reaction mixture at definite times and note the time when the blue colour reappears. You take the reciprocal of the time interval between adding the thiosulphate and seeing the blue colour reappear to be directly proportional to the rate of reaction: see Fig. 9.5.

RATE EQUATIONS

- All rate equations must be determined experimentally; they cannot be predicted from the stoichiometric chemical equations.

A rate equation is a mathematical relationship between concentration and time. There are two types of rate equation.

> *The mathematics in this topic can be difficult.*

> *Make sure you understand the chemistry first.*

- Differential rate equations express the rate of a reaction as a function of concentration
- Integrated rate equations express concentration as a function of time

These equations can be very complicated. For A-level chemistry they are restricted to those represented by the following general differential rate equation and their associated integrated rate equations:

rate $= k[A]^m[B]^n$

- k is the proportionality constant, called the rate constant, whose value depends upon temperature but not upon concentration
- m and n are the orders of reaction with respect to reactant A and reactant B
- $m + n$ is called the overall order of the reaction
- For the kinds of reaction studied in A-level courses, the orders may be 0, 1 or 2 but $m + n$ will not exceed 2

FINDING THE ORDERS OF A REACTION

As part of your practical work you could investigate the reaction of iodine with propanone by sampling the reaction mixture and titrating the iodine with aqueous sodium thiosulphate or by measuring the intensity of the red-brown colour with a colorimeter:

$$CH_3COCH_3(aq) + I_2(aq) \xrightarrow[\text{catalyst}]{\text{acid}} CH_3COCH_2I(aq) + H^+(aq) + I^-(aq)$$

To find the orders of reaction with respect to propanone, iodine and the acid catalyst, you must do several experiments to see how the concentration of each substance alters the rate.

On the one hand, you could use the method of determining how the initial rate varies with the concentration of each substance. If you measure the rate at the very start of the reaction then you will also know the concentrations of all the substances. So if you do three experiments and double the concentration of each substance in turn, you should discover how the concentration of each will affect the rate: see Fig. 9.6.

On the other hand, you could use the Ostwald isolation method of making the concentration of one substance low and the concentrations of the other two so high that their values barely change during the reaction. You repeat the procedure to isolate each substance in turn and then analyse the results to determine each order: see Fig. 9.7.

With Ostwald's method you measure the rate by measuring the change in concentration of the isolated substance with time. So in the iodination of propanone, if you keep the

	INITIAL CONCENTRATIONS/mol dm^{-3}			INITIAL RATE/10^{-5} mol dm^{-3} s^{-1}
Expt.	[iodine]	[propanone]	[HCl(aq)]	$-d[I_2(aq)]/dt$
A	0.001	0.5	1.25	1.1
B	0.002	0.5	1.25	1.1
C	0.002	1.0	1.25	2.2
D	0.002	1.0	2.50	4.4

expt A&B: rate does not depend upon [iodine]

$\Rightarrow$ $-d[I_2(aq)]/dt \propto$ [iodine]^0 and reaction is zero order w.r.t. iodine

expt B&C: rate doubles when [propanone] doubles

$\Rightarrow$ $-d[I_2(aq)]/dt \propto$ [propanone]^1 and reaction is first order w.r.t. propanone

expt C&D: rate doubles when [HCl(aq)] doubles

$\Rightarrow$ $-d[I_2(aq)]/dt \propto$ [HCl(aq)]^1 and reaction is first order w.r.t. H$_3$O$^+$(aq)

Hence $-d[I_2(aq)]/dt \propto$ [iodine]^0[propanone]^1[H$_3$O$^+$(aq)]^1

But x^0 = 1 and x^1 = x, so the differential rate equation is:

$-d[I_2(aq)]/dt = k[CH_3COCH_3(aq)][H_3O^+(aq)]$

and the iodination of propanone is a second order reaction overall

Fig. 9.6 Order of reaction from initial concentrations and rates

concentration of propanone and acid constant you find $-d[I_2(aq)]/dt$ as a measure of the rate. Since the reaction is zero order with respect to iodine, you should expect a graph of iodine concentration against time to be a straight line (gradient $= -k_o$): see Fig. 9.7.

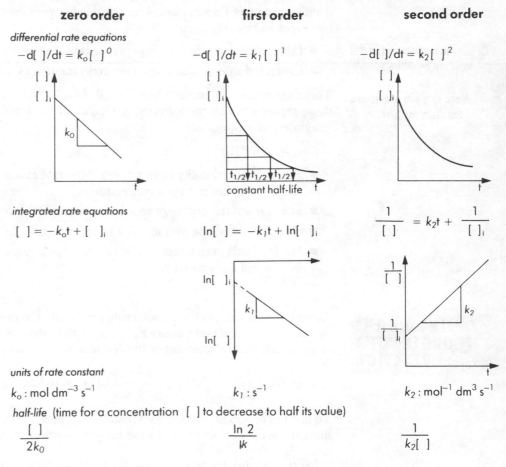

Fig. 9.7 Analysing rate equation data

The hydrolysis or inversion of sucrose to a mixture of fructose and glucose can be followed by polarimetry because the laevo rotatory power of fructose is greater than the dextro rotatory power of either glucose or sucrose:

$$C_{12}H_{22}O_{11}(aq) + H_2O(l) \rightarrow CH_2OH(CHOH)_3COCH_2OH(aq) + CH_2OH(CHOH)_4CHO(aq)$$

The reaction is first order with respect to sucrose and appears to be first order overall. So, rate of hydrolysis = k_1[sucrose] and you should expect a graph of ln[sucrose] against time to be a straight line (gradient $= -k_1$): see Fig. 9.7. However, this reaction is called a pseudo-first order reaction because there is so much water present that its concentration hardly changes.

The reaction of hydrogen and gaseous iodine to form hydrogen iodide is first order with respect to $H_2(g)$ and to $I_2(g)$, so the reaction is second order overall and the differential rate equation is:

$$\text{rate} = k_2[H_2(g)][I_2(g)]$$

If we start the reaction with the $[H_2(g)] = [I_2(g)]$ and take $-d[I_2(g)]/dt$ as a measure of the rate, then a graph of $1/[I_2(g)]$ vs time would be a straight line (gradient $= k_2$): see Fig. 9.7.

In examination questions you are often given data and asked to find the order(s) of reaction. You may be expected to use a graphical method and you may even be told which graph to plot: see Fig. 9.8.

x-AXIS	y-AXIS	ORDER
concentration	time	zero
rate	concentration	first
log concentration	time	first
rate	square of concentration	second
reciprocal concentration	time	second

Fig. 9.8 Summary of straight line graphs

The graph will usually be a straight line. But if it is a graph of [ ] vs time and it is a curve then you should read half-life values from it and expect them to be constant.

You could be given a table of data and asked to determine the order(s) of reaction 'graphically or otherwise'. If the table contains only three or four sets of values showing the effect of concentration upon the initial rate, you can deduce the order(s) by inspection: see Fig. 9.6. If, however, the table gives just one set of data containing more than four pairs of values, you probably have to plot a graph. But how do you decide which graph to plot if the question does not tell you? It can be helpful to look for half-life values.

HALF-LIFE AND FIRST ORDER REACTIONS

66 This ties in with radioactivity in Chapter 3. 99

- The half-life of a reaction is the time interval required for the concentration of a reactant to decrease to half its original value
- The half-life of a first order reaction is a constant ($= \frac{1}{2}\ln k_1$) independent of the concentration of reactant whose value is decreasing by half

The best known examples of first order reactions are the decays of radioactive isotopes: e.g.

$$^{14}_{6}C \rightarrow {}^{14}_{7}N + {}^{0}_{1}e$$

The half-lives of radioisotopes are independent of temperature and their values are listed in data books. Carbon-14 has a half-life of 5730 years and the measurement of the isotope's abundance is the basis of carbon dating.

- The variation in half-life value with concentration or time may indicate the order of the reaction. Approximate values for half-lives may be estimated by simple inspection of experimental data: see Fig. 9.9

DECOMPOSITION OF GASEOUS DINITROGEN PENTOXIDE $N_2O_5(g) \rightarrow N_2O_4(g) + \frac{1}{2}O_2(g)$									
approximate concentration values are halved	180	90				30	15	10	5
$[N_2O_5]/10^{-4}\,\text{mol dm}^{-3}$	176	125	93	71	53	39	29	16	9	5
time/10^2 s	0	6	12	18	24	30	36	48	60	72
half-life approximately constant at about 1200 s		12					12		12	
Decomposition probably first order, so plot $\ln[N_2O_5(g)]$ vs. time to confirm										

Fig. 9.9 Inspecting data for half-life values

If the data in an examination question reveal an approximately constant half-life, you could plot ln[ ] (or 2.3log[ ]) vs time. A straight line graph would confirm a first order and the gradient would give you a value for the rate constant, k_1. If the data shows an approximate half-life value changing with time, you could plot l/[ ] vs time if the half life increases and

[　] vs time if the half-life decreases. A straight-line graph would confirm a second order and a zero order respectively. And the graphs would give you a value for k_2 and k_o, respectively.

MECHANISM AND ORDER OF REACTION

You may assume that the kinetics you study at A-level concern reactions that proceed by one or more simple steps and that the overall rate of a reaction is governed by a slow step.

- A reaction mechanism is a sequence of simple steps proposed in theory to account for the overall chemical reaction that takes place
- The slowest step in a reaction mechanism is called the rate-determining step because it controls the overall rate of reaction

An extremely important feature of chemical kinetics is the information that the order(s) of a reaction may provide about the mechanism of the reaction.

- The experimentally determined rate equation may often, but not always, reveal the nature of the rate determining step

IODINATION OF PROPANONE

Any mechanism proposed for the iodination of propanone must explain the fact that you cannot increase the rate of the reaction by increasing the iodine concentration. Here is one possible mechanism (there are others):

$$(CH_3)_2CO(aq) + H_3O^+(aq) \xrightarrow{slow} (CH_3)_2COH^+(aq) + H_2O(l)$$

$$(CH_3)_2COH^+(aq) + H_2O(l) \xrightarrow{fast} CH_3{-}\underset{\underset{OH}{|}}{C}{=}CH_2(aq) + H_3O^+(aq)$$

$$CH_3{-}\underset{\underset{OH}{|}}{C}{=}CH_2(aq) + I_2(aq) \xrightarrow{fast} CH_3{-}\underset{\overset{\|}{O}}{C}{-}CH_2I(l) + HI(aq)$$

Each of these steps is called a bimolecular step because there are two reactant species involved each time. The first step is bimolecular because it involves one propanone molecule and one hydrogen ion as reactants. It is also the rate-determining step whose slowness holds up the other two faster steps. Increasing the concentration of iodine would have no effect on the rate because iodine is not involved in this slow rate-determining step; iodine is involved only in a fast step.

HYDROLYSIS OF BROMOALKANES

You could do simple test-tube reactions to discover that 2-bromo-2-methylpropane

BROMOALKANE RBr	RATE EQUATION	MECHANISM
$CH_3{-}\underset{\underset{CH_3}{\|}}{\overset{\overset{CH_3}{\|}}{C}}{-}Br$ 2-bromo-2-methyl-propane	rate α [RBr]	S_N1 (a two-step mechanism) $RBr \xrightarrow{slow} R^+$ $R^+ + OH^- \xrightarrow{fast} ROH$
$CH_3CH_2CH_2{-}\underset{\underset{H}{\|}}{\overset{\overset{H}{\|}}{C}}{-}Br$ 1-bromobutane	rate α [RBr][OH$^-$]	S_N2 (a one-step mechanism) $RBr + OH^- \xrightarrow{slow} ROH + Br^-$

Fig. 9.10 Hydrolysis of bromoalkanes

hydrolyses much faster than 1-bromobutane. The overall reaction in each case may be represented by the following equation:

$$C_4H_9Br(l) \ + \ H_2O(l) \ \rightarrow \ C_4H_9OH(l) \ + \ HBr(aq)$$

Somewhat more complicated kinetics studies suggest that the different speeds are the result of different reaction mechanisms: see Fig. 9.10.

Make sure that you do not confuse the meanings of S_N1 and S_N2. 'S' means substitution. 'N' means *nucleophilic* because a carbon atom acting as an electrophilic centre is attacked by a nucleophilic oxygen atom in a water molecule or an hydroxide ion. '1' stands for *unimolecular* and means that the rate-determining step involves only one species (one molecule of 2-bromo-2-methylpropane undergoing heterolytic fission). '2' stands for bimolecular and means that the rate-determining step involves two species (one molecule of the electrophilic 1-bromobutane and one nucleophilic hydroxide ion or water molecule). The numbers '1' and '2' do NOT refer to the number of steps in the mechanism!

ACTIVATION ENERGY (E_a) AND RATE CONSTANT (k)

If you add hydrochloric acid to aqueous sodium thiosulphate the clear, colourless solution slowly turns milky white as colloidal sulphur forms:

$$H_2S_2O_3(aq) \ \rightarrow \ H_2SO_3(aq) \ + \ S(s)$$

If you put the reaction mixture in a beaker standing on a cross marked in pencil on a piece of paper, you can time how long it takes for the liquid to become too cloudy for you to see the cross. The longer it takes, the slower is the decomposition of the thiosulphurous acid. The rate of the reaction is directly proportional to the reciprocal of the time interval because the same amount of change takes place each time. (You should compare this with the measurement of time intervals for the $HI(aq) \ + \ H_2O_2(aq)$ 'clock' reaction.)

If you repeat the same experiment under exactly the same conditions but at different temperatures you find that the rate of the reaction approximately doubles with every 10 degree rise in temperature. But for this reaction:

rate of decomposition $= k_1[H_2S_2O_3(aq)]$

But you are keeping $[H_2S_2O_3(aq)]$ constant. So the value of k_1 must increase with increasing temperature.

- The Arrhenius equation $k = Ae^{-E_a/RT}$ expresses mathematically the dependence of the rate constant upon the absolute temperature
- The Arrhenius equation may be written as $\ln k = \ln A - E_a/RT$ where R is the gas constant, A is the Arrhenius or pre-exponential factor and E_a is the activation energy for the reaction
- The activation energy for a reaction may be determined from the gradient ($= -E_a/R$) of a graph of $\ln k$ against $1/T$: see Fig. 9.11

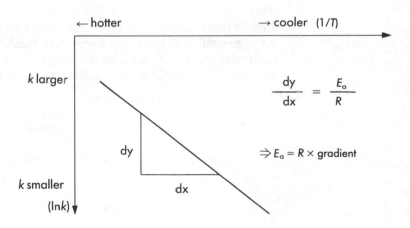

Fig. 9.11 Determining the activation energy of a reaction

THEORY OF EFFECTIVE COLLISIONS

Billions of close encounters between particles (atoms, ions and molecules) occur every second but only a small number occur as collisions with enough energy (E_a) to activate a reaction. You may think of the Arrhenius factor, A, as a measure of the billions of close encounters occurring every second between the particles. More importantly, you should regard $e^{-E_a/RT}$ as a measure of the fraction of those encounters that result in collisions causing a reaction. Most important of all, you should understand what makes the fraction large (giving a fast reaction) and what makes it small (giving a slow reaction): see Fig. 9.12.

$E_a/\text{kJ mol}^{-1}$	RELATIVE REACTION RATES		
60	1	2	5
50	60	130	240
40	4000	7000	12000
temperature/K	290	300	310

Fig. 9.12 Effect of activation energy and temperature on reaction rate

RATE OF REACTION AND TEMPERATURE

■ The higher the temperature the faster the reaction

As a rule of thumb, a ten degree rise in temperature doubles a rate of reaction. This is only an approximate rule because the increase in rate depends upon the temperature and upon the value of the activation energy, but it is worth remembering. It will remind you that quite a modest increase in temperature could produce a dramatic increase in rate: e.g. double the rate for every 10 degree rise means that a rise of 100 degrees would increase the rate by $2\times2\times2\times2\times2\times2\times2\times2\times2\times2$ (or 2^{10}) = 1024 times. A rise of 200 degrees could increase the rate by over a million times! You now have one chemical reason why chemical engineers are anxious to take heat from an exothermic reaction to control the temperature. Another important reason is the cost of a process.

■ The lower the activation energy the faster the reaction

RATE OF REACTION AND ACTIVATION ENERGY

You may consider activation energy from two points of view.

E_a AS THE ENERGY BARRIER

Figure 9.13 shows the relationship between the energies of the reactants and products, the energy change for the reaction and the activation energies for the forward and reverse reaction. Notice that the enthalpy change for a step in a reaction is related to the difference between its activation energies in the forward and reverse directions: see Fig. 9.13. These reaction profiles show the activation energies as barriers that must be surmounted before reactants can turn into products and before products can reform into reactants. You may think of molecules that have reached the top of the barrier as being in a transition state. For some simple cases you can imagine these molecules to have formed an activated complex in which the old bonds have partly broken and the new bonds have partly formed: see Fig. 9.14. From this point of view the activation energy could be related to the difference between the partly broken and partly formed bond energies. You should not attempt to relate the activation energy to the energy required to break any particular bond. As a general rule, however, since bond breaking requires energy you should expect reactions involving the breaking of strong bonds to have high activation energies and to be very slow. This would in part explain why many organic reactions are slow.

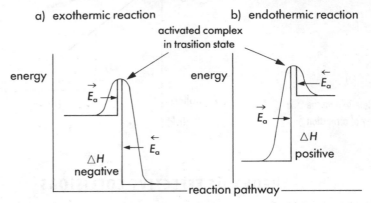

Fig. 9.13 Reaction profiles

$\overrightarrow{E_a}$ = activation energy of forward $(\overrightarrow{E_a})$ and reverse $(\overleftarrow{E_a})$ reaction

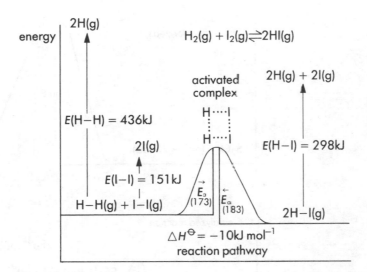

Fig. 9.14 Formation of an activated complex in a transition state

■ Breaking strong covalent bonds requires energy and may lead to high activation energies resulting in slow reactions

Figure 9.13(a) shows an exothermic reaction and figure 9.13(b) shows an endothermic reaction. Notice that the activation energy for the endothermic direction of any chemical change must always be greater than the activation energy for the exothermic direction.

E_a AND THE DISTRIBUTION OF ENERGY

Figure 9.15 indicates how energy is distributed between a collection of molecules and shows how the proportion of molecules (with enough energy to react) varies with temperature. All you need to know about the function f(E,n) represented on the y-axis is that it makes the total area under the curve proportional to the total number of molecules and the area under any portion of the curve proportional to the number of molecules with energies in that range. Three important points follow from this.

As the temperature rises

1) the peak of the curve moves to the right so the mean value of the function f(E,n) and, therefore, the mean energy of the molecules increases
2) the curve flattens so the total area under it and, therefore, the total number of molecules remains constant
3) the area under the curve to the right of E_a and, therefore, the number of molecules colliding with enough energy to cause a reaction roughly doubles for every ten degree rise in temperature.

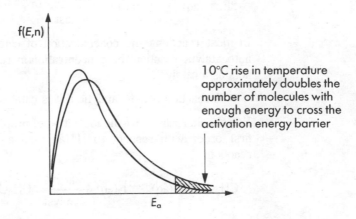

Fig. 9.15 Maxwell-Boltzmann distribution of energies

■ Catalysts speed up reactions but are not consumed by reactions and therefore do not appear as reactants in the overall equations
■ Catalysts provide alternative reaction pathways with activation energies lower than those of the uncatalysed reactions: see Fig. 9.16
■ Catalysts speed up the rate of attainment of equilibrium for a reversible reaction without altering its composition at equilibrium

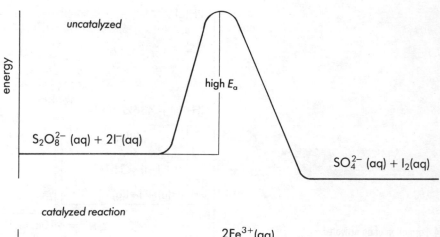

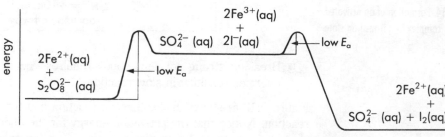

Fig. 9.16 Catalysts provide an
alternative pathway with lower E_a

- Catalysts lower by the same amount the activation energies of the forward and reverse reactions of a reversible reaction

- An homogeneous catalyst takes part in a reaction and an increase in its concentration will increase the speed of the rate determining step

The oxidation of iodide anions by peroxodisulphate(VI) anions is slow:

$$S_2O_8^{2-}(aq) \ + \ 2I^-(aq) \quad \xrightarrow[\Rightarrow \text{ slow}]{\text{high } E_a} \quad 2SO_4^{2-}(aq) \ + \ I_2(aq)$$

Iron(II) or iron(III) cations catalyse the reaction by providing an alternative pathway:

$$S_2O_8^{2-}(aq) \ + \ 2Fe^{2+}(aq) \quad \xrightarrow[\Rightarrow \text{ fast}]{\text{low } E_a} \quad 2SO_4^{2-}(aq) \ + \ 2Fe^{3+}(aq)$$

$$2Fe^{3+}(aq) \ + \ 2I^-(aq) \quad \xrightarrow[\Rightarrow \text{ fast}]{\text{low } E_a} \quad 2Fe^{2+}(aq) \ + \ I_2(aq)$$

In most reactions the concentration of catalyst remains constant with time but in an autocatalytic reaction the concentration increases with time because one of the products acts as a catalyst.

- A reaction is autocatalytic if it is catalysed by one of its products

You should realise that the iodination of propanone is autocatalytic because the reaction is first order with respect to $H^+(aq)$ and aqueous hydrogen ions are produced in the reaction:

$$CH_3COCH_3(aq) \ + \ I_2(aq) \quad \xrightarrow[\text{catalyst}]{\text{acid}} \quad CH_3COCH_2I(aq) \ + \ \underset{\uparrow}{H^+(aq)} \ + \ I^-(aq)$$

product acts as catalyst

- An heterogeneous catalyst provides a surface onto which reactants are adsorbed and from which products of the reaction are desorbed

- Many industrially important heterogeneous catalysts are d-block transition metals that use their 3d and 4s subshells to adsorb reactant molecules and lower activation energies by weakening bonds in the adsorbed molecules

EXAMINATION QUESTIONS AND OUTLINE ANSWERS

Q1 a) i) Using the same set of axes draw **two** curves to show the distribution of energies in the same sample of gas molecules at different temperatures T_1 and T_2, where T_2 is the higher temperature. Label the axes and label each curve. What is the significance of the area under the curves? *(6)*

ii) By reference to the curves explain why an increase in temperature increases the rate of a gaseous reaction. *(4)*

b) i) Explain why the addition of a catalyst increases the rate of a reaction. *(3)*

ii) Write an equation for a reaction which is catalysed by a transition **metal**. Name the catalyst. *(3)*

c) The data below refer to a reaction between **X** and **Y**.

Experiment	Initial concentration/ mol dm^{-3}		Initial rate of reaction/ mol dm^{-3} s^{-1}
	X	**Y**	
1	0.25	0.25	1.0×10^{-2}
2	0.50	0.25	4.0×10^{-2}
3	0.50	0.50	8.0×10^{-2}

i) Deduce the order of reaction with respect to both **X** and **Y**. Write a rate expression for the reaction. *(3)*

ii) Calculate a value for the rate constant. Include units in your answer. *(3)*

iii) Calculate the initial rate of reaction using an initial concentration of 0.80 mol dm^{-3} for both **X** and **Y**. *(3)*

(AEB 1988)

Q2 a) Define and briefly explain the meaning of **each** of the following terms:

rate of reaction;   rate constant (coefficient);   order of reaction. *(5)*

b) In three different experiments (A, B and C), the following initial rates of reaction were measured for the alkaline hydrolysis of 1-bromobutane at 60 °C.

$$C_4H_9Br + OH^- = C_4H_9OH + Br^-$$

Experiment	*Initial concentration/mol dm^{-3}*		*Initial rate of decrease of OH$^-$ concentration/mol dm^{-3} s^{-1}*
	C_4H_9Br	OH^-	
A	0.1	0.1	1×10^{-5}
B	0.2	0.2	4×10^{-5}
C	0.2	0.05	1×10^{-5}

i) Calculate the order of reaction with respect to
  1) 1-bromobutane,
  2) the hydroxide ion.
  Explain your reasoning. *(4)*

ii) Give the equation for the reaction. *(1)*

iii) Give the overall order of the reaction. *(1)*

iv) Calculate
  1) the numerical value of the rate constant,
  2) the units for the rate constant. *(2)*

v) Describe the mechanism of this reaction and explain the importance of the kinetic data in determining the mechanism. *(4)*

c) Give briefly the essential experimental details of a method of determining the rate of this reaction **or** of any **one** other reaction of your choice.   *(3)*

(WJEC 1988)

Outline answers

Q1

(a) (i)(ii) See Fig 9·15 and text page 129

(b) (i)(ii) See Fig 9·16 and text page 129

(c) (i) expt 1 → 2 doubles [X] and quadruples rate
⇒ second order for X
expt 2 → 3 doubles [Y] and doubles rate
⇒ first order for Y
⇒ rate = K[X]²[Y]

(ii) for expt 1 $\dfrac{1.0\times10^{-2}}{(\text{mol dm}^{-3}\text{ s}^{-1})} = K\dfrac{(0.25)^2(0.25)}{(\text{mol dm}^{-3})^2(\text{mol dm}^{-3})}$
⇒ K = 0.64 mol⁻² dm⁶ s⁻¹

(iii) initial rate = 0.64 × (0.80)² × (0.80)
= 32.8 mol dm⁻³ s⁻¹

Q2

(a) See chapter 9

(b) (i) (1) First order
(2) First order

(ii) $-\dfrac{d[OH^-]}{dt} = K[C_4H_9Br][OH^-]$

(iii) second order overall

(iv) (1) 1 × 10⁻³
(2) mol⁻¹ dm³ s⁻¹

(v) S_{N2} See Fig. 16·3.

(c) See chapter 9

EXAMINATION QUESTION AND TUTOR'S ANSWER

Q3   Nitrogen(II) oxide reacts with chlorine to form nitrosyl chloride according to the equation

$2NO + Cl_2 \rightleftharpoons 2NOCl$

The progress of the reaction may be followed by measuring the change in the total pressure of the mixture with time at 300 K. The initial rate of increase of the concentration of NOCl was obtained by this means.

	Initial concentrations /mol dm^{-3}	/mol dm^{-3}	Initial rate of increase of NOCl /mol dm^{-3} s^{-1}
Experiment	[Cl$_2$]	[NO]	
1	0.10	0.10	0.0001
2	0.10	0.20	0.0004
3	0.10	0.30	0.0009
4	0.20	0.10	0.0002
5	0.30	0.10	0.0003

a) i) What is the order of the reaction with respect to NO?

Second

ii) What is the order of the reaction with respect to Cl$_2$?

First

iii) What is the overall order of the reaction?

Third   (3)

b) i) Write a rate equation for the reaction.

$$\frac{d[NOCl]}{dt} = k[NO]^2[Cl_2]$$

ii) Calculate the value of the rate constant k, and give its units, using the results of ONE of the experiments.

$$rate = 0.0001 \; mol \, dm^{-3} \, s^{-1}$$
$$when \; [Cl_2] = 0.10 \quad mol \, dm^{-3}$$
$$[NO] = 0.10 \quad mol \, dm^{-3}$$
$$\Rightarrow \quad 0.0001 = k(0.10)^2(0.10)$$
$$\Rightarrow \quad k = \frac{0.0001}{(0.10)^2(0.10)}$$
$$= 0.1$$

Answer $k = 0.1 \; mol^2 \, dm^{-6} \, s^{-1}$ (6)

c) The production of nitrosyl chloride from NO and Cl$_2$ is catalysed by the presence of charcoal.

With the aid of a reaction pathway diagram explain the function of the catalyst.

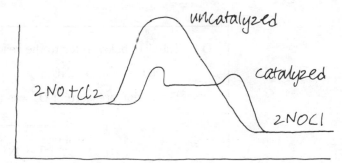

The charcoal lets the reaction take place at its surface — it is a heterogeneous catalyst (solid but reactants are gases) — to give an alternative route with lower activation energy. (4)

d) A molecule of nitrosyl chloride is polarised as follows:

$$\overset{\delta-}{Cl}-\overset{\delta+}{NO}$$

Nitrosyl chloride will add to double bonds in a similar way to hydrogen bromide which is also polarised.

$$\overset{\delta-}{Br}-\overset{\delta+}{H}$$

i) Draw the structure of the product from the reaction of propene with nitrosyl chloride.

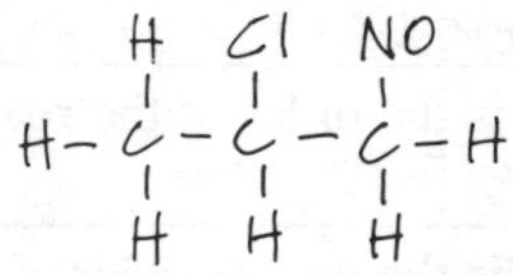

ii) Suggest a mechanism for the reaction of NOCl with ethene.

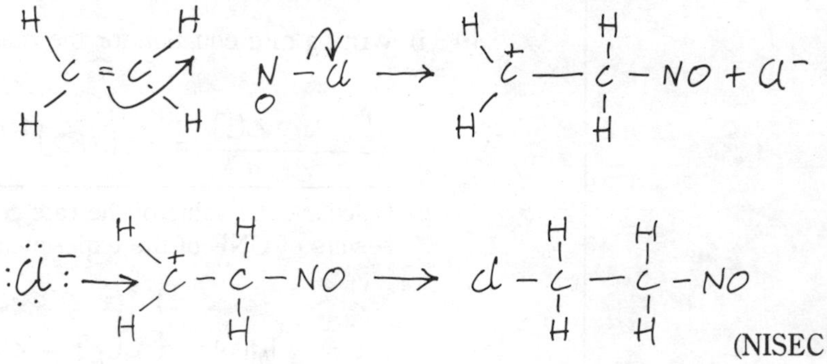

(5)

(NISEC 1988)

STUDENT'S ANSWER WITH EXAMINER'S COMMENTS

Q4 The data below refer to the reaction

$$2NO + O_2 \rightarrow 2NO_2$$

The partial pressure of O_2 was the same for each experiment.

Initial rate /N m^{-2} s^{-1}	p^2_{NO} /N^2 m^{-4}
1.70	0.010
6.80	0.040
27.2	0.16
61.2	0.36
108	0.64
170	1.00

a) Plot the data.

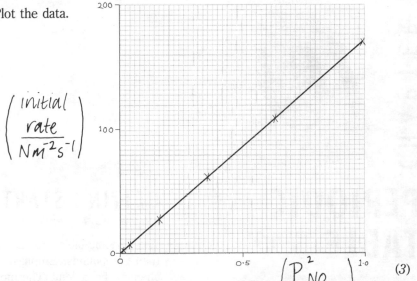

$$\left(\frac{initial\ rate}{Nm^{-2}s^{-1}}\right)$$

(3)

Good. Always include units on the axes.

b) What is the order of reaction with respect to NO? $\left(\dfrac{P_{NO}^{2}}{N^{2}m^{-4}}\right)$
Justify your answer.

First order because the graph is a straight line ✗

_____

_____ (2)

No. The rate is proportional to the *square* of the partial pressure of NO, so it is second order.

c) When the partial pressure of O_2 was doubled to a new constant value the gradient of the graph in a) doubled. What is the order with respect to O_2? Explain your answer.

First order because the partial pressure is like concentration for a gas. The reaction is first order if the rate doubles when you double concentration. ✓

(2)

d) Give the rate equation for this reaction.
What are the units of the rate constant? $(Nm^{-2})^2\ (Nm^{-2})$

rate = k P_{NO}^{2} × P_{O_2} ✓

units are s^{-1} ✗

_____ (2)

$Nm^{-2}\,s^{-1}$

Units of k must balance with the units of the other terms: i.e. $s^{-1}(Nm^{-2})^{-2} \Rightarrow$ $N^{-2}m^{4}s^{-1}$

e) To calculate the activation energy of another gas phase reaction, $2N_2O \rightarrow 2N_2 + O_2$, the reaction was monitored at various temperatures, and a graph of $\ln k$ against $1/T$ (in K^{-1}) was plotted (k is the rate constant for the reaction). The gradient of the graph had a numerical value of -2.95×10^4. The Arrhenius equation may be expressed in the form

$\ln k = \ln A - \dfrac{E_A}{RT}$. Calculate the activation energy for this reaction,

stating the units.

$(R = 8.31\,J\,K^{-1}\,mol^{-1})$

gradient is $-\dfrac{EA}{R} = -2.95 \times 10^4$
$\Rightarrow EA = 2.95 \times 10^4 \times 8.31\,Jk^{-1}\,mol^{-1}$
$= 245(145)\,Jk^{-1}\,mol^{-1}$

(4)

(*Total 13 marks*)
(ULSEB 1988)

Too many insignificant figures will lose a mark.

No. This would be the units of entropy change. Energy units are $J\,mol^{-1}$ or $245\,kJ\,mol^{-1}$

CHAPTER

10

PERIODIC TABLE

GETTING STARTED

The periodic table provides a framework for studying the elements and their compounds. Emphasis is upon patterns in physical properties and chemical behaviour. The most important features are a) the similarities shown by a group and by a transition series, b) the trends within a group and across a period and c) the anomalous behaviour of the elements from lithium to fluorine in the second period. Reactions classified as acid-base and redox may be related to the bonding and structure of the reactants and products.

ESSENTIAL PRINCIPLES

The periodic table is extremely important. You must study it regularly and very thoroughly. All syllabuses require a knowledge and understanding of the broad patterns in the physical and chemical properties of the elements and their compounds down groups and across a period. You will also have to know the chemistry of certain groups and specific elements and their compounds in detail. Check your syllabus to find out precisely which groups, elements and types of compound you need to study.

The Royal Society of Chemistry (Burlington House, Piccadilly, LONDON W1V 0BN) and Time Life International Books have published an illustrated periodic table in the form of a large wall-chart. It is excellent. It is in colour with detailed notes and photographs of each element. Many of my students have their own. Do get yourself a copy. You can order it through bookshops or directly from The Distribution Centre, Blackhorse Road, Letchworth, Hertfordshire, SG6 1HN and save over half the cost if you order ten copies. Put it on your wall at home. Each morning read about one element and think about it when you are having your breakfast. Stick to this medium-term plan until you have covered all the elements in your syllabus.

> It will help to have your own illustrated periodic table.

THE PERIODIC TABLE

- The periodic table is an arrangement of all the elements in increasing atomic number order to form four areas called s-, p-, d- and f-blocks
- The s-block elements are divided vertically into two groups and the p-block elements are divided vertically into six groups
- The d-block elements are divided horizontally into three rows usually called the transition series of elements
- The f-block elements are divided horizontally into two rows usually called the lanthanoids (lanthanides) and actinoids (actinides)

Fig. 10.1 Periodic classification of the elements

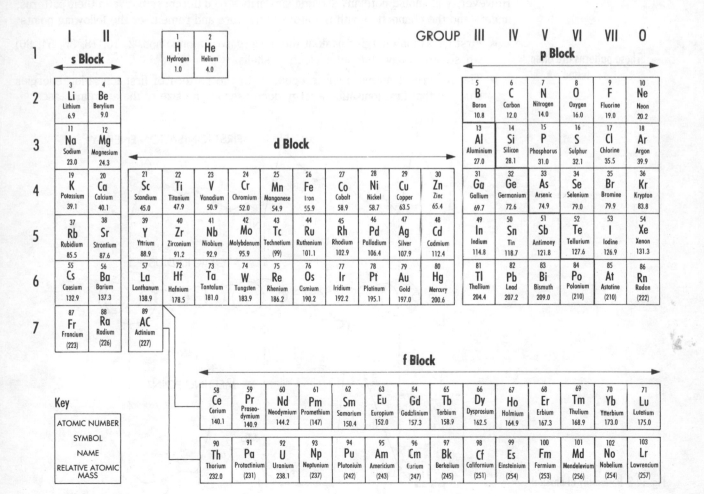

PERIODICITY AND THE PERIODIC TABLE

■ Periodicity is the recurrence of similar properties at regular intervals when elements are arranged in increasing atomic number order

You need to understand what we mean by periodicity but you do not need to learn a formal definition by heart. The word 'periodic' means 'occurring at regular intervals.' You will be familiar with time intervals counted in seconds, minutes, hours, days, etc: e.g., a birthday occurs regularly at constant 52-week intervals (years). And you may know about musical intervals counted in notes: e.g., a 'C' on the piano keyboard occurs regularly at constant 8-note intervals (octaves). In the 'periodic' table the intervals are counted in elements but the size of the interval is NOT constant. The number of elements in each interval changes from 2 to 32 in the following sequence: 2, 8, 8, 18, 18, 32. For example, similar properties shown by the family of alkali metals recur at these intervals:

alkali metal:	(hydrogen)	lithium	sodium	potassium	rubidium	caesium	francium
atomic no.:	1	3	11	19	37	55	87
interval:		2	8	8	18	18	32

PROPERTIES OF ELEMENTS

We usually show the periodic variation of certain physical properties with atomic number graphically: see Fig. 10.2. Notice that we connect the points only to show the pattern. Bar or spike charts serve the same purpose but plotting and joining points is quicker than drawing bars. Your school or college may have computer programs that you can use to plot and display a variety of properties against atomic number even in a 'three-dimensional representation'. First ionisation energy, atomic radius and melting point must be considered for the first twenty elements at least. Electronegativity, atomic volume and boiling point may also be included, together with the standard molar enthalpy changes of melting and boiling.

You will not be expected to remember actual values or very fine details of these graphs. However, you should note any striking similarities or differences between their patterns, understand the connection with the atomic structure and remember the following points:

These patterns are often tested.

■ First ionisation energies peak at the noble gases (atomic nos. 2, 10, 18, 36, 54, 86) whose atoms have full outer electron shells

■ The pattern of atomic radii is opposite to the pattern of first ionisation energies showing that first ionisation energy decreases as the size of the atom increases

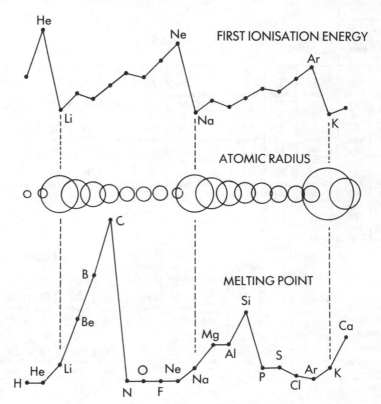

Fig. 10.2 Periodicity of physical properties

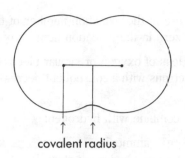 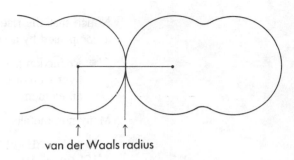

Fig. 10.3 Atomic radii   covalent radius   van der Waals radius

You may need to distinguish between the covalent radius (half the distance between the nuclei of two atoms in the same molecule) and the van der Waals radius (half the distance between the nuclei of two atoms in adjacent molecules): see Fig. 10.3.

- Melting and boiling points peak at group 4 (atomic nos. 6, 14, 32, 50, 82) headed by carbon with its giant covalent diamond structure

- The pattern of standard molar enthalpy changes of melting and boiling is similar to the pattern of melting and boiling points showing that melting and boiling increases as the enthalpy change increases

STRUCTURE AND BONDING

You can reveal pattern in the structure and bonding of s-block and p-block elements by colour-coding or shading a 'shortened form' of the periodic table: see Fig. 10.4. You should mentally associate giant metallic or covalent structures with the elements in groups 1 to 4 and the appropriate parts of the histograms of physical properties against atomic number. You should also mentally associate simple molecular structures with the elements in group 5 and beyond. You should compare these patterns with those for the properties of the compounds of the elements.

giant covalent structures

Li	Be	B	C	N_2	O_2	F_2	Ne
Na	Mg	Al	Si	P_4	S_8	Cl_2	Ar
K	Ca	Ga	Ge	As	Se	Br_2	Kr
Rb	Sr	In	Sn	Sb	Te	I_2	Xe
Cs	Ba	Tl	Pb	Bi	Po	At	Rn

simple covalent molecules

giant metallic structures

Fig. 10.4 Structure and bonding in the s- and p-block elements

The d-block elements are metallic and rather similar to each other
The f-block elements are even more alike and are not needed for A-level

REDOX REACTIONS

- Oxidation may involve gain of oxygen or similar electronegative non-metals, loss of hydrogen or loss of electrons with a consequent increase in the oxidation number of an element

Many elements burn in oxygen or fluorine to form their highest oxidation state oxides or fluorides: e.g.

$$K(s) + O_2(g) \longrightarrow KO_2(s) \quad \text{potassium superoxide}$$
$$S_8(s) + 24F_2(g) \longrightarrow 8SF_6(g) \quad \text{sulphur hexafluoride}$$

Notice that the rise above zero in the oxidation number of the potassium and sulphur is accompanied by a fall below zero in the oxidation number of the oxygen and fluorine

■ Reduction may involve loss of oxygen or similar electronegative non-metals, gain of hydrogen or gain of electrons with a consequent decrease in the oxidation number of an element

Many non-metallic elements combine with hydrogen:

$N_2(g) + 3H_2(g) \rightleftharpoons 2NH_3(g)$ ammonia
$\frac{1}{2}Cl_2(g) + \frac{1}{2}H_2(g) \longrightarrow HCl(g)$ hydrogen chloride

Notice that the fall below zero in the oxidation number of the nitrogen and chlorine is accompanied by a rise above zero in the oxidation number of the hydrogen.

■ Redox is the simultaneous occurrence of reduction and oxidation

You should be able to recognise a redox reaction by assigning oxidation numbers to each elemental species in an equation and finding a change in oxidation number for at least one species. Apply the oxidation number rules carefully (see Chapter 13) to avoid pitfalls: e.g. the combination of lithium with hydrogen is a redox reaction but the lithium is being oxidised (loss of electrons) and the hydrogen reduced (gain of electrons)!

$Li(s) + \frac{1}{2}H_2(g) \longrightarrow Li^+H^-(s)$ lithium hydride

Lithium is a more powerful reducing agent than hydrogen. This, as you will see, is what you would expect if you place hydrogen at the top of group 1 in the periodic table.

■ Metallic elements tend to be reducing agents and show an increase in oxidation number in redox reactions

■ Non-metallic elements tend to be oxidising agents and show a decrease in oxidation number in redox reactions: see Fig. 10.5

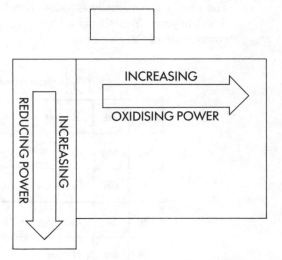

Fig. 10.5 Trend in redox reactivity of the elements

This fits with the broad pattern established in GCSE work that the reactivity of the alkali metals increases with increasing atomic number DOWN the group and that the reactivity of the halogens increases with decreasing atomic number UP the group.

PROPERTIES OF COMPOUNDS

The character of the elements is reflected in the chemical properties of their compounds. You need to study the patterns in properties of the oxides, chlorides and hydrides of the elements usually in the period from sodium to argon.

OXIDATION NUMBERS

Every element can be assigned an oxidation number according to a set of rules: see Chapter 13. For uncombined elements the oxidation number is 0. For combined elements the number may vary from −4 up to +7.

■ The maximum oxidation numbers of the first twenty elements exhibit periodicity: see Fig. 10.6

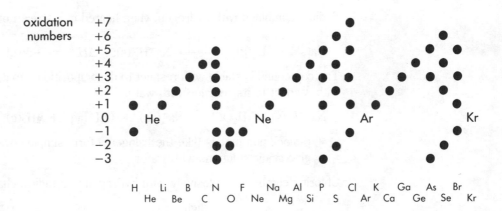

oxidation numbers

Fig. 10.6 Periodicity of oxidation numbers

OXIDES

You should remember from your GCSE work that magnesium forms a basic oxide, i.e. an oxide that reacts with acids to form salts and water:

$$MgO(s) + 2H^+(aq) \longrightarrow Mg^{2+}(aq) + H_2O(l)$$

and that sulphur forms acidic oxides, i.e. an oxide that reacts with bases to form salts and water:

$$SO_2(g) + 2OH^-(aq) \longrightarrow SO_3^{2-}(aq) + H_2O(l)$$
$$SO_3(s) + 2OH^-(aq) \longrightarrow SO_4^{2-}(aq) + H_2O(l)$$

In between, aluminium forms an amphoteric oxide, i.e. an oxide that can act as an acid and a base:

$$Al_2O_3(s) + 6H^+(aq) \longrightarrow 2Al^{3+}(aq) + H_2O(l)$$
$$Al_2O_3(s) + 2OH^-(aq) \longrightarrow 2AlO_2^-(aq) + H_2O(l)$$

- Metallic elements tend to form basic oxides which, if highly ionic, form hydroxides with water
- Non-metallic elements tend to form acidic oxides which are covalent and produce oxoacids with water: see Fig. 10.7
- Basic oxides have giant structures that are often ionic whereas acidic oxides have simple molecular structures that are covalent

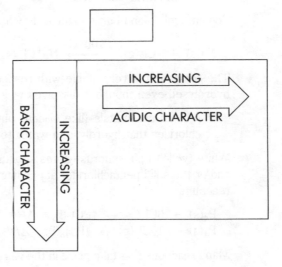

Fig. 10.7 Trend in acid-base character of oxides

HYDRIDES

You will know something about ammonia (NH_3), hydrogen chloride (HCl), methane (CH_4) and water (H_2O) from your GCSE work. These are all hydrides of non-metals. You now need to know something about hydrides of metals: see Chapter 11.

- s-block metals form ionic hydrides containing the hydride anion H^-, in a giant structure, that react with water to give a base and hydrogen

Sodium combines with hydrogen when heated in a stream of the dry gas:

$$Na(s) + \tfrac{1}{2}H_2(g) \xrightarrow{100\,°C} Na^+H^-(s); \quad \Delta H^\ominus = -56\,kJ\,mol^{-1}$$

The compound is stable with respect to decomposition into its elements but very unstable with respect to its reaction with water:

$$Na^+H^-(s) + H_2O(l) \rightarrow Na^+(aq) + OH^-(aq) + \tfrac{1}{2}H_2(g); \quad \Delta H^\ominus = -554\,kJ\,mol^{-1}$$

- p-block non-metals (like the halogens) form simple covalent molecular hydrides that give acidic solutions with water

Chlorine combines explosively with hydrogen in a radical chain mechanism:

$$\tfrac{1}{2}H_2(g) + \tfrac{1}{2}Cl_2(g) \xrightarrow[\text{flame}]{\text{light or}} HCl(g); \quad \Delta H^\ominus = -92\,kJ\,mol^{-1}$$

The compound is stable with respect to decomposition into its elements but 'dissolves' rapidly in water where it completely ionises:

$$HCl(g) + H_2O(l) \rightarrow H_3O^+(aq) + Cl^-(aq); \quad \Delta H^\ominus = -167\,kJ\,mol^{-1}$$

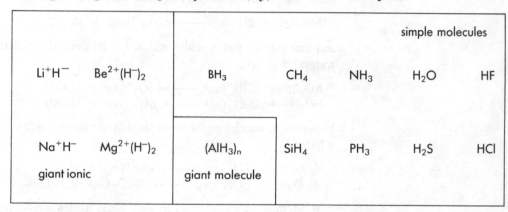

Fig. 10.8 Trend in properties of hydrides

CHLORIDES

The pattern in the structure and bonding of the chlorides is similar to the patterns for the oxides and hydrides.

- s-block metals form ionic chlorides that have giant structures and dissolve in water to form neutral solutions

Sodium ignites and burns vigorously when melted in chlorine gas:

$$Na(s) + \tfrac{1}{2}Cl_2(g) \xrightarrow{100\,°C} Na^+Cl^-(s); \quad \Delta H^\ominus = -411\,kJ\,mol^{-1}$$

The compound is very stable with respect to decomposition into its elements and is not hydrolysed even though it dissolves well in water.

- p-block non-metals (like silicon, phosphorus and sulphur) form covalent molecular chlorides that hydrolyse in water to give acidic solutions

White (yellow) phosphorus ignites spontaneously in chlorine to form the liquid trichloride and/or the solid pentachloride, depending upon the conditions and relative amounts of the reactants;

$$P_4(s) + 6Cl_2(g) \rightarrow 4PCl_3(l); \quad \Delta H^\ominus = -1279\,kJ\,mol^{-1}$$
$$P_4(s) + 10Cl_2(g) \rightarrow 4PCl_5(l); \quad \Delta H^\ominus = -1774\,kJ\,mol^{-1}$$

Many reactions that take place in the gas phase produce a flame that gives out light as well as heat because they are exothermic enough to vaporise the reactants and product(s). Many elements burn in chlorine because the halogen is so reactive and the chloride so volatile. (Note that the volatility of the chlorides is behind the use of concentrated hydrochloric acid in the flame test.)

- When a metal having two chlorides burns in chlorine the compound formed is often covalent with the metal in its higher oxidation state, e.g.:

$$2Fe(s) + 3Cl_2(g) \rightarrow 2FeCl_3(s) \text{ [iron(III) chloride]}$$

Hydrolysis of chlorides

- In general the tendency to hydrolyse increases as the chlorides become more covalent in character: see Fig. 10.9

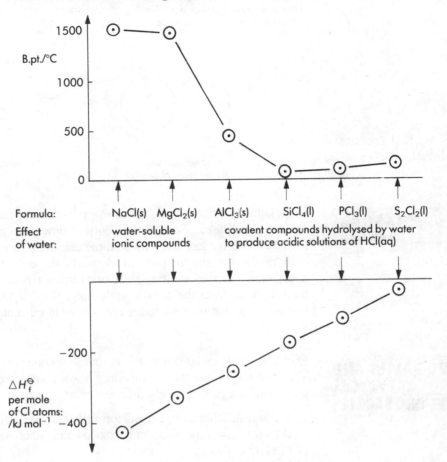

Fig. 10.9 Trend in properties of chlorides

PATTERNS IN BONDING

- The covalent character of the oxides, hydrides and chlorides increases with increasing atomic number of the elements from left to right across a period of the periodic table

- The ionic character of the oxides, hydrides and chlorides increases with increasing atomic number of the elements from top to bottom down a group of the periodic table: see Fig. 10.10

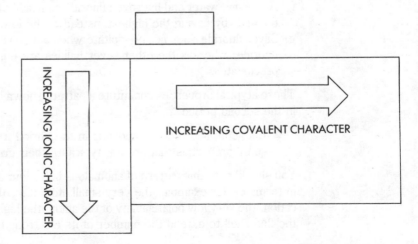

Fig. 10.10 Pattern in covalent-ionic character of the oxides, chlorides and hydrides of the elements

These patterns in bonding can be discussed in terms of the polarising power of cations and the polarisability of anions: see Fig. 10.11. You can think of a cation distorting an anion by attracting its negative charge.

The greater the charge density (amount of positive charge in a given volume) of the cation, the greater is the distortion or polarisation of the anion and the less ionic (more covalent) is the character of the bonding: see Chapter 4.

Fajans' rules:

covalent character of the bonding increases
or ionic character of the bonding decreases
if the cation radius decreases
if the anion radius increases
if the charge on the ions increase

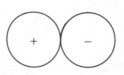

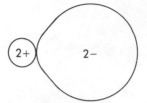

ionic bonding
no covalent character

small cation polarises large anion
some covalent character

Fig. 10.11 Polarisation of polarisable anion by polarising cation

DIAGONAL RELATIONSHIPS

From left to right across the period (Li to F) metallic character decreases and non-metallic character increases. From top to bottom down the groups (Li to Cs, Be to Ba, etc.) metallic character increases and non-metallic character decreases. Taken together, these two trends give rise to pairs of elements (Li and Mg, Be and Al, B and Si, etc.) in neighbouring groups showing similarities in their properties; e.g. lithium and magnesium burn in air to form the normal oxide only, their carbonates decompose on heating in a bunsen flame and their chlorides are soluble in ethanol and slowly hydrolysed by water.

ANOMALIES AND ATYPICAL PROPERTIES

Each element from lithium to fluorine in the second period is the first member of its group. These elements and their compounds show some properties that break the patterns of group similarities and trends. For example:

a) If you heat lithium in air it will form the normal oxide (Li_2O) and lithium nitride (Li_3N): the other alkali metals form peroxides and superoxides and do not form their nitrides directly in this way

b) You can decompose lithium carbonate or nitrate completely to lithium oxide in a bunsen flame: the same is not true for the other alkali metal carbonates and nitrates

c) Beryllium does not react with water or steam, whereas magnesium and the other metals in group 2 will react to form hydrogen

d) Boron forms a stable trichloride (BCl_3) in which the boron atom has only 6 electrons in its outer shell, whereas aluminium forms the dimer Al_2Cl_6 to give 8 electrons in the outer shell of Al

e) Tetrachloromethane (CCl_4) does not react even with hot water, whereas the tetra-chlorides of the other group 4 elements are all readily hydrolysed

f) Ammonia, water and hydrogen fluoride contradict the trend in the hydrides of groups 5, 6 and 7, by having the highest instead of the lowest boiling points

g) Silver fluoride does not precipitate when aqueous silver nitrate is added to an aqueous fluoride, whereas the other silver halides are only sparingly soluble and readily form precipitates.

These atypical properties constitute a pattern shown by the elements and their compounds in the second period.

■ Elements and their compounds in the period from lithium to fluorine inclusive have some properties that are not typical of their group

You should note this pattern of anomalous behaviour and be able to suggest an explanation in terms of, for example, the very small atom size, the very high polarising power of the cation, the very low polarisability of the anion, the high electronegativity and the inability of the 2nd shell to extend the number of its electrons beyond eight.

FACTS THAT BREAK THE PATTERN

The construction of a periodic table has played an important part in the development of our theories of atomic structure and bonding. These theories in turn help us to 'explain' the similarities and trends in a group and across a period. Even the atypical properties associated with the second period elements (Li to F) may be explained and regarded as a pattern. However, this periodic classification of the elements has its limitations.

You should realise that the patterns of chemical behaviour in the periodic table are not perfect. There are inconsistencies and 'breaks' in the patterns. Here are some examples. You should keep your eyes open for others.

a) You can see that the alkali metals become more reactive with increasing atomic number by the way a small piece behaves when placed onto water in a large glass trough; see Chapter 11. The reactions become more vigorous and give out more heat. If we use the standard molar enthalpy changes of formation of the ionic alkali metal halides and the covalent hydrogen halides as a quantitative measure of reactivity we find that some of these values do NOT fit nicely the broad pattern of increasing reactivity DOWN the alkali metal group and UP the halogen group! (see Fig. 10.12)

b) As a general rule the more reactive an element, the more stable are its compounds. The reactivity of the halogens increases from iodine to fluorine as the atomic number decreases. So we should expect chlorine compounds to be more stable than iodine compounds. They usually are. However, white silver chloride and off-white silver bromide rapidly darken in sunlight but silver iodide remains yellow and does not photodecompose!

c) Although mercury is a dense metal it is also a liquid! It becomes a solid at $-39\,°C$.

↓ A B →	HYDROGEN	FLUORINE	CHLORINE	BROMINE	IODINE
Hydrogen	0	−271	−92	−36	+27
Lithium	−91	−616	−409	−351	−270
Sodium	−56	−574	−411	−361	−288
Potassium	−58	−567	−437	−394	−328
Rubidium	−52	−558	−435	−395	−334
Caesium	−54	−554	−443	−406	−347

e.g. $Na(s) + \frac{1}{2}H_2(g) \rightarrow NaH(s);\quad \Delta H^{\ominus} = -56\,kJ\,mol^{-1}$

Fig. 10.12 Standard molar enthalpy changes of formation/ $kJ\,mol^{-1}$ of AB

SAMPLE QUESTION AND ANSWER

Q1 This question concerns the patterns and trends in the properties of the elements and their compounds. The letters shown in the outline of the periodic table below are used to indicate the positions but NOT the symbols of seven elements. Use these letters when answering the questions.

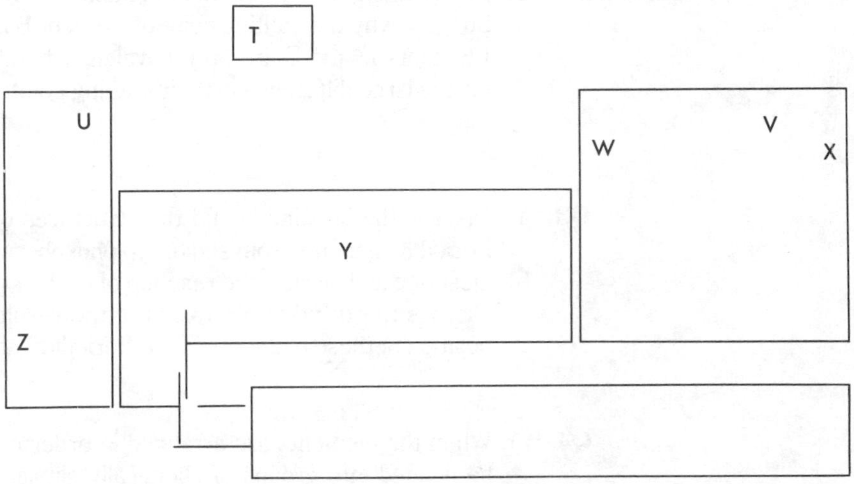

a) Give the letter(s) of the element(s) in the

   i) s-block; U Z   ii) p-block V W X   iii) d-block Y

b) Give the letter of an element with an oxide that is

   i) basic: Z   ii) acidic: V   iii) amphoteric: W

c) Give the letter of the element that has the
   i) lowest first ionisation energy: __Z__
   ii) highest electronegativity index: __V__
   iii) greatest range of oxidation states in its compounds: __Y__

d) Using the letters T and U as the symbols, write formula and state the type of bonding in the compound they form when heated together.
   formula: __UT_2__   type of bonding: Ionic $U^{2+}(T^-)_2$

e) Using the letters V and W as the symbols, predict the formula and shape of the molecule formed when they react together.
   formula: __WV_3__   shape of molecule: trigonal planar

EXAMINATION QUESTIONS AND OUTLINE ANSWERS

Q2

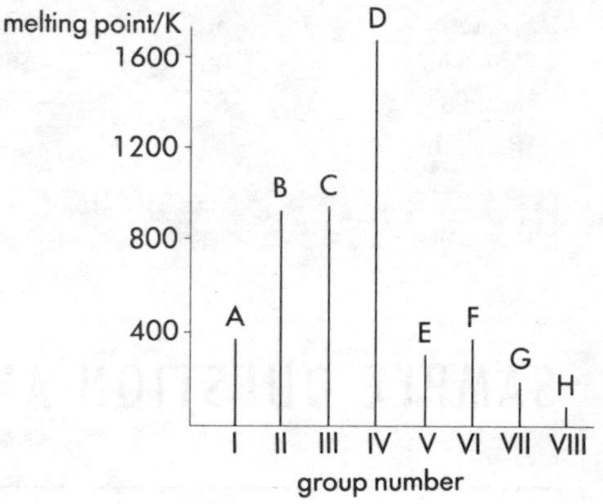

The graph shows the melting points for the elements across a Period in the Periodic Table.

a) Identify the Period represented by the graph.   *(1)*
b) The bonding in both elements A and B is metallic.
   Suggest why the melting point of element B is higher than that of element A.   *(1)*
c) Elements D and E are both covalently bonded. In terms of structure, account for the large difference in their melting points.   *(1)*
                                                                                    (3)
                                                                            (SEB 1988)

Q3 a) Discuss the bonding in and the structures of the chlorides of the
        Period 3 elements from sodium to phosphorus inclusive.   *(14)*
     b) Describe and discuss the reaction of each of these chlorides with water.   *(11)*
     c) Discuss the trend in pH values of the resulting solutions in relation to the
        position of these elements in the Periodic Table.   *(5)*
                                                                        (JMB (As specimen))

Q4 a) When the elements are arranged in order of increasing atomic number, they can
        be divided into groups of chemically similar elements. Explain **briefly** how this
        can be explained in terms of the electronic configuration of the elements.   *(4)*
        State, and as far as possible explain, the changes which occur in (i) the atomic
        radius, and (ii) the first ionisation energy of the elements in one of the first two
        periods of the Periodic Table, that is lithium to neon or sodium to argon.   *(8)*
     b) Discuss the changes in bonding and properties of the oxides of the elements
        in the second period, that is sodium to argon.   *(8)*
                                                                            (ODLE 1988)

Q5 a) This question concerns the hydrides of the elements carbon, nitrogen, oxygen and fluorine in Period 2 of the Periodic Table and the corresponding hydrides of the elements silicon, phosphorus, sulphur and chlorine in Period 3.

Discuss the variation in the following properties of these hydrides with increasing Atomic Number (i) within a period, and (ii) within a group of the Periodic Table:

1) the ionic/covalent nature of the bonding;   (4)
2) the shapes of the molecules;   (4)
3) the acid/base properties of the molecules.   (4)

Give explanations for the trends which you describe.   (3)

b) Describe the nature of the bonding in sodium hydride.

Explain how the electrolysis of molten sodium hydride supports the existence of this bonding. Write ionic equations for the reactions which occur at the electrodes stating which reaction is an oxidation and which one is a reduction.   (5)

(WJEC 1988)

Q6 Give the formulae of the chlorides (if any) of the elements in the third period of the Periodic Table (sodium to argon). Describe and account for the bonding in these chlorides and their reaction (if any) with water.   (12)

Draw a 'dot-and-cross' diagram to show the electronic structure of a molecule of boron trifluoride, BF_3. Use the electron pair repulsion theory to predict the shape of this molecule.   (4)

State and explain how you would expect boron trifluoride to react with ammonia. Draw a dot-and-cross diagram to show the electronic structure of the product of this reaction.   (4)

(UCLES 1988)

Outline answers

Q2

(a) period 3 (Na – Ar)

(b) atoms provide more electrons to hold the ions together because atom B has 2 outer (valence) electrons but A has only 1 outer electron.

(c) D has a giant covalent structure
    E has a simple molecular structure

Q3, Q4, Q5(a)

These are typical questions on this topic. Study chapter 10 thoroughly. Be prepared to use charts and graphs in your answer. Make it attractive for the examiner to look at, as well as easy to read.

Q5(b)

(b) Ionic bonding Na^+ $H:^-$
Electrolysis of molten sodium hydride gives hydrogen at the ANODE (and NOT the cathode):-

$$2H^- \longrightarrow H_2 + 2e^-$$
oxidation at anode

$$2Na^+ + 2e^- \longrightarrow 2Na$$
reduction at cathode

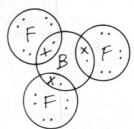

Q6 'dot-and-cross' diagram

no lone pair

Trigonal planar molecule – mutual repulsion of 3 bonded pairs of electrons

STUDENT'S ANSWER WITH EXAMINER'S COMMENTS

Q7 The following graph shows the first ionisation energy of twelve elements plotted against their atomic numbers. The elements in order are represented by the letters A to L with their atomic numbers increasing each time by one. You are not expected to identify the elements and the letter A does not represent hydrogen. Use these letters A to L when answering the questions below.

first ionisation energy

A B C D E F G H I J K L

a) i) Give the letter of a green gas used in water purification.
H ✓ (1)

ii) State the number of electrons in the outer energy level of its atom in the ground state.
7 ✓ (1)

b) i) Give the letters of two unreactive gases whose atoms have eight electrons in their outer energy levels.
A and I ✓ (1)

ii) State one industrial use for such gases.
filling light bulbs ✓ (1)

> **J, K also correct**

c) i) Give the letters of two elements in the s-Block of the periodic table.
B and C ✓ (1)

> **B is an alkali metal in group 1 so count from there to E.**

ii) Give the letter of one element in Group IV of the periodic table.
D ✗ (1)

d) i) State the significance of the drop in ionisation energy from element A to B and I to J.
In A, I, the electron shells are full, so the electron in B or J is further away from the nucleus ✓ (1)

ii) State the significance of the drop in ionisation energy from element C to D and K to L.
Elements C and K have full s-shells ✓

> **Better to say s-orbitals or s-subshells.**

(1)

(AEB AS-level specimen)

CHAPTER 11

s-BLOCK ELEMENTS AND HYDROGEN

THE s-BLOCK ELEMENTS

THE COMPOUNDS OF s-BLOCK ELEMENTS

ANOMALOUS BEHAVIOUR OF LITHIUM AND BERYLLIUM

HYDROGEN

GETTING STARTED

Group 1 (alkali metals) and group 2 (alkaline earth metals) constitute the s-block elements. The emphasis is upon characteristic features of the s-block elements and their compounds, patterns in physical properties and chemical behaviour within each group, and similarities and differences between the two groups. The anomalous properties of lithium and beryllium and the diagonal relationship with magnesium and aluminium are mentioned. Hydrogen is unique and treated separately but it has some features in common with the alkali metals.

ESSENTIAL PRINCIPLES

THE s-BLOCK ELEMENTS

These elements and their compounds have much in common because their properties are governed by s-orbital electrons in the outermost shell of their atoms.

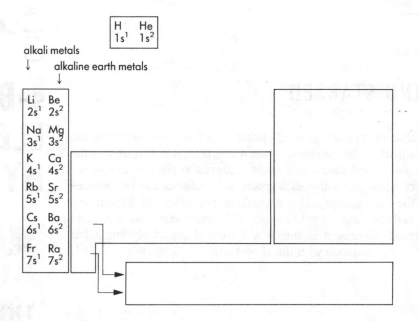

Fig. 11.1 s-block elements and hydrogen

PHYSICAL PROPERTIES

- The s-block elements are all ductile, malleable, electrically and thermally conducting metals, softer than most transition metals with lower densities, melting points, boiling points and standard enthalpy changes of melting and of boiling

The metallic bonding is weaker in the s-block elements than in most transition metals and in both groups its strength decreases as the atomic number and size of the atom increases down the group.

- The alkali metals are softer and have lower densities, melting points, boiling points and standard enthalpy changes of melting and of boiling than the alkaline earth metals in the same period

The trends in physical properties within each group and the differences between the two groups may be explained in terms of the strengths of the metallic bonding and the structures of the elements: see Chapter 3.

- Hardness, density, standard enthalpy change of melting and of boiling all decrease down the groups with increasing atomic number
- All the alkali metals have the same body-centred cubic structure and their melting and boiling points decrease down the group from Li-Cs whilst the alkaline earth metals have different structures and their melting and boiling points vary irregularly in the group

> Alkali metals have body-centred structures, NOT close-packed ones.

SYMBOL	STRUCTURE	m.pt/°C	b.pt/°C	SYMBOL	STRUCTURE	m.pt/°C	b.pt/°C
Li	BCC	181	1342	Be	HCP	1278	2970
Na	BCC	98	883	Mg	HCP	649	1107
K	BCC	63	760	Ca	FCC	839	1484
Rb	BCC	39	686	Sr	FCC	769	1384
Cs	BCC	29	669	Ba	BCC	725	1640

BCC body-centred cubic; HCP hexagonal close packed; FCC face-centred cubic

Fig. 11.2 Physical properties of s-block elements

CHEMICAL PROPERTIES

The s-block metals act as powerful reducing agents by losing their outer s-subshell electrons to form very stable cations with the same electronic configurations as their corresponding noble gases: e.g.

$$Na(2.8.1) \rightarrow Na^+(2.8) + e^- \quad Mg(2.8.2.) \rightarrow Mg^{2+}(2.8) \quad [Ne\ 2.8]$$

The fixed oxidation states (+1 for group 1 and +2 for group 2) in the compounds of the s-block elements can be rationalised by considering the appropriate Born-Haber energy cycles: see Chapter 6.

■ Alkali metals are more reactive than their corresponding alkaline earth metals

REACTION WITH NON-METALS

■ The s-block metals burn vigorously in air to form normal oxides, peroxides or superoxides depending upon their reactivity

GROUP 1	GROUP 2
normal oxides (O^{2-})	normal oxides (O^{2-})
$4Li(s) + O_2(g) \rightarrow 2Li_2O(s)$ $4Na(s) + O_2(g) \rightarrow 2Na_2O(s)$ sodium forms a mixture of the normal oxide and the peroxide on burning in air	$2Be(s) + O_2(g) \rightarrow 2BeO(s)$ $2Mg(s) + O_2(g) \rightarrow 2MgO(s)$ $2Ca(s) + O_2(g) \rightarrow 2CaO(s)$ $2Sr(s) + O_2(g) \rightarrow 2SrO(s)$ $2Ba(s) + O_2(g) \rightarrow 2BaO(s)$
peroxides ($^-O{-}O^-$)	peroxides ($^-O{-}O^-$)
$2Na(s) + O_2(g) \rightarrow Na_2O_2(s)$	$Ba(s) + O_2(g) \rightarrow BaO_2(s)$
superoxides (O_2^-)	superoxides (O_2^-)
$K(s) + O_2(g) \rightarrow KO_2(s)$ $Rb(s) + O_2(g) \rightarrow RbO_2(s)$ $Cs(s) + O_2(g) \rightarrow CsO_2(s)$	there are stable superoxides of Na, Ca, Sr and Ca but they are not formed by burning in air

Fig. 11.3 s-block oxides formed by burning the elements in air

■ The s-block metals react, often violently, with halogens and with sulphur to form halides and sulphides

$$2Na(s) + Cl_2(g) \xrightarrow{heat} 2NaCl(s)$$

$$Mg(s) + S(s) \xrightarrow{heat} MgS(s)$$

Sodium, in the UK, and magnesium, in the rest of the world, are used as reducing agents in the industrial extraction of titanium from $TiCl_4$:

$$4Na + TiCl_4 \xrightarrow{1000\,°C} 4NaCl + Ti$$

■ The s-block metals combine, on heating, with hydrogen to form ionic hydrides that react readily with water

$$Ca(s) + H_2(g) \xrightarrow{heat} CaH_2(s)$$
$$CaH_2(s) + 2H_2O(l) \longrightarrow Ca(OH)_2(s) + 2H_2(g)$$

REACTION WITH WATER

■ The s-block metals react with water to form hydrogen and hydroxides (or oxides) that are strongly basic

$$2Na(s) + 2H_2O(l) \longrightarrow 2Na^+(aq) + 2OH^-(aq) + H_2(g)$$

$$Mg(s) + H_2O(g) \xrightarrow{heat} MgO(s) + H_2(g)$$

The s-block metal cations are so stable that they do not normally discharge at the cathode during electrolysis of their aqueous solutions; the less stable aqueous hydrogen ion discharges instead. Consequently, these very electropositive metals must be extracted by electrolysis of their molten anhydrous salts.

THE COMPOUNDS OF s-BLOCK ELEMENTS

The compounds important for A-level chemistry are the oxides, hydroxides, hydrides, halides, carbonates, hydrogencarbonates and nitrates.

PHYSICAL PROPERTIES

- The s-block metal compounds are usually colourless or white ionic solids unless the anion includes a transition metal, e.g. Cr or Mn

There are patterns in the solubilities of the s-block compounds: e.g.

> These patterns are important. Learn them.

All nitrates are soluble in water
All s-block chlorides, bromides and iodides are soluble
All group 1 hydroxides are soluble
All group 2 carbonates are insoluble

But they are not always simple or consistent: e.g.

Down group 2 (increasing atomic number from Mg to Ba)
the solubility of the hydroxides increases but
the solubility of the sulphates decreases

These patterns are difficult to explain because solubility depends upon a number of factors, including the lattice energy of the crystal and the hydration energy of the ions: see Chapter 6.

The s-block compounds usually have high melting points typical of ionic crystals. The chlorides are said to be the easiest to vaporise, so this is why you use concentrated hydrochloric when conducting a flame test: see Fig. 11.4.

GROUP 1	FLAME COLOUR	VIA BLUE GLASS	GROUP 2	FLAME COLOUR	VIA BLUE GLASS
lithium	carmine	violet	beryllium	–	–
sodium	yellow	invisible	magnesium	invisible	–
potassium	lilac	purple red	calcium	yellowish red	violet
rubidium	lilac	purple red	strontium	scarlet red	violet
caesium	lilac	purple red	barium	yellowish green	–

Fig. 11.4 Flame colours of the s-block chlorides

CHEMICAL PROPERTIES

Thermal decomposition

- The normal oxides, chlorides and sulphates do not decompose when heated in a bunsen flame

- The alkali metal compounds are generally more stable to heat than the compounds of their corresponding alkaline earth metals

The polarising power of the group 2 cations is greater than that of their corresponding group 1 cations because they are smaller and have twice the charge. Consequently, the carbonates and nitrates of the group 2 metals decompose into the oxides more readily than those of group 1.

- In a bunsen flame the group 1 nitrates decompose mostly into oxygen and the nitrite whereas the group 2 nitrates decompose into oxygen, nitrogen dioxide and the alkaline earth metal oxide

$$2KNO_3(s) \xrightarrow{\text{heat}} 2KNO_2(s) + O_2(g)$$

$$2Ca(NO_3)_2(s) \xrightarrow{\text{heat}} 2CaO(s) + 2NO_2(g) + O_2(g)$$

- In a bunsen flame the group 2 carbonates decompose into the oxide and carbon dioxide whereas the group 1 carbonates (except Li_2CO_3) do not decompose

$$CaCO_3(s) \xrightarrow{\text{heat}} CaO(s) + CO_2(g)$$

Only the group 1 metals (except lithium) form a solid hydrogencarbonate.

- Aqueous and solid hydrogencarbonates decompose readily into the carbonate, carbon dioxide and water

$$2NaHCO_3(s) \xrightarrow{\text{heat}} Na_2CO_3(s) + H_2O(l) + CO_2(g)$$

$$Ca^{2+}(aq) + 2HCO_3^-(aq) \xrightarrow{\text{heat}} CaCO_3(s) + H_2O(l) + CO_2(g)$$

Electrical decomposition

- The molten salts and molten group 1 hydroxides conduct direct current electrolytically with the discharge of cations at the negative electrode (cathode) forming the s-block metal

$$Na^+ + e^- \xrightarrow[\text{molten } Na^+Cl^-]{\text{at the cathode}} Na$$

Metallic sodium is manufactured by electrolysis of molten sodium chloride to which calcium chloride and/or barium is added to lower the melting point. The applied voltage is high enough to discharge sodium ions but not the more stable barium and calcium ions.

- The aqueous salts and hydroxides conduct direct current electrolytically with the discharge of hydrogen ions at the negative electrode (cathode) forming hydrogen gas

$$2H^+(aq) + 2e^- \xrightarrow{\text{at the cathode}} H_2(g)$$

As the hydrogen ions discharge and their concentration near the cathode falls, more water molecules ionise and the concentration of hydroxide ions increases to maintain the equilibrium:

$$H_2O(l) \rightleftharpoons H^+(aq) + OH^-(aq)$$

The result is that aqueous sodium hydroxide forms near the cathode. In industrial processes the cathode and anode compartments are separated by a porous diaphragm (or, more recently, a semipermeable membrane) to prevent recombination of the products. The extremely important industrial electrolysis of aqueous sodium chloride (brine) may be summarised by:

$$2NaCl(aq) + 2H_2O(l) \xrightarrow{\text{D.C. electricity}} \underset{\text{[at anode]}}{Cl_2(g)} + \underset{\text{[————at cathode————]}}{H_2(g) + 2NaOH(aq)}$$

Sodium ions are discharged at a mercury cathode if the applied voltage is not too high because gases like hydrogen require an additional applied voltage (called the overvoltage) before they can be produced at smooth, shiny, silvery metal surfaces. This is the principle behind the electrolytic production of high quality sodium hydroxide by the Castner-Kellner process. However, this method is now gradually being replaced by electrolysis of brine in the environmentally safer diaphragm or membrane cells.

CHEMICAL ATTACK

Reaction with water

When an s-block compound dissolves in (or is attacked by) water, the resulting solution contains the aqueous s-block cation. From the point of view of the cation, the process may be represented by:

$$Na^+(s) \longrightarrow Na^+(aq)$$

Whether the compound reacts or just dissolves will depend upon the anion. The behaviour of the anion will also determine the pH of the resulting solution.

- The s-block hydrides and the hydroxides, oxides and carbonates of the group 1 metals (except for lithium) give alkaline solutions with water: see Fig. 11.5

Reaction with aqueous acid

You can think of these reactions as being the same as those for water (see Fig. 11.5) but

NaH	hydride	$H^-(s) + H_2O(l) \rightarrow OH^-(aq) + H_2(g)$
NaOH	hydroxide	$OH^-(s) (+ \text{ water}) \rightarrow OH^-(aq)$
Na_2O	oxide	$O^{2-}(s) + H_2O(l) \rightarrow 2OH^-(aq)$
Na_2O_2	peroxide	$O_2^{2-}(s) + H_2O(l) \rightarrow 2OH^-(aq) + \frac{1}{2}O_2(g)$
KO_2	superoxide	$2O_2^-(s) + H_2O(l) \rightarrow 2OH^-(aq) + 1\frac{1}{2}O_2(g)$
Na_2CO_3	carbonate	$CO_3^{2-}(s) + H_2O(l) \rightarrow OH^-(aq) + HCO_3^-(aq)$

Fig. 11.5 Alkali metal compounds give alkaline solutions in water

followed by the $OH^-(aq)$ ions combining with $H^+(aq)$ ions (from the acid) to form water molecules: e.g.

hydride ion	+ water:	$H^-(s) + H_2O(l) \rightarrow OH^-(aq) + H_2(g)$		[A]
hydroxide ion	+ acid:	$H^+(aq) + OH^-(aq) \rightarrow H_2O(l)$		[B]

$$\Rightarrow \text{ hydride ion } + \text{ aqueous acid: } \quad H^-(s) + H^+(aq) \rightarrow H_2(g) \qquad [C]$$

Notice that you get equation [C] by adding together the left-hand sides and the right-hand sides of [A] and [B] and cancelling the $H_2O(l)$ and $OH^-(aq)$ that now appear on both sides of the equation.

If you do this for the carbonate reaction, you would get:

$$CO_3^{2-}(s) + H^+(aq) \rightarrow HCO_3^-(aq) \qquad [X]$$

But the hydrogencarbonate ion is a base and would react further:

$$H^+(aq) + HCO_3^-(aq) \rightarrow H_2CO_3(aq) \rightarrow H_2O(l) + CO_2(g) \qquad [Y]$$

So the effect of excess aqueous acid upon a carbonate (add [X] to [Y] and cancel the $HCO_3^-(aq)$ to get [Z]) would be:

$$CO_3^{2-}(s) + 2H^+(aq) \rightarrow H_2CO_3(aq) \rightarrow H_2O(l) + CO_2(g) \qquad [Z]$$

- All carbonates react with aqueous acids that are stronger than carbonic acid (H_2CO_3) to give water and carbon dioxide gas

Reaction with involatile anhydrous acid

- Concentrated phosphoric (or sulphuric) acid displaces the hydrogen halides from the s-block metal halides: see Chapter 12

ANOMALOUS BEHAVIOUR OF LITHIUM AND BERYLLIUM

Lithium and beryllium are the first members of their groups, so some of the properties of these two elements (and their compounds) will be exceptions to the general patterns described above: see Fig. 11.6.

LITHIUM	BERYLLIUM
atypical property	atypical property
Li does not form alums LiF insoluble in water	Be does not react with water or steam Be does not combine **directly** with $H_2(g)$
atypical properties showing diagonal relationship with Mg	atypical properties showing diagonal relationship with Al
combines directly with nitrogen $6Li(s) + N_2(g) \rightarrow 2Li_3N(s)$ $3Mg(s) + N_2(g) \rightarrow Mg_3N_2(s)$	oxide is covalent and amphoteric $BeO(s) + 2HCl(aq) \rightarrow BeCl_2(aq) + H_2O(l)$ $Al_2O_3(s) + 6HCl(aq)$ $\qquad\qquad \rightarrow 2AlCl_3(aq) + 3H_2O(l)$
burns to normal oxide in air $4Li(s) + O_2(g) \rightarrow 2Li_2O(s)$ $2Mg(s) + O_2(g) \rightarrow 2MgO(s)$	$BeO(s) + 2NaOH(aq)$ $\qquad\qquad \rightarrow Na_2BeO_2(aq) + H_2O(l)$ $Al_2O_3(s) + 6NaOH(aq)$ $\qquad\qquad \rightarrow 2NaAlO_2(aq) + 3H_2O(l)$
hydroxide and carbonate will decompose in a bunsen flame $2LiOH(s) \rightarrow Li_2O(s) + H_2O(l)$ $Mg(OH)_2(s) \rightarrow MgO(s) + H_2O(l)$ $Li_2CO_3(s) \rightarrow Li_2O(s) + CO_2(g)$ $MgCO_3(s) \rightarrow MgO(s) + CO_2(g)$	

Fig. 11.6 Anomalies and diagonal relationships of lithium and beryllium

| HYDROGEN |

In some versions of the periodic table hydrogen and helium are 'boxed' side-by-side in a prominent position at the top. This arrangement should remind you that the chemistry of hydrogen is unique and governed by the first electron shell with its 1s orbital having a maximum of 2 electrons.

| H 1s^1 hydrogen | He 1s^2 helium |

In other versions the hydrogen is placed above group 1 and the helium above group 0. Helium clearly belongs with neon and argon in being a completely unreactive (inert) noble gas. But what has hydrogen, the most abundant element in the universe, got in common with the alkali metals?

PHYSICAL PROPERTIES

It is a colourless, odourless, tasteless diatomic gas (b.pt. $-252\,°C$; m.pt. $-259\,°C$) whose density is about 15 times less than that of air. It is very slightly soluble in water, ethanol and ethoxyethane. It is clearly not similar to the alkali metals in its physical properties!

CHEMICAL PROPERTIES

In most of its compounds hydrogen is covalently bonded. And here is another way in which hydrogen differs from the alkali metals! They can reduce hydrogen to the hydride anion in which it has an oxidation state of -1 and the same electronic configuration as helium:

$$Na\ (2.8.1) \rightarrow Na^+\ (2.8) + e^-\ [Ne\ 2.8] \qquad\qquad e^- + H\ (1) \rightarrow H^-\ (2)\ [He\ 2]$$

However, in most of its compounds hydrogen is in an oxidation state of $+1$. Now if the atom were simply to lose its electron (as in a mass spectrometer) the resulting cation would actually be just the proton that constitutes the nucleus. It would be so minute that it would be an intensely polarising cation and an extremely powerful Lewis acid, readily accepting a lone pair of electrons from any Lewis base donor atom. Therefore in test tube reactions the formation of hydrogen ions requires a Lewis base to donate an electron-pair; see chapter 7.

$$H_2O:(l) + H-Cl(g) \rightarrow H_2O:H^+(aq) + :Cl^-(aq)$$
$$H_3N:(g) + H-Cl(g) \rightarrow H_3N:H^+Cl^-(s)$$

Consequently, even as a positive ion, hydrogen has a full outer (first) shell with the same electron configuration as the lithium cation and the helium atom.

- In most of its reactions hydrogen acts as a reducing agent and is oxidised from an oxidation state of 0 to $+1$

You may find the following definition of reduction in your textbook:

- Reduction is the removal of oxygen (or halogen) the addition of hydrogen, the gain of electrons or a decrease in oxidation number

REACTION WITH NON-METALS

$$2H_2(g) + O_2(g) \xrightarrow[\text{light}]{\text{flame}} 2H_2O(l)$$

Hydrogen = water producer (Greek: *hydor* – water; *gennaein* – to produce).

$$3H_2(g) + N_2(g) \underset{\text{iron catalyst}}{\overset{450\,°C\ 200\,atm}{\rightleftharpoons}} 2NH_3(g)$$

Industrial synthesis of ammonia by the Haber process

REACTION WITH OXIDES

$$3H_2(g) + WO_3(s) \xrightarrow{850\,°C} W(s) + 3H_2O(l)$$

Industrial extraction of high quality tungsten for light bulb filaments

$$2H_2(g) + CO(g) \xrightarrow[\text{ZnO/Cr}_2O_3 \text{ catalyst}]{400\,°C \ 300\,atm} CH_3OH(g)$$

$$3H_2(g) + CO_2(g) \xrightarrow[\text{Cu based catalyst}]{250\,°C \ 70\,atm} CH_3OH(g) + H_2O(g)$$

Industrial production of methanol from synthesis gas (mixture CO, CO$_2$ & H$_2$)

REACTION WITH UNSATURATED ORGANIC COMPOUNDS

$$C_6H_6(l) + 3H_2(g) \xrightarrow[\text{Ni catalyst}]{150\,°C} C_6H_{12}(l)$$

hydrogenation of benzene

Industrial production of cyclohexane for manufacture of nylon-6

$$>C=C< + H_2(g) \xrightarrow[\text{Ni catalyst}]{150\,°C} \begin{array}{c} | \quad | \\ -C-C- \\ | \quad | \\ H \quad H \end{array}$$

Hardening of (unsaturated) oils to (saturated) fats

Industrial production of margarine

SIMILARITY WITH ALKALI METALS

- Hydrogen, like lithium, forms a slightly charged positive ion with the same electron configuration as He
- In most of its compounds hydrogen, like all the alkali metals, has an oxidation number of +1
- In most of its reactions hydrogen, like all the alkali metals, acts as a reducing agent
- Hydrogen, like all the alkali metals, reacts with halogens to form binary halides whose aqueous solutions are good electrolytic conductors

EXAMINATION QUESTION AND OUTLINE ANSWER

Q1 Discuss the group properties of the elements in Groups I (Li-Cs) and II (Be-Ba) of the Periodic Table by considering
    a) trends in the physical properties of the elements and of their ions as the groups are descended.   *(8)*
    b) the reactions of the elements with water and the nature of the products formed.   *(8)*
    c) the properties of their carbonates and hydrogencarbonates.   *(8)*
    d) the abnormalities by the first element and its compounds in each group.   *(6)*
                                                       (JMB 1988)

Outline answer

Q1 (a) This is a typical question on the patterns and trends in the s-block groups. See chapter 11 page 150

(b) Chlorides have less covalent character because the cation size increases and the ion is less strongly polarizing

---

(c) Lattice energies are more negative for the doubly charged cations. Hence hydration energy is less likely to provide the energy needed to break up the solid lattice.

(d) Thermal stability of carbonates increases with atomic number (Mg → Ba) because cation becomes less strongly polarizing.

EXAMINATION QUESTION AND TUTOR'S ANSWER

Q2 This question concerns the alkaline earth elements in Group 2 of the Periodic Table.

a) The following is a table of metallic and ionic radii.

Element	Metallic Radius /nm	Ionic Radius M^{2+}/nm
Beryllium	0.112	0.030
Magnesium	0.160	0.065
Calcium	0.197	0.094
Strontium	0.215	0.110
Barium	0.221	0.134

How do you account for the fact that

i) for all elements the ionic radius is smaller than the metallic radius

They form cations by losing their outer shell valence electrons: Mg 2.8.2 → Mg^{2+} 2.8.

ii) the metallic radius increases from beryllium to barium

They have more inner shells of electrons to shield the outer electrons from the nuclear charge

iii) the radius of the K^+ ion is greater than the radius of the Ca^{2+} ion although both ions have the same electronic configuration?

The charge on the nucleus of the Ca^{2+} ion is higher than that on the K^+ nucleus - so more attraction   (4)

b) Explain why $BeCl_2(g)$ is a covalent and linear molecule but $MgCl_2(s)$ is ionic.

The Be^{2+} would be too strongly polarizing so it shares electrons Cl-Be-Cl that mutually repel. Mg^{2+} is less polarizing so it transfers electrons to give Cl^- ions   (3)

c) State with reasons which of the ions in the table has the least exothermic molar enthalpy of hydration.

Ba^{2+} it is the largest so it has the smallest charge density and attracts the water molecules the least strongly

(3)

d) Magnesium sulphate is very soluble in water whereas barium sulphate is not. Suggest what factors are relevant in accounting for this.

Hydration energy of Mg^{2+} (and SO_4^{2-}) provides energy needed to break down the lattice. Barium sulphate has too low a hydration energy.

66 These are good
answers 99

(3)

(*Total 13 marks*)

(ULSEB 1988)

STUDENT'S ANSWER WITH EXAMINER'S COMMENTS

Q3 a) Write an equation for the reaction between lithium and water.

66 Include the state symbols
e.g. Li(s); LiOH(aq) 99

$2Li + 2H_2O \rightarrow 2LiOH + H_2$

(1)

b) Give two examples which show that compounds of lithium differ in their properties from those formed by other Group I elements.

Example 1 Lithium carbonate decomposes on heating ✓

Example 2 Lithium nitrate decomposes into Lithium oxide on heating ✓

(2)

c) Give one simple test which could be used to show that beryllium chloride has covalent character and state the result you would expect to obtain.

66 All solid compounds are
non-conductors. Test the *molten*
compound. 99

Test See if it conducts

Result non-conductor

(2)

d) Given that beryllium hydroxide is amphoteric, state what you would observe if aqueous sodium hydroxide was slowly added, with shaking, to an aqueous solution of a beryllium salt, until present in excess. Write ionic equations for the reactions occurring.

66 White solid. 99

Observations the hydroxide would precipitate and then dissolve again

66 State symbols are
essential. 99

Equations $BeCl_2 + 2NaOH \rightarrow Be(OH)_2 (s) + 2NaCl$

$Be(OH)_2 + 2NaOH \rightarrow Na_2BeO_2 (aq) + 2H_2O$

(4)

e) State which Group II carbonate is the least stable to heat and give a reason for your answer.

Carbonate Magnesium carbonate

Reason Mg^{2+} ion polarizes the CO_3^{2-} ion

66 Beryllium carbonate
because Be^{2+} is even more
strongly polarising than
Mg^{2+} 99

(2)

(*Total 11 marks*)

(JMB 1988)

PRACTICE QUESTION

Q4 a) i) State **three** chemical properties which you associate with the elements of the
          s-block (Groups I and II of the Periodic Table) or their compounds. Give **two**
          reasons why the elements of the s-block have these typical properties.   *(5)*
       ii) Why do the first members of each Group have some properties which
          are not typical of the s-block generally? Give an example.   *(3)*

   b) Explain why the chlorides of the Group II metals are less extensively
       hydrolysed with increasing atomic number of the metal.   *(4)*

   c) Almost all salts of the Group I metals are soluble in water, whereas many salts of
       Group II metals (e.g. carbonates, some sulphates) are insoluble. Suggest reasons
       for this in terms of the enthalpy changes involved.   *(4)*

   d) What gradation in the thermal stabilities of Group II carbonates (Mg to Ba)
       would you expect to occur and why?   *(4)*

                                                                    (OCSEB 1988)

p-BLOCK ELEMENTS

THE p-BLOCK ELEMENTS

GROUP III: ALUMINIUM

GROUP IV: CARBON, SILICON, TIN, LEAD

GROUP V: NITROGEN, PHOSPHOROUS

MANUFACTURE OF AMMONIA

MANUFACTURE OF NITRIC ACID

GROUP VI: OXYGEN, SULPHUR

WATER

GROUP VII: THE HALOGENS

GROUP 0: NOBLE GASES

GETTING STARTED

The p-block consists of the halogens, the noble gases and the elements of groups 3 to 6 inclusive. The emphasis is upon trends across the block and within selected groups. Chemical behaviour varies widely with the reactivity of the metals, metalloids and non-metals and the comparative lack of reactivity of the noble gases. Similarities within a group are shown best by the halogens and nobles gases. Trends within a group are shown most dramatically by the group 4 elements carbon to lead. The chemistry of individual elements is related to their group properties and their position in the p-block: see Fig. 12.1.

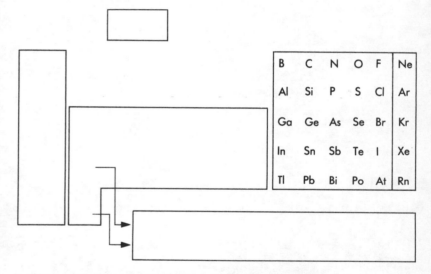

Fig. 12.1 p-Block elements (groups 3,4,5,6, halogens and noble gases)

ESSENTIAL PRINCIPLES

THE p-BLOCK ELEMENTS

These groups of elements and their compounds have little in common. This is quite obvious from the way chemical behaviour changes across the block from left to right in period 3 from aluminium (reactive metal) to chlorine (reactive non-metal) and argon (inert noble gas). You should study the trends in the physical and chemical properties of the elements: see Chapter 10. You should also study the trends in the properties of the chlorides, hydrides and oxides: see Fig. 12.2.

You may well get questions about the characteristics of the s-block and of the d-block because of the similarities of the elements and their compounds within each block. You are much less likely to get questions about the p-block in general. Instead you are more likely to be asked about particular groups and about individual elements.

Chlorine, bromine and iodine may already be familiar to you from your GCSE work. For A-level you will have to study in more detail the halogens and perhaps the noble gases as a group showing marked similarities with some trends in chemical behaviour. You will probably also need to study at least one other p-block group where the emphasis would be on marked trends with some similarities in chemical behaviour. Such trends are most marked in group 4 (C – Pb) but your syllabus could specify group 5 (N – Bi) or 6 (O – Te).

Most syllabuses require a knowledge and understanding of the chemistry of particular p-block elements. Those selected serve to illustrate important characteristics of their group as well as the variety of chemical behaviour to be found in inorganic chemistry. Check your syllabus carefully to find out which elements (together with at least their hydrides, oxides and chlorides) you need to study (see Fig. 12.3).

Make sure you relate the chemistry of these elements and their compounds not only to the group they are in but also to their position in the p-block. In the following sections I have concentrated on the most popular and important elements.

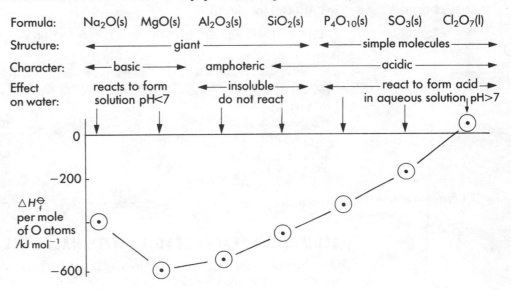

Fig. 12.2 Trend in properties of oxides

Fig. 12.3 Individual p-block elements to be studied in detail

NOMENCLATURE

The examination boards have published a joint statement on the naming of chemicals and the use of chemical (and mathematical) symbols and units. Some boards (e.g. ULSEB) reproduce this in their syllabus. Some boards (e.g. JMB) publish a leaflet on nomenclature, units and conventions they will use in their question papers. All boards refer to the book Chemical Nomenclature, Symbols and Terminology for use in school science. This has been published by the Association for Science Education (ASE). Your teacher or lecturer will probably have a copy. You do not need the book but it might benefit you to refer to it occasionally.

■ The main aim of the examination boards is to make their question papers clear and unambiguous

Systematic names give you information about the structure of a compound and they can be derived logically. But some names are not logical and systematic because the structure is complicated (e.g. nylon) or uncertain (e.g. bleaching powder) or because they are widely adopted traditional names (e.g. ammonia). You should learn the most common trivial and semi-systematic names but use systematic nomenclature as far as possible. However, in your examinations you should not be penalised for using any meaningful name unless the question clearly requires a systematic one.

GROUP III: ALUMINIUM

Aluminium is an important element with an amphoteric oxide.

You need to study the chemistry of aluminium in detail. You may need to know a little about the anomalous behaviour of boron but you do not have to study gallium, indium and thallium or the group as a whole.

Aluminium is high in the electrochemical series and classed as a reactive metal that can form covalent and ionic compounds (in which its oxidation number is +3) whose aqueous solutions contain the hexaaquaaluminium(III) ion $[Al(H_2O)_6]^{3+}(aq)$ or hydrated aluminium ion $Al^{3+}(aq)$ for short:

$$Pt[H_2(g)]|2H^+(aq)\vdots Al^{3+}(aq)|Al(s) \qquad E^\ominus = -1.66\,V$$

	BORON	ALUMINIUM
electron configuration	B $1s^2 2s^2 2p^1$	Al $1s^2 2s^2 3s^2 3p^1$
covalent radius/nm	0.09	0.13
cationic radius/nm	(0.012)	0.053
character	non-metallic	metallic
melting point/°C	2300	660
$\Delta H^\ominus_{melting}$/kJ mol^{-1}	22.18	10.67
electrical conduction	insulator	conductor
structure	giant covalent network	cubic close-packed metal

Fig. 12.4 Properties of boron and aluminium

ALUMINIUM REACTS READILY WITH HALOGENS

One drop of water placed on the apex of a conical heap of a dry mixture of aluminium powder and iodine powder catalyses an exciting 'volcanic' reaction. You may have seen this demonstrated in a fume cupboard. The reaction is complicated but may involve the formation of AlI_3 followed by its hydrolysis to Al_2O_3 and HI. Decomposition of the HI could yield the purple clouds of iodine vapour and the flames of hydrogen burning in the air. A fine spray of water makes the 'dying embers' flare up and become white hot!

In contrast to the reaction with iodine, aluminium spontaneously ignites and burns vigorously in fluorine gas to form aluminium fluoride:

$$2Al(s) + 3F_2(g) \rightarrow 2AlF_3(s) \text{ [or } Al^{3+}(F^-)_3 \text{ to show ionic structure]}$$

You can see this dangerous reaction demonstrated on the video 'The Elements Organised'; produced and broadcast for the Open University. As part of your practical work you may well prepare aluminium chloride in the laboratory by burning aluminium in a stream of dry chlorine:

$$2Al(s) + 3Cl_2(g) \rightarrow 2AlCl_3(s) \quad \text{[or } Al_2Cl_6(g) \text{ to show covalent molecule]}$$

The simple Al^{3+} cation is small and so strongly polarising that aluminium chloride which sublimes at 178 °C is essentially covalent and even aluminium fluoride which sublimes at 1291 °C has some covalent character.

In BCl_3 and $AlCl_3$ the molecules are trigonal planar because the group 3 atom has only six outer shell electrons as three shared pairs: see Fig. 12.5.

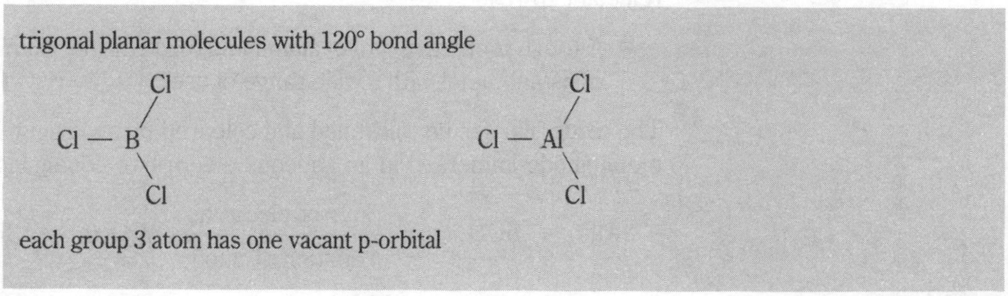

trigonal planar molecules with 120° bond angle

each group 3 atom has one vacant p-orbital

Fig. 12.5 Shape and structure of group 3 chlorides

The trihalides act as Lewis acids because the B and Al atoms can accept a pair of electrons to form a dative bond. So, for example, as the catalyst in the Friedel-Crafts organic reactions, aluminium chloride promotes the formation of the electrophile: see Fig. 12.6.

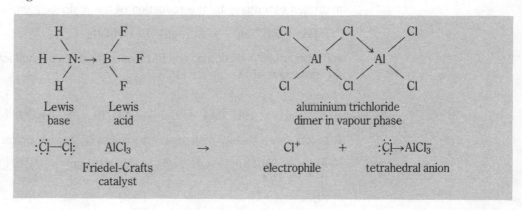

Lewis base   Lewis acid   aluminium trichloride dimer in vapour phase

$:\ddot{Cl}-\ddot{Cl}:$ $AlCl_3$ → Cl^+ + $:\ddot{Cl}{\rightarrow}AlCl_3^-$

Friedel-Crafts catalyst   electrophile   tetrahedral anion

Fig. 12.6 Formation of dative bonds

Both covalent trichlorides react readily with water. For boron trichloride the net effect is to replace the halogen atoms by hydroxyl groups to produce boric acid $B(OH)_3$. This trioxoboric(III) acid is a weak acid that partially ionises to give hydrogen ions and borate ions (tetrahydroxoborate(III) ion):

$$BCl_3(l) + 3H_2O(l) \rightarrow B(OH)_3(aq) + 3H^+(aq) + 3Cl^-(aq)$$
$$B(OH)_3(aq) + H_2O(l) \rightleftharpoons H^+(aq) + B(OH)_4^-(aq)$$

Aluminium trichloride 'fumes' HCl in moist air but 'dissolves' in water to give a solution containing hydrated aluminium ions and chloride ions. The polarizing effect of the aluminium cation on the attached water molecules facilitates the loss of a proton. The pK_a of the hexaaquaaluminium(III) ion is 5.0 and the solution is acidic:

$$AlCl_3(s) + 6H_2O(l) \rightarrow [Al(H_2O)_6]^{3+}(aq) + 3Cl^-(aq)$$
$$[Al(H_2O)_6]^{3+}(aq) \rightleftharpoons H^+(aq) + [Al(H_2O)_5OH]^{3+}(aq)$$

Six water molecules can form dative bonds with the aluminium cation because the aluminium atom can extend its outer electron shell beyond eight by using the 3d sub-shell. Boron cannot do this.

■ Aluminium (but not boron) atoms can extend their outer electron shell

Boron forms various hydrides, diborane $B_2H_6(g)$ is the simplest, but aluminium forms only one; a complex polymeric $AlH_3(s)$. Their tetrahydrido- derivatives $NaBH_4$ and $LiAlH_4$ contain the tetrahedral BH_4^- anion and are powerful nucleophilic reducing agents used in organic chemistry to convert, for example, a carboxylic acid to a primary alcohol. The hydrides and their derivatives are readily hydrolysed and dangerously flammable so they must be kept out of contact with water and air.

Aluminium powder forms its oxide when it burns in air:

$$2Al(s) + 1\tfrac{1}{2}O_2(g) \rightarrow Al_2O_3(s) \qquad \Delta H^\ominus = -1676\,\text{kJ}\,\text{mol}^{-1}$$

and when it violently displaces iron from iron(III) oxide:

$$2Al(s) + Fe_2O_3(s) \rightarrow Al_2O_3(s) + 2Fe(s) \qquad \Delta H^\ominus = -852\,\text{kJ}\,\text{mol}^{-1}$$

As the oxide of a reactive element, aluminium oxide is of course very stable with respect to decomposition into its elements. It is a refractory oxide with a melting point above 2000 °C.

It also forms a very tough film (about 10 nm thick) on the aluminium surface. This oxide layer so well protects the metal from chemical attack that you will normally find aluminium reluctant to react.

■ A tough protective film of aluminium oxide makes aluminium a good corrosion resistant metal with a wide range of uses

The oxide film can be thickened and coloured by anodising i.e. electrolytic oxidation of the metal anode immersed in an aqueous electrolyte containing water-soluble dyes:

$$2Al(s) + 6OH^-(aq) \xrightarrow[\text{oxidation at anode}]{\text{loss of electrons}} Al_2O_3(s) + 3H_2O(l) + 6e^-$$

■ Reactions of aluminium are energetically feasible but kinetically hindered by an oxide layer shielding the metal from direct attack

The amphoteric oxide layer is attacked much more readily by alkalis than by acids. So aluminium reacts with moderately concentrated aqueous hydrochloric acid to form hydrogen essentially by the reduction of the hydrogen ion:

$$Al(s) + 3H^+(aq) \rightarrow Al^{3+}(aq) + 1\tfrac{1}{2}H_2(g)$$

and with aqueous nitric acid and nitrates to form ammonium ions or ammonia essentially by the reduction of the nitrate ion:

$$Al(s) \rightarrow Al^{3+}(aq) + 3e^- \qquad\qquad \text{[oxidation no. } 0 \rightarrow +3\text{]}$$
$$NO_3^-(aq) + 10H^+(aq) + 8e^- \rightarrow NH_4^+(aq) + 3H_2O(l) \text{ [oxidation no. } +5 \rightarrow -3\text{]}$$

It reacts more readily with aqueous sodium hydroxide or carbonate:

$$Al(s) + 3H_2O(l) + OH^-(aq) \rightarrow Al(OH)_4^-(aq) + 1\tfrac{1}{2}H_2(g)$$
$$\uparrow \qquad\quad \uparrow \qquad\qquad\qquad \uparrow \qquad\qquad\quad \uparrow$$
$$[0] \qquad [+1] \qquad\qquad\qquad [+3] \qquad\qquad [0] \quad \text{[oxidation states]}$$

You may regard this as the reduction of the hydrogen in water and the oxidation of aluminium under alkaline conditions. This reaction is the reason why aluminium saucepans carry the warning to use no soda!

ELECTROLYTIC EXTRACTION OF ALUMINIUM

Aluminium is manufactured by the electrolysis of aluminium oxide obtained from bauxite (see below) in molten cryolite (Na_3AlF_6) containing some calcium fluoride to lower the melting point to about 950 °C: see Fig. 12.7.

Note the following points:

1 We cannot use an **aqueous** electrolyte because hydrogen ions (not aluminium ions) would be discharged at the cathode
2 We must remove Fe_2O_3 and TiO_2 impurities from the Al_2O_3 in the bauxite *before* electrolysis otherwise iron and titanium would also be formed at the cathode and contaminate the aluminium

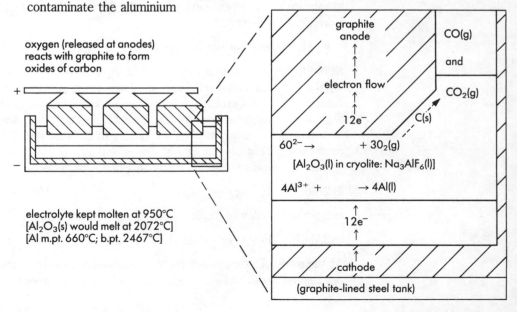

oxygen (released at anodes) reacts with graphite to form oxides of carbon

electrolyte kept molten at 950°C
[Al_2O_3(s) would melt at 2072°C]
[Al m.pt. 660°C; b.pt. 2467°C]

graphite anode

CO(g)

and

CO₂(g)

electron flow

12e⁻   C(s)

$6O^{2-} \rightarrow \qquad + 3O_2(g)$

[Al_2O_3(l) in cryolite: Na_3AlF_6(l)]

$4Al^{3+} + \qquad \rightarrow 4Al(l)$

12e⁻

cathode

(graphite-lined steel tank)

Fig. 12.7 Extraction of aluminium

The purification of the bauxite ore depends upon the fact that Al_2O_3 is amphoteric and the oxide impurities are not:

$$Al_2O_3/Fe_2O_3/TiO_2(s) + 2NaOH(aq) + 3H_2O(l)$$

$$\xrightarrow{150\,°C,\ 4\,atm} 2NaAl(OH)_4(aq) + Fe_2O_3/TiO_2(s)$$

The insoluble oxides of iron and titanium are removed, the hot sodium aluminate solution cooled and seeded with pure alumina (Al_2O_3) to precipitate the trihydrate which is vacuum filtered and then dehydrated:

$$Al_2O_3.3H_2O(s) \xrightarrow{1000\,°C} Al_2O_3(s) + 3H_2O(l)$$

Major industrial uses of aluminium oxide are in the glass, paper and pottery manufacture, in water treatment, as a refractory, an abrasive and insulator, and as a catalyst promoter and support in the chemical and petroleum industries.

GROUP IV: CARBON, SILICON, TIN, LEAD

You need to study the trends from the non-metals carbon and silicon to metals tin and lead. You will need to learn some detailed chemistry for these four elements but you can ignore the metalloid germanium.

GROUP CHARACTERISTICS

All the group 4 elements can form covalent compounds in which the oxidation state of the element is +4. They can also have an oxidation number of +2 in some compounds: see Fig. 12.8. Remember that you should expect carbon and its compounds to show some atypical behaviour as the first member of the group. For example, CO_2 is a simple molecule whilst the oxides of the other elements are giant structures: see Fig. 12.9.

	CARBON	SILICON	TIN	LEAD
electron outer shell	$2s^2 2p^2$	$3s^2 3p^2$	$5s^2 5p^2$	$6s^2 6p^2$
max. oxidation no.	+4	+4	+4	+4
highest hydride	CH_4	SiH_4	SnH_4	PbH_4
highest oxide	CO_2	SiO_2	SnO_2	PbO_2
highest chloride	CCl_4	$SiCl_4$	$SnCl_4$	$PbCl_4$
monoxide (ox. no. +2)	CO	SiO	SnO	PbO

Fig. 12.8 Common features in group 4

> The change from non-metal to metal is most striking in this group.

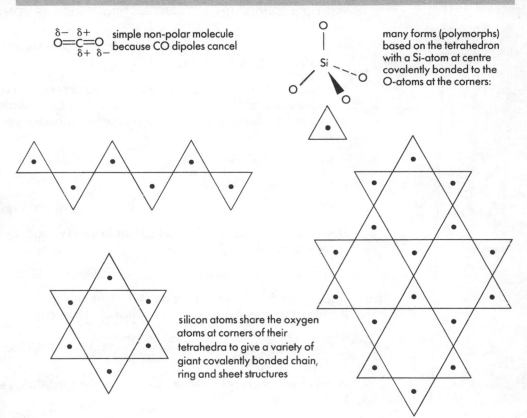

Fig. 12.9 Structures of carbon dioxide and silica

GROUP TRENDS

Hydrides

The range and stability of the hydrides decreases with atomic number down the group: see Fig. 12.10.

carbon	alkanes:	C_nH_{2n+2}, alkenes C_nH_{2n}, etc . . .	very stable hydrocarbons
silicon	silanes:	Si_nH_{2n+2} (n = 1 to 8 inc.)	hydrolyse and ignite
tin	stannanes:	SnH_4 and Sn_2H_6	very unstable
lead	plumbane:	PbH_4	only traces detected

Fig. 12.10 Hydrides of group 4

Oxides

The elements form oxides when heated in oxygen: see Fig. 12.11.

Fig. 12.11 Reaction of group 4 elements with oxygen

$C(s)$ +	$O_2(g) \rightarrow CO_2(g)$;	$\Delta H^\ominus = -394\,kJ\,mol^{-1}$	carbon dioxide
$Si(s)$ +	$O_2(g) \rightarrow SiO_2(s)$;	$\Delta H^\ominus = -910\,kJ\,mol^{-1}$	silicon(IV) oxide
$Sn(s)$ +	$O_2(g) \rightarrow SnO_2(s)$;	$\Delta H^\ominus = -581\,kJ\,mol^{-1}$	tin(IV) oxide
$Pb(s)$ +	$\frac{1}{2}O_2(g) \rightarrow PbO(s)$;	$\Delta H^\ominus = -217\,kJ\,mol^{-1}$	lead(II) oxide

The first two (non-metal) oxides are acidic: e.g.

$$CO_2(g) + Ca(OH)_2(aq) \rightarrow CaCO_3(s) + H_2O(l)$$

[carbon dioxide turns limewater milky (ppt calcium carbonate)]

$$SiO_2(s) + CaO(s) \xrightarrow[\text{blast furnace}]{900\,°C\ in\ the} CaSiO_3(l)$$

earthy impurities   slag

The oxides of tin and lead are amphoteric and react with acids and alkalis but the lead oxides are more basic than those of tin: e.g.

$$SnO_2(s) + 6HCl(aq) \xrightarrow{conc.\ acid} 2H_3O^+(aq) + SnCl_6^{2-}(aq)$$

hexachlorostannate(IV) ion

$$SnO_2(s) + 2NaOH(aq) \xrightarrow{conc.\ alkali} 2Na^+(aq) + [Sn(OH)_6]^{2-}(aq)$$

hexaaquastannate(IV) ion

$$PbO(s) + 2HCl(aq) \rightarrow PbCl_2(s) + H_2O(l)$$

The lead(II) chloride is soluble in hot water but insoluble in cold water.

Notice that lead(IV) oxide cannot be formed by direct combination of the elements. In the laboratory, using moderately concentrated nitric acid, you could oxidise tin to tin(IV) oxide:

$$Sn(s) + 4HNO_3(aq) \rightarrow SnO_2(s) + 4NO_2(g) + 2H_2O(l)$$

but lead to only lead(II) ions:

$$Pb(s) + 2H^+(aq) + 2HNO_3(aq) \rightarrow Pb^{2+}(aq) + 2NO_2(g) + 2H_2O(l)$$

However, you can oxidise the lead(II) ions to lead(IV) oxide by warming the lead(II) nitrate solution with chlorate(I) solution:

$$Pb^{2+}(aq) + ClO^-(aq) + OH^-(aq) \rightarrow PbO_2(s) + H^+(aq) + Cl^-(aq)$$

Lead(IV) oxide is formed in the car battery by anodic oxidation of lead(II) sulphate during the electrolytic recharging process (when electrical energy is being stored):

$$PbSO_4(s) + H_2O(l) \xrightarrow[\text{being stored}]{electrical\ energy} PbO_2(s) + 2H^+(aq) + SO_4^{2-}(aq) + 4e^-$$

When electrical energy is being taken out of the car battery, the oxidation number of the lead falls from +4 to +2.

Halides

All four chlorides XCl_4 are liquids whose simple covalent molecules have a tetrahedral structure. All except CCl_4 hydrolyse to the oxide XO_2 and HCl: see Fig. 12.12.

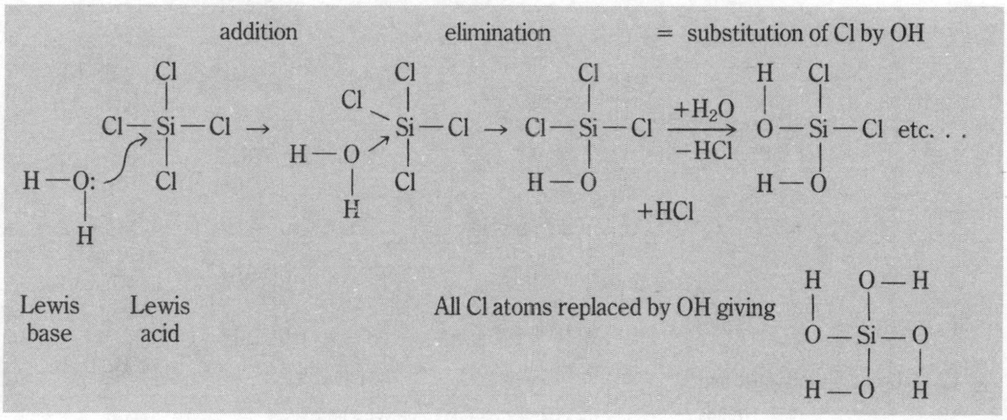

Fig. 12.12 Hydrolysis of silicon(IV) chloride

$$SiCl_4(l) \; + \; 4H_2O(l) \rightarrow 4HCl(aq) \; + \; Si(OH)_4(s) \; \text{or} \; SiO_2.2H_2O(s)$$

If you calculate the standard enthalpy change of reaction, you find the hydrolysis of tetrachloromethane to be exothermic (energetically feasible):

$$CCl_4(l) \; + \; 2H_2O(l) \rightarrow CO_2(g) \; + \; 4HCl(g)$$

$\Delta H_f^{\ominus}/\text{kJ}$ $\quad -130 \quad\quad 2 \times (-286) \quad -394 \quad 4 \times (-92)$

$\Delta H_{\text{hydrolysis}}^{\ominus} = \{4 \times (-92) + (-394)\} - \{(-130) + 2 \times (-286)\}$

$\quad\quad\quad\quad = -60 \, \text{kJ} \, \text{mol}^{-1}$

Tetrachloromethane does NOT hydrolyse because it is kinetically hindered. The carbon atom (unlike silicon) cannot extend its electron shell beyond eight. So a strong C—Cl bond must break before a C—O bond can form. This means a very high activation energy barrier and an extremely slow reaction. In the hydrolysis of $SiCl_4$ some of the energy released by the formation of a strong Si—O bond can provide the energy absorbed by the breaking of a fairly weak Si—Cl bond.

Tin(II) chloride 'dissolves' in water to produce an acidic solution by the dissociation of its hydrated cation. Lead(II) chloride is soluble in hot water but precipitates as a white solid when the solution is cooled. You should regard these two chlorides as essentially ionic structures with some covalent character.

Most compounds of carbon and of silicon are covalent. In their ionic compounds the elements are usually covalently bonded as part of the anion: e.g.

CO_3^{2-} carbonate $\quad\quad HCO_3^{-}$ hydrogencarbonate $\quad\quad SiO_3^{2-}$ silicate

Carbon and silicon do not form 'simple' cations; they would be too strongly polarising. By contrast, tin and lead can react with aqueous acids to form Sn^{2+} and Pb^{2+}:

$$Sn(s) \; + \; 2HCl(aq) \rightarrow Sn^{2+}(aq) \; + \; 2Cl^{-}(aq) \; + \; H_2(g)$$

Tin(II) chloride (stannous chloride) has a wide range of industrial uses as a reducing agent. As well as its function in lead-acid storage batteries, lead(IV) oxide (lead dioxide) has a variety of industrial uses as an oxidising agent.

■ As the atomic number increases down the group the stability of the +2 oxidation state increases and that of the +4 oxidation state decreases

GROUP V: NITROGEN, PHOSPHORUS

You need to study the chemistry of the first two elements in some detail. You should know that the character of the elements and their compounds changes from non-metallic to metallic with increasing atomic number down the group but otherwise you do not need to study arsenic, antimony or bismuth.

GROUP CHARACTERISTICS AND TRENDS

The group 5 elements can have oxidation numbers of +3 and +5 in their compounds but the stability of the +5 oxidation state decreases with increasing atomic number down the group: see Fig. 12.13. Remember that as the first member of the group, nitrogen (and its compounds) will show some atypical behaviour. For example, ammonia has an

ELECTRONIC CONFIGURATION	b.pt /°C	N_p	HYDRIDES	OXIDES AND TYPE		CHLORIDES
N [He]$2s^2 2p^3$	−196	3.0	NH_3 N_2H_4	N_2O_5 N_2O_4 NO_2 N_2O_3 NO N_2O	acidic acidic acidic acidic neutral neutral	NCl_3
P [Ne]$3s^2 3p^3$	280	2.1	PH_3 P_2H_4	P_4O_{10} P_4O_6	acidic acidic	$PCl_4^+ PCl_6^-$ PCl_3
As [Ar]$3d^{10}4s^2 4P^3$	sub	2.0	AsH_3	As_4O_{10} As_4O_6	acidic acidic	$AsCl_3$
Sb [Kr]$4d^{10}5s^2 5p^3$	1750	1.9	SbH_3	Sb_2O_5 Sb_2O_3	acidic amphoteric	$SbCl_3$
Bi [Xe]$4f^{14}5d^{10}6s^2 6p^3$	1833	1.9	BiH_3	Bi_2O_5 Bi_2O_3	acidic basic	$BiCl_3$

Fig. 12.13 Group 5 similarities and trends

N_p = Pauling's electronegativity index

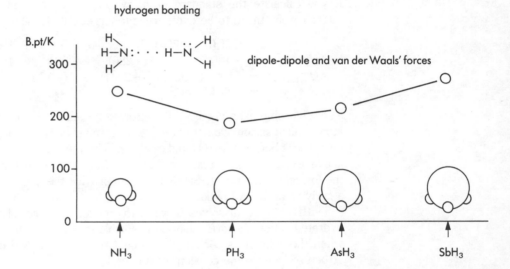

Fig. 12.14 Boiling points of the group 5 hydrides

exceptionally high boiling point owing to intermolecular hydrogen bonding caused by the high electronegativity of the nitrogen atom: see Fig. 12.14.

NITROGEN AND PHOSPHORUS

■ Nitrogen is unreactive and exists as a diatomic gas whereas phosphorus is reactive and has at least three solid allotropes: see Fig. 12.15

RED PHOSPHORUS	WHITE PHOSPHORUS
violet-red amorphous powder	transparent yellowish brittle waxy solid
density $2.34\,g\,cm^{-3}$	density $1.82\,g\,cm^{-3}$
sublimes at 416 °C	melts under water at 44.1 °C
giant covalent molecules	simple covalent molecules (P_4)
insoluble in organic solvents	soluble in organic solvents
does not react with NaOH(aq)	reacts with NaOH(aq) → PH_3(g) + NaH_2PO_4(aq)

Fig. 12.15 Two allotropes of phosphorus

The lack of reactivity of the nitrogen molecule (N_2) is explained by the triple bond which is almost as short as a C—H bond but more than twice as strong. The reactivity of the phosphorus molecule (P_4) is explained by the single bond which is almost as weak as a F—F bond and by the small P—P—P bond angle causing strain in the molecule: see Fig. 12.16. Both elements are typical non-metals. They form acidic oxides, volatile hydrides and covalent chlorides. Their oxidation numbers range from −3 to +5, with nitrogen being remarkable in having stable compounds for each oxidation state: see Fig. 12.17.

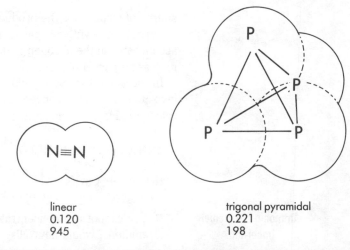

Fig. 12.16 Molecules of nitrogen and phosphorus	shape of molecule	linear	trigonal pyramidal
	bond length/nm	0.120	0.221
	bond strength/kJ mol^{-1}	945	198

Ox. no.	NAMES AND FORMULAE
+5	nitric acid (HNO$_3$); nitrate ion (NO$_3^-$); dinitrogen pentoxide (N$_2$O$_5$)
+4	nitrogen dioxide (NO$_2$); dinitrogen tetroxide (N$_2$O$_4$)
+3	nitrous acid (HNO$_2$); nitrite ion (NO$_2^-$); dinitrogen trioxide (N$_2$O$_3$)
+2	nitrogen monoxide (NO)
+1	dinitrogen oxide (N$_2$O)
0	nitrogen
−1	hydroxylamine (NH$_2$OH)
−2	hydrazine (NH$_2$NH$_2$)
−3	ammonia (NH$_3$); ammonium ion (NH$_4^+$)

Fig. 12.17 Oxidation states of nitrogen

HYDRIDES

Both elements form hydrides of formula XH$_3$ and X$_2$H$_4$: see Fig. 12.18.

You will meet hydrazine (N$_2$H$_4$) and its derivatives in your study of organic chemistry but ammonia is by far the most important hydride of group 5.

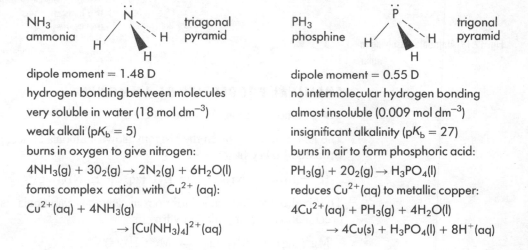

NH$_3$ ammonia	triagonal pyramid	PH$_3$ phosphine	trigonal pyramid
dipole moment = 1.48 D		dipole moment = 0.55 D	
hydrogen bonding between molecules		no intermolecular hydrogen bonding	
very soluble in water (18 mol dm^{-3})		almost insoluble (0.009 mol dm^{-3})	
weak alkali (pK_b = 5)		insignificant alkalinity (pK_b = 27)	
burns in oxygen to give nitrogen:		burns in air to form phosphoric acid:	

4NH$_3$(g) + 3O$_2$(g) → 2N$_2$(g) + 6H$_2$O(l)

PH$_3$(g) + 2O$_2$(g) → H$_3$PO$_4$(l)

forms complex cation with Cu^{2+}(aq):

reduces Cu^{2+}(aq) to metallic copper:

Cu^{2+}(aq) + 4NH$_3$(g)

→ [Cu(NH$_3$)$_4$]$^{2+}$(aq)

4Cu^{2+}(aq) + PH$_3$(g) + 4H$_2$O(l)

→ 4Cu(s) + H$_3$PO$_4$(l) + 8H$^+$(aq)

Fig. 12.18 Comparison of the trihydrides of nitrogen and phosphorus

MANUFACTURE OF AMMONIA

Ammonia is manufactured on a very large scale by the Haber process:

$$N_2(g) + 3H_2(g) \xrightleftharpoons[\substack{\text{catalyst of iron} \\ \text{promoted by KOH}}]{450\,°C\ 200\,atm} 2NH_3(g); \quad \Delta H^\ominus = -92\,kJ\,mol^{-1}$$

You should be able to explain the optimum conditions in terms of the law of chemical equilibrium and the principles of reaction kinetics: see Chapters 7 and 9. When the gases are passed over the heterogeneous catalyst, a conversion of about 15% is achieved. The ammonia is liquefied by cooling and the 85% unreacted synthesis gas is recycled.

You should know that the synthesis gas (H$_2$:N$_2$ 3:1 by mol) is made chiefly from air (N$_2$:O$_2$ 4:1 by mol), natural gas (CH$_4$) and some naphtha (a petroleum fraction) after sulphur compounds have been removed (to avoid poisoning the catalysts). This is one source of sulphur for the production of sulphuric acid: see Chapter 12. You do not need to know how the synthesis gas is made. The process is a source of atmospheric argon which accumulates in the nitrogen-hydrogen mixture. The argon is regularly removed because its increasing partial pressure would adversely affect the yield of ammonia.

In the laboratory you could make dry ammonia by warming an ammonium salt with an involatile base (or a concentrated solution of a stronger alkali) and passing the gas over calcium oxide (or soda-lime) to dry it:

$$2NH_4Cl(s) + Ca(OH)_2(s) \xrightarrow{heat} CaCl_2(s) + 2H_2O(l) + 2NH_3(g)$$

$$H_2O(g) + CaO(s) \xrightarrow[\text{ammonia gas}]{\text{to dry the}} Ca(OH)_2(s)$$

> **Ammonia is extremely important.**

- You cannot use concentrated sulphuric or calcium chloride as drying agents because ammonia would neutralise the acid and would form a complex with the calcium ion

PHYSICAL PROPERTIES OF AMMONIA

You should know from your GCSE work that ammonia is a colourless gas with a distinctive smell. And you should reason that the gas must be less dense than air by comparing the relative molecular mass of ammonia ($NH_3 = 17$ because $14 + 3 \times 1 = 17$) with that of air ($N_2 : O_2$ $4 : 1 \approx 29$ because $\frac{4}{5}$ of 28 $+ \frac{1}{5}$ of 32 $= 28.8$). You should also know that the gas dissolves very fast and very well in water to give an alkaline solution. For A-level you need to explain the solubility in terms of properties of the molecules: see Fig. 12.19.

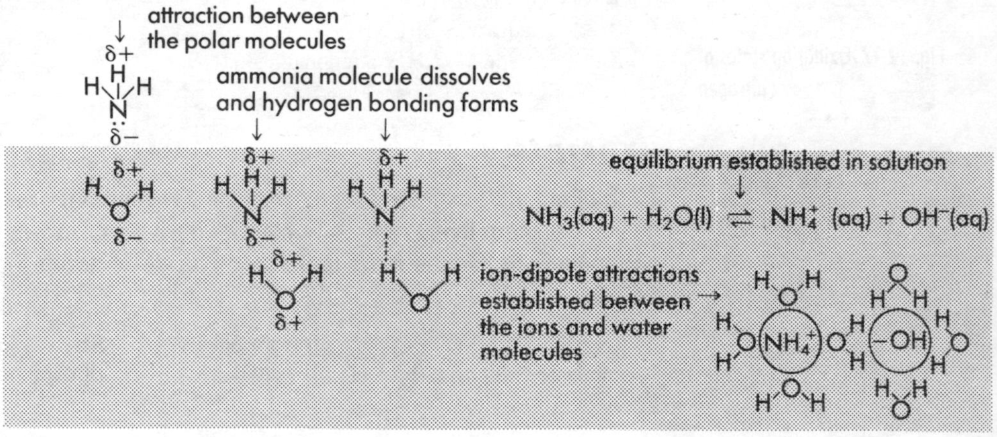

Fig. 12.19 Ammonia dissolving in water

CHEMICAL PROPERTIES OF AMMONIA

- Ammonia can act as a Lewis base

The lone electron-pair on the N-atom allows ammonia to accept a proton and act as a weak Bronsted-Lowry base:

$$HO\text{—}H(l) + :NH_3(aq) \rightleftharpoons HO^-(aq) + H:NH_3^+(aq)$$

The hydroxide ion concentration in aqueous ammonia is usually enough to cause precipitation of many metal hydroxides: e.g.

$$Mg^{2+}(aq) + 2OH^-(aq) \rightarrow Mg(OH)_2(s)$$

If, however, you add some ammonium chloride to the aqueous ammonia, the $OH^-(aq)$ concentration decreases (according to Le Chatelier's principle and the law of chemical equilibrium) and some metal hydroxides do not precipitate.

The lone electron-pair on the N-atom allows ammonia to form a dative bond and act as a ligand in complex formation: see Chapter 13. If you add excess ammonia to some aqueous metal salts the precipitated metal hydroxide redissolves: e.g.

$$Cu(OH)_2(s) + 4NH_3(aq) \rightarrow [Cu(NH_3)_4]^{2+}(aq) + 2OH^-(aq)$$

The lone pair also enables ammonia to attack electrophilic centres and act as a nucleophile in organic chemical reactions: see Chapter 16. For example,

$$RCH_2Cl + NH_3 \rightarrow RCH_2NH_2 + HCl \quad (or \ RCH_2NH_3^+Cl^-)$$

- Ammonia can undergo redox reactions

Very powerful reducing agents can react with ammonia: e.g.

$$2Na(s) \ + \ 2NH_3(g) \ \xrightarrow{\text{heat}} \ 2NaNH_2(s) \ + \ H_2(g)$$

$$[0] \qquad\qquad [+1] \qquad\qquad\quad [+1] \qquad\quad [0] \text{ [oxidation states]}$$

The reducing agent sodium is oxidised (ox. no. rises from 0 to +1) and the hydrogen (not the nitrogen) in the ammonia is reduced (ox. no. falls from +1 to 0). The nitrogen remains in an oxidation state of -3.

In many of its reactions ammonia acts as the reducing agent and the oxidation number of the nitrogen rises from -3: e.g.

$$2NH_3(g) \ + \ 3CuO(s) \ \xrightarrow{\text{heat}} \ 3Cu(s) \ + \ N_2(g) \ + \ 3H_2O(l)$$

You should realise that the 'decomposition' of some ammonium compounds are actually redox reactions in which nitrogen in the ammonium cation is being oxidised (from -3) by the anion. You may have seen the 'vulcano' produced as ammonium dichromate(VI) decomposes:

$$(NH_4)_2Cr_2O_7(s) \ \xrightarrow{\text{heat}} \ N_2(g) \ + \ Cr_2O_3(s) \ + \ 4H_2O(l); \quad \Delta H^\ominus \ = \ -1116\,\text{kJ}\,\text{mol}^{-1}$$

$$[-3]\,[+6] \qquad\qquad\qquad\quad [0] \qquad\quad [+3] \qquad\qquad\qquad \text{[oxidation states]}$$

A cone of green chromium(III) oxide powder develops and a shower of red-hot sparks fly out of the crater in the middle. Care is needed with this demonstration which must be performed in a fume cupboard because dichromate dust is extremely harmful.

■ Ammonium nitrate can 'decompose' violently:

$$NH_4^+NO_3^-(s) \ \xrightarrow{\text{heat}} \ N_2O(g) \ + \ 2H_2O(l)); \quad \Delta H^\ominus \ = \ -124\,\text{kJ}\,\text{mol}^{-1}$$

$$[-3]\,[+5] \qquad\qquad\quad [+1] \qquad\quad \text{[reverse of disproportionation]}$$

$$\text{dinitrogen (nitrous) oxide}$$
$$\text{[laughing gas]}$$

Ammonium nitrate is a component of explosives used for rock-blasting in quarries. It is also a fertiliser that must be stored in a cool place out of direct sunlight. If you were to heat a small sample in a test tube you should observe the last tiny portion 'explode with a pop', see a trace of brown fumes in the tube and witness a beautiful white smoke ring curl through the air! Solid ammonium nitrite does not exist. If dilute solutions of ammonium chloride and sodium nitrite are mixed and gently warmed the aqueous ammonium nitrite 'decomposes' rapidly:

$$NH_4^+(aq) \ + \ NO_2^-(aq) \ \xrightarrow{\text{heat}} \ N_2(g) \ + \ 2H_2O(l)$$

$$[-3] \qquad\qquad [+3] \qquad\qquad\quad [0]$$

This was a method of making 'chemically pure' nitrogen (i.e. uncontaminated by argon and other noble gases). The discovery that its density was slightly less than that of nitrogen extracted from the air was a clue to the existence of the unreactive noble gases.

MANUFACTURE OF NITRIC ACID

Ammonia (unlike phosphine, PH_3) does not burn in air but it will burn in oxygen or react with oxygen in the presence of an heterogeneous transition metal catalyst. The oxidation of ammonia to nitric acid is an extremely important large scale industrial process:

$$4NH_3(g) \ + \ 5O_2(g) \ \xrightarrow[\text{Pt/Rh catalyst}]{900\,^\circ\text{C} \ 5\,\text{atm}} \ 4NO(g) \ + \ 6H_2O(g); \quad \Delta H^\ominus \ = \ -994\,\text{kJ}\,\text{mol}^{-1}$$

Here again you should be able to explain the optimum conditions in terms of the law of chemical equilibrium and the principles of reaction kinetics: see Chapters 7 and 9. The exact composition and pressure of the ammonia/air gas mixture, the flow rate through the platinum-rhodium gauze catalyst, and the temperature are all carefully monitored and controlled to minimise unwanted reactions. For example, we do not want the formation and explosive decomposition of ammonium nitrate! We do not want ammonia oxidised to nitrogen by other redox reactions such as:

$$4NH_3(g) + 6NO(g) \rightarrow 5N_2(g) + 6H_2O(g) \quad \text{[if flow rate too fast]}$$
$$2NO(g) \rightarrow N_2(g) + O_2(g) \quad \text{[if flow rate too slow]}$$
$$4NH_3(g) + 3O_2(g) \rightarrow 2N_2(g) + 6H_2O(g) \quad \text{[if catalyst is too hot]}$$

The gaseous products (nitrogen monoxide and water) are compressed, cooled and mixed with air to achieve spontaneous oxidation of nitrogen monoxide to dioxide and hydration to nitric acid:

$$2NO(g) + O_2(g) \rightarrow 2NO_2(g); \quad \Delta H^\ominus = -114\,kJ\,mol^{-1}$$
$$3NO_2(g) + H_2O(l) \rightarrow 2HNO_3(aq) + NO(g)$$

OXIDES

All the group 5 elements form pentoxides (X_2O_5) that are acidic and trioxides (X_2O_3) that become less acidic down the group from N_2O_3 to Bi_2O_3: see Fig. 12.13. You may be asked to compare the trioxides and pentoxides of nitrogen and phosphorus: see Fig. 12.20.

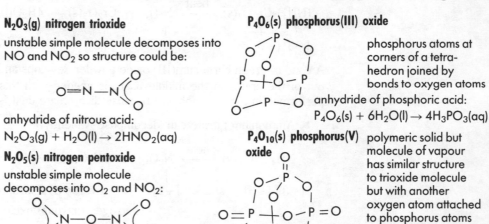

$N_2O_3(g)$ nitrogen trioxide

unstable simple molecule decomposes into NO and NO_2 so structure could be:

anhydride of nitrous acid:
$N_2O_3(g) + H_2O(l) \rightarrow 2HNO_2(aq)$

$N_2O_5(s)$ nitrogen pentoxide

unstable simple molecule decomposes into O_2 and NO_2:

anhydride of nitric acid:
$N_2O_5(g) + H_2O(l) \rightarrow 2HNO_3(aq)$

$P_4O_6(s)$ phosphorus(III) oxide

phosphorus atoms at corners of a tetra-hedron joined by bonds to oxygen atoms

anhydride of phosphoric acid:
$P_4O_6(s) + 6H_2O(l) \rightarrow 4H_3PO_3(aq)$

$P_4O_{10}(s)$ phosphorus(V) oxide

polymeric solid but molecule of vapour has similar structure to trioxide molecule but with another oxygen atom attached to phosphorus atoms forming tetrahedra

anhydride of phosphoric acid:
$P_4O_{10}(s) + 6H_2O(l) \rightarrow 4H_3PO_4(aq)$

Fig. 12.20 Comparing the oxides of nitrogen and phosphorus

N_2O_4, NO_2 and NO, and to a lesser extent N_2O, are more important than the anhydrides of nitrous acid and nitric acid. Their structures are not easy to represent because they contain delocalised electrons. And two of them, NO_2 and NO, have an odd number of valence electrons: see Fig. 12.21. They are often formed in redox reactions that you could do in test tubes in the laboratory: e.g. heating metal nitrates or reacting elements with nitric acid. If you add some dilute hydrochloric acid to solid sodium nitrite in a test tube, the liquid becomes cold, turns blue and gives off a colourless gas (NO). A pale brown gas appears higher up in the tube which feels warm at that point: see Fig. 12.22. If you warm the top of the tube the gas becomes a darker brown. You should be able to explain this effect by applying the law of chemical equilibrium to the following reversible reaction:

$$N_2O_4(g) \rightleftharpoons 2NO_2(g); \quad \Delta H^\ominus = 58\,kJ\,mol^{-1}$$
[colourless]   [dark brown]

dinitrogen tetraoxide	nitrogen dioxide	nitrogen oxide	dinitrogen oxide

Fig. 12.21 Structures of nitrogen oxides

The actual structures contain delocalised electrons. They may be regarded as resonance hybrids of the pairs of structures represented above

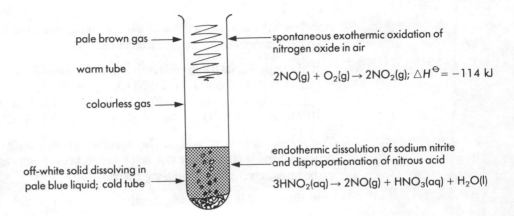

Fig. 12.22 Reaction of sodium nitrite and hydrochloric acid

pale brown gas

warm tube

colourless gas

off-white solid dissolving in pale blue liquid; cold tube

spontaneous exothermic oxidation of nitrogen oxide in air

$2NO(g) + O_2(g) \rightarrow 2NO_2(g); \Delta H^\ominus = -114$ kJ

endothermic dissolution of sodium nitrite and disproportionation of nitrous acid

$3HNO_2(aq) \rightarrow 2NO(g) + HNO_3(aq) + H_2O(l)$

OXOACIDS AND SALTS

Nitrogen has a wider range of oxides than phosphorus but phosphorus has a wider range of oxoacids than nitrogen: see Fig. 12.23.

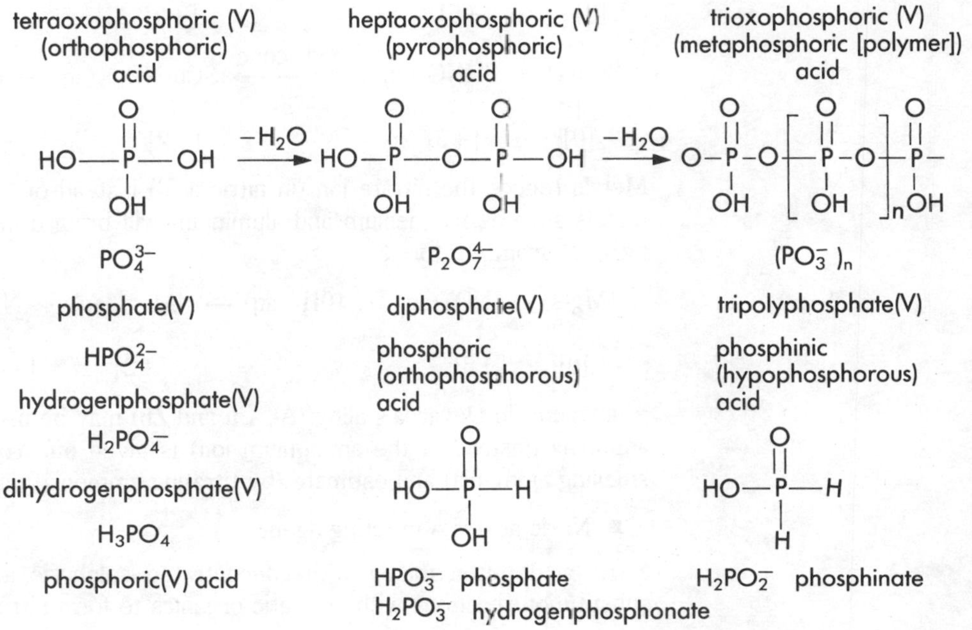

tetraoxophosphoric (V) (orthophosphoric) acid

heptaoxophosphoric (V) (pyrophosphoric) acid

trioxophosphoric (V) (metaphosphoric [polymer]) acid

PO_4^{3-}

phosphate(V)

HPO_4^{2-}

hydrogenphosphate(V)

$H_2PO_4^-$

dihydrogenphosphate(V)

H_3PO_4

phosphoric(V) acid

$P_2O_7^{4-}$

diphosphate(V)

phosphoric (orthophosphorous) acid

HPO_3^- phosphate
$H_2PO_3^-$ hydrogenphosphonate

$(PO_3^-)_n$

tripolyphosphate(V)

phosphinic (hypophosphorous) acid

$H_2PO_2^-$ phosphinate

Fig. 12.23 Some phosphorus oxoacids and oxoanions

The three most important acids are nitric, nitrous and phosphoric(V) acid.

Nitric acid

- Nitric acid is a strong monoprotic (monobasic) acid

It completely ionises in dilute aqueous solution to give a trigonal planar nitrate ion with three identical nitrogen-oxygen bonds involving delocalised electrons and a trigonal pyramidal oxonium ion: see Fig. 12.24.

$HNO_3(l) + H_2O(l) \rightarrow NO_3^-(aq) + H_3O^(aq)$

- Almost all nitrates are very soluble in water

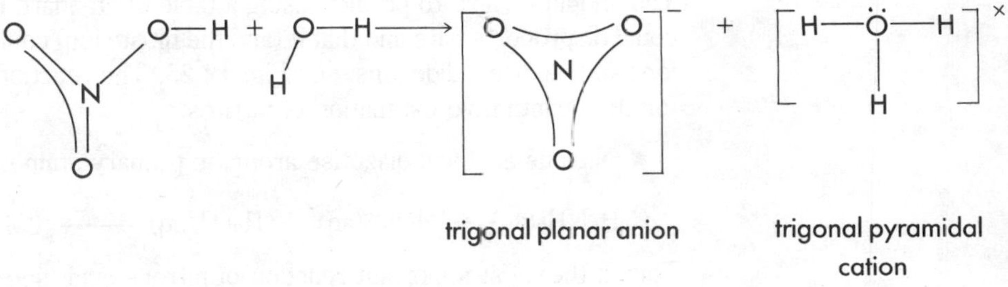

Fig. 12.24 Nitric acid completely ionises in water

trigonal planar anion

trigonal pyramidal cation

Aqueous nitric acid neutralises alkalis, dissolves basic oxides and liberates carbon dioxide from carbonates and hydrogencarbonates:

$$HNO_3(aq) + NaOH(aq) \rightarrow NaNO_3(aq) + H_2O(l)$$
$$2HNO_3(aq) + CuO(s) \rightarrow Cu(NO_3)_2(aq) + H_2O(l)$$
$$2HNO_3(aq) + ZnCO_3(s) \rightarrow Zn(NO_3)_2(aq) + H_2O(l) + CO_2(g)$$
$$HNO_3(aq) + KHCO_3(s) \rightarrow KNO_3(aq) + H_2O(l) + CO_2(g)$$

You should use nitric acid in the aqueous silver nitrate test (for halides) and in the aqueous barium nitrate test (for sulphate) to prevent unwanted precipitates (hydroxides, carbonates, etc.) giving confusing results.

- Nitric acid is an oxidising agent

You could be asked to predict the outcome of reactions involving nitric acid or nitrates using a table of selected standard redox potentials: see the questions at the end of this chapter. Although reduction of the nitrogen in the nitric acid to dinitrogen oxide seems most energetically feasible, it is often kinetically hindered. Nitrogen oxide and nitrogen dioxide are the most common reduction products: e.g.

$$\underset{[0]}{Cu(s)} + \underset{[+5]}{4HNO_3(aq)} \xrightarrow[\text{acid}]{\text{concentrated}} \underset{[+2]}{Cu(NO_3)_2(aq)} + 2H_2O(l) + \underset{[+4]}{2NO_2(g)}$$

$$\underset{[0]}{3Cu(s)} + \underset{[+5]}{8HNO_3(aq)} \xrightarrow[\text{acid}]{\text{mod. conc}} \underset{[+2]}{3Cu(NO_3)_2(aq)} + 4H_2O(l) + \underset{[+2]}{2NO(g)}$$

Metals reduce the nitrate ion (in nitric acid) instead of the hydrogen ion. Very reactive metals such as magnesium and aluminium will bring down the oxidation number of the nitrogen from +5 to −3:

$$\underset{[0]}{4Mg(s)} + \underset{[+5]}{NO_3^-(aq)} + 10H^+(aq) \rightarrow \underset{[+2]}{4Mg^{2+}(aq)} + \underset{[+4]}{NH_4^+(aq)} + 3H_2O(l)$$

Aluminium (in Devarda's alloy: Al, Cu and Zn) may be used in alkaline conditions so that ammonia (instead of the ammonium ion) is given off. You could use this to detect (by smelling ammonia) and estimate (by titrating ammonia) nitrates.

- Nitric acid is a nitrating agent

Nitric acid (often mixed with concentrated sulphuric acid) brings about electrophilic substitution reactions with aromatic organics to form nitro compounds: see chapter 15.

Nitrous acid

- Nitrous acid is a weak unstable monoprotic (monobasic) acid

It is formed by adding hydrochloric acid to aqueous sodium nitrite:

$$H_3O^+(aq) + NO_2^-(aq) \rightarrow H-O-N=O(aq) + H_2O(l)$$

The pale blue colour of the solution may be due to the presence of N_2O_3.

- Nitrous acid can act as an oxidising agent and a reducing agent

Nitrous acid readily disproportionates into nitrogen oxide and nitric acid: see Fig. 12.22.

$$\underset{[+3]}{3HNO_2(aq)} \rightarrow \underset{[+2]}{2NO(g)} + \underset{[+5]}{HNO_3(aq)} + H_2O(l) \quad \text{[oxidation states]}$$

You should be able to predict, using a table of standard redox potentials, that nitrous acid could disproportionate and that it (and the nitrite ion) could reduce acidified manganate(VII) ions and oxidise iodide ions: see Fig. 12.25. The reaction with manganate(VII) is the basis for the quantitative estimation of nitrites.

- Nitrous acid will diazotise aromatic primary amines

$$C_6H_5NH_2(aq) + HNO_2(aq) + H_3O^+(aq) \xrightarrow{5\,°C} C_6H_5N_2^+(aq) + 3H_2O(l)$$

This is the most important reaction of nitrous acid: see Chapter 20.

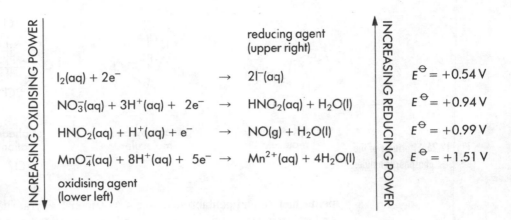

Fig. 12.25 Predicting redox reactions of nitrous acid

Phosphoric(V) acid

■ Phosphoric(V) acid is a triprotic (tribasic) acid

In aqueous solution H_3PO_4 is a somewhat weak acid that ionises as follows:

$$H_3PO_4(aq) + H_2O(l) \rightarrow H_3O^+(aq) + H_2PO_4^-(aq)$$
acid ($pK_a = 2.1$)   conjugate base

If a strong alkali (e.g. NaOH) is added to the solution the dihydrogenphosphate(V) ion (a weak acid) donates a proton:

$$H_2PO_4^-(aq) + H_2O(l) \rightarrow H_3O(aq) + HPO_4^{2-}(aq)$$
acid ($pK_a = 7.2$)   conjugate base

And in its turn the hydrogenphosphate(V) ion (a very weak acid) will donate a proton:

$$HPO_4^{2-}(aq) + H_2O(l) \rightarrow H_3O^+(aq) + PO_4^{3-}(aq)$$
acid ($pK_a = 12.4$)   conjugate base

Notice that because the hydrogenphosphate(V) ion is such a weak acid, the phosphate(V) ion is a very strong base. Consequently aqueous trisodium phosphate has a very high pH and is dangerously caustic!

An aqueous solution containing equal amounts, by mole, of sodium dihydrogen-phosphate(V) (NaH_2PO_4) and disodium hydrogenphosphate(V) (Na_2HPO_4) has a pH of 7.2. It is an effective buffer because the solution contains an acid, $H_2PO_4^-(aq)$, and a base, $HPO_4^{2-}(aq)$ in equal amounts. The acid can deal with any alkalis added and the base can deal with any acids added.

■ Phosphoric(V) acid is an involatile non-oxidising acid

If you warm an alkali metal halide with phosphoric(V) acid, the involatile acid displaces the more volatile hydrogen halide:

$$H_3PO_4(l) + KBr(s) \xrightarrow{\text{heat}} KH_2PO_4(s) + HBr(g)$$

■ Phosphoric(V) acid is a good dehydrating agent

If you warm cyclohexanol with phosphoric(V) acid, you can obtain cyclohexene in reasonable yield: see Chapter 15.

$$C_6H_{11}OH(l) \xrightarrow[-H_2O]{H_3PO_4 \text{ heat}} C_6H_{10}(l)$$

HALIDES

Nitrogen trichloride is a yellow oil and nitrogen triiodide is a black solid. They both have trigonal pyramidal molecules. They are dangerously explosive and not important for A-level chemistry.

■ Nitrogen cannot extend its outer shell beyond eight electrons so it does not form pentahalides

The phosphorus halides are more important than the nitrogen halides. The phosphorus trihalide molecules are trigonal pyramids: see Fig. 12.26. In the vapour state, phosphorus

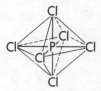

trigonal pyramidal molecule	trigonal bipyramidal molecule	tetrahedral cation	octahedral anion
PCl_3	PCl_5	PCl_4^+	PCl_6^-

Fig. 12.26 Structures of the phosphorus halides

pentachloride and pentabromide exist as trigonal bipyramidal molecules whose degree of endothermic dissociation increases with increasing temperature: see Chapter 7.

$$PCl_5(g) \rightleftharpoons PCl_3(g) + Cl_2(g)$$

In the solid state the structures are more complicated:

$PCl_4^+PCl_6^-$ tetrachlorophosphonium hexachlorophosphate(V)
$PBr_4^+Br^-$ tetrabromophosphonium bromide

Phosphorus pentaiodide does not exist. Phosphorus triiodide is very unstable and usually prepared in situ by reacting iodine with red phosphorus. For example, to make 1-iodobutane from butan-1-ol you could put red phosphorus in a flask and cover it with some of the alcohol. You could dissolve the iodine in the rest of the alcohol and add the solution to the flask a little at a time (because the reaction is very exothermic):

$$3I_2(\text{in butanol}) + 6C_4H_9OH(l) + 2P(\text{red}) \rightarrow 6C_4H_9I(l) + 2H_3PO_3(l)$$

The phosphorus chlorides PCl_5 and PCl_3 are readily hydrolysed to phosphoric(V) acid and phosphonic acid respectively: e.g.

$$PCl_3(l) + 3H_2O(l) \rightarrow H_3PO_3(l) + 3HCl(g)$$
$$PCl_5(s) + C_2H_5OH(l) \rightarrow C_2H_5Cl(l) + POCl_3(l) + HCl(g)$$

They may used to test for (and replace by Cl) OH groups in many organic compounds. You can identify misty HCl fumes with ammonia (gives white smoke of ammonium chloride, NH_4Cl).

GROUP VI: OXYGEN, SULPHUR

Oxygen and sulphur are two very important elements that you have already met in your GCSE work. You now need to study the chemistry of these two elements and their compounds in more detail. You do not need to study the other elements (Se and Te) or the group characteristics.

OXYGEN

Liquid oxygen is pale blue in colour and paramagnetic i.e. weakly attracted by a magnet. Paramagnetism is due to the presence of one or more unpaired electrons. But a diatomic oxygen molecule (O_2) has an even number (12) of valence electrons. It follows that if we represent the structure of the molecule by $O{=}O$ we should not assume that each O-atom has two pairs of non-bonded electrons. Oxygen exists as two gaseous allotropes: see Fig. 12.27.

Oxides

An oxide is a compound of an element with oxygen. The electronegativity of oxygen is second only to that of fluorine and the atom is small enough to stabilise many elements in their highest oxidation state. An oxygen atom can gain two electrons to form the oxide ion O^{2-} which results in ionic structures (e.g. $Mg^{2+}O^{2-}$) with high lattice energies and, consequently, very exothermic enthalpy changes of formation. It can also share two electrons to form two covalent bonds —O— which results in simple molecules e.g. CO_2 or in giant covalent structures e.g. SiO_2.

Most elements form one or more oxides. This large variety of oxides can be classified

oxygen (dioxygen) O_2

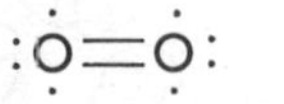

colourless odourless gas

$$:\!\overset{..}{O}\!=\!\overset{..}{O}\!:$$

b.pt. $-183°C$

produced by fractional distillation of liquid air

very reactive oxidant combines with most elements to form oxides and supports combusion of organics to carbon dioxide and water

used in steel making

ozone (trioxygen) O_3

pale blue pungent gas

b.pt. $-112°C$

produced by silent electric discharge through air

extremely reactive oxidant attacks organic compounds and cleaves C=C bonds to form ozonides that hydrolyse to aldehydes and ketones

used for air/water purification

Fig. 12.27 Gaseous allotropes of oxygen

in various ways. You should understand how oxides can be classed as acidic, amphoteric or basic and be able to explain the patterns of these oxides in the periodic table: see Chapter 10.

■ Oxides of metals in a high oxidation state tend to be acidic and oxides of metals in a low oxidation state tend to be basic

Some oxides e.g. carbon monoxide CO, nitrogen oxide NO and dinitrogen oxide N_2O are classified as neutral because they do not react with acids or with bases. Some are called mixed oxides because they behave like two separate oxides: e.g. Pb_3O_4 reacts like $PbO_2.2PbO$. An acidic oxide that dissolves in water to form one acid may be called an acid anhydride: e.g.

$$SO_2(g) + H_2O(l) \rightarrow H_2SO_3(aq)$$

If two acids are formed, the oxide may be called a mixed anhydride: e.g.

$$N_2O_4(g) + H_2O(l) \rightarrow HNO_2(aq) + HNO_3(aq)$$

dinitrogen tetroxide   nitrous acid   nitric acid

Peroxides

These are compounds whose structure contains two oxygen atoms that are in an oxidation state of -1 and joined by a covalent bond: e.g.

hydrogen peroxide (H—O—O—H)   sodium peroxide ($Na^{+-}O$—O^-Na^+)

Pure hydrogen peroxide is a pale blue liquid: m.pt. $-0.4°C$; b.pt. $150°C$: Trace impurities catalyse its decomposition and make it a dangerous explosion and fire risk. You may have used dilute aqueous hydrogen peroxide and manganese(IV) oxide as a catalyst to prepare some oxygen in the laboratory. The hydrogen peroxide disproportionates:

$$\text{H—O—O—H(aq)} \xrightarrow[\text{catalyst}]{MnO_2(s)} H_2O(l) + \tfrac{1}{2}O_2(g)$$
$$\quad\;[-1\;-1] \qquad\qquad\qquad [-2] \quad\;\; [0] \qquad \text{[oxidation states]}$$

Hydrogen peroxide and the peroxides are powerful oxidising agents with a wide range of uses that depend upon their oxidising properties. Organic peroxides are used as initiators in polymerisation reactions because they readily decompose into free radicals: see Chapter 15. A very powerful oxidising agent can force hydrogen peroxide to act as a reducing agent: e.g.

$$2MnO_4^-(aq) + 5H_2O_2(aq) \rightarrow 2Mn^{2+}(aq) + 8H_2O(l) + 5O_2(g)$$
$$\quad[+7] \qquad\quad\; [-1] \qquad\qquad\; [+2] \qquad\qquad\qquad\quad [0]$$

WATER

This is probably the most important oxide. You will already know a great deal about it. Its physical and chemical properties are governed by the structure of the molecule: see Fig. 12.28. Remember that as the first member of the group oxygen (and its compounds) will behave atypically. Water has an exceptionally high boiling point owing to inter-molecular hydrogen bonding caused by the high electronegativity of the oxygen atom: see Fig. 12.29. Water is exceptional even among the three atypical hydrides [NH_3 ($M_r = 17$; b.pt. $-33°C$), H_2O ($M_r = 18$; b.pt. $100°C$) and HF ($M_r = 20$; b.pt. $20°C$] because it

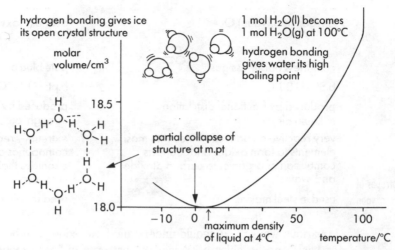

Fig. 12.28 Structure and physical properties of water

Open crystal structure makes ice less dense than liquid water at its m.pt. This makes water unusual in that it expands on freezing—hence burst water pipes in winter—and the ice floats on the liquid—hence ice-skating.

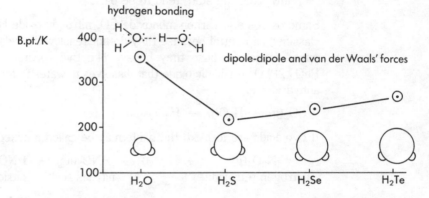

Fig. 12.29 Boiling points of the group 6 hydrides

boils at a higher temperature than hydrogen fluoride. All three hydrides are hydrogen bonded but whereas NH_3 and HF form only one hydrogen bond per molecule, water forms two hydrogen bonds per molecule.

- ■ Water can act as a polar solvent, a catalyst, an acid, a base, a ligand, a nucleophile, an oxidising agent and a reducing agent

WATER AS A SOLVENT

Many ionic solids readily dissolve in water because the polar water molecules are attracted by the ions which become hydrated and disperse into the water. The water molecules cause the attractive forces between the ions to diminish as they penetrate the crystal lattice. The hydration energy given out compensates for the lattice energy taken in. The difference between these two energy terms is small and represents the solution energy: see Fig. 12.30. Covalent substances may dissolve in water if their molecules are polar and/or able to form hydrogen bonds with water molecules (e.g. lower alcohols and carbohydrates) or if their molecules can ionise (e.g. lower amines and carboxylic acids).

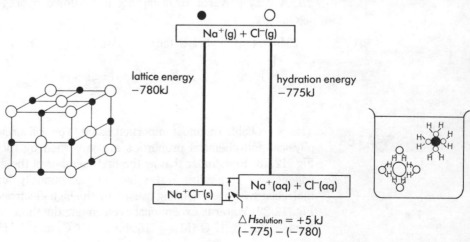

Fig. 12.30 Energy changes upon dissolving

WATER AS A CATALYST

From 1888 to 1912, H. B. Baker published (in the Transactions of the Chemical Society) the results of many experiments in which he investigated the effect of trace amounts of moisture upon various physical and chemical properties. For example, he found that the boiling point of dinitrogen tetraoxide (data book value 22 °C) to be 69 °C after he dried the gas with phosphorus(V) oxide for one year! He also found that a mixture of hydrogen and oxygen (in the ratio 2:1 by volume) dried with phosphorus(V) oxide for ten days would not explode when small sparks were passed through it or silver was melted in it! You can read more about this in T. M. Lowry's Historical Introduction to Chemistry, published by MacMillan & Co in 1915.

You may already know that dry iron does not rust, dry chlorine does not bleach or react with metals and dry hydrogen chloride dissolved in dry methylbenzene does not react with calcium carbonate. We may not fully appreciate the widespread importance of water as a catalyst because it is so difficult to remove completely all traces of moisture from our chemicals and apparatus!

WATER AS AN AMPHOTERIC OXIDE

A water molecule can act as an acid by donating a proton and as a base by accepting a proton:

$$H_2O(l) + H_2O(l) \rightleftharpoons H_3O^+(aq) + OH^-(aq)$$

Pure water is 'neutral' because the hydrogen ion concentration $[H_3O^+(aq)]$ is equal to the hydroxide ion concentration $[OH^-(aq)]$. At 25 °C, pure water has a pH = 7 because $K_w = [H_3O^+(aq)][OH^-(aq)] = 10^{-14} \, mol^2 \, dm^{-6}$ at that temperature: see Chapter 7.

WATER AS A LIGAND

Many crystalline salts are hydrated: e.g. cobalt(II) chloride-6-water and copper(II) sulphate-5-water. Some or all of the water molecules are attached to the cation by dative covalent bonds formed by the O-atom donating a lone electron-pair: see Chapter 13. The colour change when the anhydrous salt becomes hydrated is the basis of testing for water with cobalt chloride paper:

$$Co^{2+}(Cl^-)_2(s) + 6H_2O(l) \rightarrow [Co(H_2O)_6]^{2+}(Cl^-)_2(s)$$
$$\text{blue} \qquad\qquad\qquad\qquad\qquad \text{pink}$$

We use anhydrous calcium chloride, magnesium chlorate(VII) and sodium sulphate as drying agents in organic chemistry because they readily form hydrates but do not readily dissolve in organic liquids.

WATER AS A NUCLEOPHILE

Hydrolysis of inorganic and organic halogen compounds may involve nucleophilic attack by water molecules: see Chapter 12.

$$2H_2O(l) + SiCl_4(l) \rightarrow SiO_2(s) + 4HCl(g)$$
$$H_2O(l) + CH_3COCl(l) \rightarrow CH_3CO_2H(l) + HCl(g)$$

WATER AS A REDOX AGENT

Whether water is reduced or oxidised depends upon the reagents and conditions. For example, sodium metal or methane gas can reduce the H-atoms in water to hydrogen:

$$2Na(s) + 2H_2O(l) \rightarrow 2NaOH(aq) + H_2(g)$$

$$CH_4(g) + H_2O(l) \xrightarrow[\text{Ni catalyst}]{900\,°C \ 30\,atm} CO(g) + 3H_2(g)$$

And fluorine can oxidise the O-atoms in water to oxygen:

$$F_2(g) + 2H_2O(l) \rightarrow 4HF(aq) + O_2(g)$$

Reduction (at the cathode) and oxidation (at the anode) occurs simultaneously at inert (platinum) electrodes during the electrolysis of (acidified) water:

$$2H_2O(l) \xrightarrow{\text{D.C. electricity}} \underset{\text{at cathode}}{2H_2(g)} + \underset{\text{at anode}}{O_2(g)}$$

SULPHUR

Solid sulphur is yellow and exists as rhombic and monoclinic crystals. These two allotropes consist of S_8 (ring) molecules that are packed in different ways in their crystal lattices. Below 96 °C rhombic sulphur is the stable allotrope. Monoclinic sulphur is the stable form between 96 °C and 119 °C, the melting point.

If you have heated molten sulphur in a test tube, you will have seen the liquid darken and become almost black at its boiling point. You may also have been surprised to see that the liquid becomes more viscous at first but then become less viscous as the temperature nears its boiling point. An explanation is that the S_8 rings break open and join together to form extremely long chain molecules; so the liquid becomes very viscous. At higher temperatures these long chains break into smaller ones; so the liquid becomes less viscous. If you pour the dark liquid sulphur rapidly into cold water, a rubbery mass of plastic sulphur is formed. This gradually changes back to a yellow solid as the chains break and the S_8 molecules reform.

Reaction with elements

Metals can occur as sulphide ores (e.g. PbS – galena; ZnS – zinc blende) and can reduce sulphur to sulphides: e.g.

$$8Fe(s) + S_8(s) \xrightarrow{\text{heat}} 8FeS(s) \text{ iron(II) sulphide}$$
$$\underset{[0]}{\uparrow} \qquad \underset{[-2]}{\uparrow} \qquad \text{[oxidation states]}$$

Chlorine, fluorine and oxygen will oxidise sulphur by combining directly with it: e.g.

$$4Cl_2(g) + S_8(s) \rightarrow 4S_2Cl_2(g) \qquad \text{disulphur dichloride}$$
$$\underset{[+1]}{\uparrow}$$
$$8O_2(g) + S_8(s) \rightarrow 8SO_2(g) \qquad \text{sulphur dioxide}$$
$$\underset{[+4]}{\uparrow}$$
$$24F_2(g) + S_8(s) \rightarrow 8SF_6(g) \qquad \text{sulphur hexafluoride}$$
$$\underset{[+6]}{\uparrow}$$

Sulphur hexafluoride gas is so stable that it is widely used as an electrical insulator.

Hydrogen sulphide

In contrast to water (H_2O), hydrogen sulphide (H_2S) is a dangerously flammable, very poisonous gas with a characteristic smell of 'rotten eggs'. In a plentiful supply of air it burns to water and sulphur dioxide, or to sulphur if there is not enough air:

$$2H_2S(g) + 3O_2(g) \xrightarrow{\text{excess air}} 2H_2O(l) + 2SO_2(g)$$
$$2H_2S(g) + O_2(g) \xrightarrow{\text{insufficient air}} 2H_2O(l) + 2S(s)$$

The gas is oxidised to water and sulphur as it reduces acidified aqueous dichromate(VI) ions to chromium(III) ions. You can use the colour change from orange to green and the appearance of the yellow precipitate as a test for hydrogen sulphide gas.

It is slightly soluble in water and behaves as a very weak diprotic (dibasic) acid:

$$H_2S(aq) + H_2O(l) = H_3O^+(aq) + HS^-(aq) \qquad pK_a = 7.1$$
$$HS^-(aq) + H_2O(l) = H_3O^+(aq) + S^{2-}(aq) \qquad pK_a = 12.9$$

If you add hydrogen sulphide (or its aqueous solution) to aqueous salts of d- and p-block metals, you should observe precipitates of insoluble sulphides. A filter paper moistened with aqueous lead(II) ethanoate turns black in hydrogen sulphide. You can use this as a fairly sensitive test for the gas:

$$H_2S(g) + Pb^{2+}(aq) \rightarrow 2H^+(aq) + PbS(s) \quad \text{lead(II) sulphide [black]}$$

Oxides

Sulphur dioxide and sulphur trioxide are acidic oxides that form sulphurous acid and sulphuric acid with water. Sulphur dioxide is produced by burning sulphur (or fossil fuels containing sulphur compounds) in air or oxygen. It is also produced when sulphide ores are 'roasted' in air as a first step in the extraction of some metals:

$$2ZnS(s) + 2O_2(g) \xrightarrow{\text{heat}} 2ZnO(s) + SO_2(g)$$

Strong reducing agents can reduce (the S-atom in) sulphur dioxide to sulphur: e.g.

$$2H_2S(g) + SO_2(g) \xrightarrow[\text{as catalyst}]{\text{moisture}} 3S(s) + 2H_2O(l)$$
$$\uparrow \qquad\qquad \uparrow$$
$$[+4] \qquad\qquad [0] \qquad \text{[oxidation states]}$$

This reaction does not take place if the gases are dry. If water is added to the dry gaseous mixture in a jar, sulphur deposits on the moist parts of the glass.

Sulphur dioxide usually acts as a reducing agent. This explains its use as an anti-oxidant in food preservation. It also explains its use in water purification for removing excess chlorine:

$$SO_2(aq) + Cl_2(aq) + 2H_2O(l) \rightarrow H_2SO_4(aq) + 2HCl(aq)$$
$$\uparrow \qquad \uparrow \qquad\qquad\qquad \uparrow \qquad\qquad \uparrow$$
$$[+4] \quad [0] \qquad\qquad\qquad [+6] \qquad\quad [-1]$$

Sulphur trioxide is produced by the oxidation of sulphur dioxide using vanadium(V) oxide (promoted by potassium sulphate and supported by silicon oxide) as the heterogeneous catalyst:

$$2SO_2(g) + O_2(g) \xrightarrow[V_2O_5]{417\,°C} 2SO_3(g); \quad \Delta H^\ominus = -197\,\text{kJmol}^{-1}$$

This is the essential reaction in the Contact Process for the production of sulphuric acid.

OXOACIDS AND SALTS

Sulphurous acid and sulphites

Sulphurous acid is formed when sulphur dioxide 'dissolves' in water. It is a weak diprotic (dibasic) acid that exists only in aqueous solution:

$$SO_2(g) + H_2O(l) = H_2SO_3(aq)$$
$$H_2SO_3(aq) = H^+(aq) + HSO_3^-(aq) \qquad pK_a = 1.8$$
$$HSO_3^-(aq) = H^+(aq) + SO_3^{2-}(aq) \qquad pK_a = 7.2$$

Sulphites are formed when sulphur dioxide 'dissolves' in aqueous alkali:

$$SO_2(g) + 2OH^-(aq) = SO_3^{2-}(aq) + H_2O(l)$$

Sulphur dioxide is released when sulphites are acidified. You should compare this with the action of acid on carbonates:

$$2H^+(aq) + SO_3^{2-}(aq) \rightarrow H_2SO_3(aq) \rightarrow H_2O(l) + SO_2(g)$$

Sulphurous acid and sulphites are readily oxidised to sulphuric acid and sulphates: e.g.

$$Cr_2O_7^{2-}(aq) + 2H^+(aq) + 3SO_2(g) \rightarrow 2Cr^{3+}(aq) + H_2O(l) + 3SO_4^{2-}(aq)$$
$$\uparrow \qquad\qquad\qquad\qquad \uparrow \qquad\quad \uparrow \qquad\qquad\qquad \uparrow$$
$$[+6] \qquad\qquad\qquad\qquad [+4] \qquad [+3] \qquad\qquad\qquad [+6]$$

The colour change of the aqueous dichromate(VI) from orange-yellow to green is used as a test for sulphur dioxide and other gaseous reducing agents.

$$2MnO_4^-(aq) + 6H^+(aq) + 5SO_3^{2-}(aq) \rightarrow 2Mn^{2+}(aq) + 3H_2O(l) + 5SO_4^{2-}(aq)$$
$$\uparrow \qquad\qquad\qquad\qquad \uparrow \qquad\qquad\qquad \uparrow \qquad\qquad\qquad \uparrow$$
$$[+7] \qquad\qquad\qquad\qquad [+4] \qquad\qquad [+2] \qquad\qquad\qquad [+6]$$

The colour change of the aqueous manganate(VII) from purple to colourless is used to indicate the end-point when you titrate sulphite solutions.

Thiosulphurous acid and thiosulphates

Sodium sulphite is reduced to sodium thiosulphate by boiling the aqueous solution with sulphur:

$$Na_2SO_3(aq) \ + \ S(s) \ \xrightarrow{\text{heat}} \ Na_2S_2O_3(aq)$$
$$\uparrow \qquad\qquad \uparrow \qquad\qquad\qquad \uparrow$$
$$[+4] \qquad\quad [0] \qquad\qquad\quad [+2]$$

You will use the quantitative reaction of aqueous thiosulphate with aqueous iodine to carry out titrations as part of your practical work in analytical chemistry:

$$2S_2O_3^{2-}(aq) \ + \ I_2(aq) \ \rightarrow \ 2I^-(aq) \ + \ S_4O_6^{2-}(aq) \qquad \text{[tetrathionate]}$$
$$\uparrow \qquad\qquad\quad \uparrow \qquad\qquad \uparrow \qquad\qquad \uparrow$$
$$[+2] \qquad\qquad [0] \qquad\quad [-1] \qquad\quad [+2\tfrac{1}{2}]$$

Thiosulphurous acid is an unstable weak acid. So when you add hydrochloric acid to aqueous sodium thiosulphate, the thiosulphurous acid that is formed will disproportionate:

$$2H^+(aq) \ + \ S_2O_3^{2-}(aq) \ \rightarrow \ H_2S_2O_3(aq) \ \rightarrow \ H_2O(l) \ + \ SO_2(g) \ + \ S(s)$$
$$\uparrow \qquad\qquad\qquad\qquad\qquad\qquad\qquad \uparrow \qquad\quad \uparrow$$
$$[+2] \qquad\qquad\qquad\qquad\qquad\qquad [+4] \qquad [0]$$

SULPHURIC ACID AND SULPHATES

Sulphur trioxide reacts violently with water to produce a stable mist of sulphuric acid:

$$H_2O(l) \ + \ SO_3(s) \ \rightarrow \ H_2SO_4(l)$$

In the very large scale industrial manufacture of sulphuric acid, the sulphur trioxide produced in the Contact Process is 'dissolved' in 98% sulphuric acid until the 2% water is converted to sulphuric acid. This 100% acid is mixed with water to produce more 98% acid and the process repeated.

- Never add water to concentrated sulphuric acid
- Always add concentrated sulphuric acid to water in small amounts at a time with stirring to mix well and dissipate the heat

Pure sulphuric acid is a colourless, syrupy liquid. It is not very volatile (b.pt. 338 °C) so it can be used to prepare hydrogen chloride in the laboratory:

$$NaCl(s) \ + \ H_2SO_4(l) \ \rightarrow \ NaHSO_4(s) \ + \ HCl(g)$$

- Sulphuric acid can act as an acid, a catalyst, a dehydrating agent, an oxidising agent and a sulphonating agent

Sulphuric acid as an acid

Sulphuric acid is a strong diprotic (dibasic) acid. In water it ionises completely into hydrogen ions and hydrogensulphate ions:

$$H_2SO_4(l) \ + \ H_2O(l) \ \rightarrow \ H_3O^+(aq) \ + \ HSO_4^-(aq)$$

The hydrogensulphate ion is a weak acid so its ionisation is usually incomplete:

$$HSO_4^-(aq) \ + \ H_2O(l) \ = \ H_3O^+(aq) \ + \ SO_4^{2-}(aq) \quad pK_a = 2.0$$

Aqueous sulphuric acid shows the expected behaviour of neutralising alkalis and liberating CO_2 from carbonates. However, it gives a precipitate with aqueous barium hydroxide because barium sulphate is extremely insoluble:

$$H_2SO_4(l) \ + \ Ba(OH)_2(aq) \ \rightarrow \ BaSO_4(s) \ + \ 2H_2O(l)$$

And with a lump of marble the reaction subsides because a film of fairly insoluble calcium sulphate forms on the surface of the calcium carbonate and protects it from further attack:

$$H_2SO_4(l) \ + \ CaCO_3(s) \ \rightarrow \ CaSO_4(s) \ + \ H_2O(l) \ + \ CO_2(g)$$

The industrial production of ammonium sulphate fertiliser depends upon the acid-base reaction of sulphuric acid with ammonia:

$$H_2SO_4(l) + 2NH_3(g) \rightarrow (NH_4)_2SO_4(s)$$

Sulphuric acid as a catalyst

Sulphuric acid catalyses the esterification of carboxylic acids with alcohols: see Chapter 17.

$$CH_3CO_2H(l) + C_2H_5OH(l) \xrightarrow[\text{catalyst}]{H_2SO_4} CH_3CO_2C_2H_5(l) + H_2O(l)$$

Sulphuric acid as a dehydrating agent

You can use concentrated sulphuric acid to dry gases (e.g. CO_2, HCl but *not* NH_3) and to dehydrate alcohols to form alkenes or ethers: see Chapter 17.

$$C_2H_5OH(l) \xrightarrow[-H_2O]{\text{conc. } H_2SO_4} CH_2{=}CH_2(g) \text{ or } C_2H_5OC_2H_5(l)$$

Sulphuric acid as an oxidising agent

Although the *less volatile* sulphuric acid will readily *displace* the *more volatile* hydrogen iodide from potassium iodide it will also even more readily oxidise the halide to iodine:

$$H_2SO_4(l) + KI(s) \rightarrow KHSO_4(s) + HI(g)$$
$$8HI(g) + H_2SO_4(l) \rightarrow 4I_2(s) + H_2S(g) + 4H_2O(l)$$
$$\qquad\uparrow \qquad\quad \uparrow \qquad\quad \uparrow \qquad\quad \uparrow$$
$$\quad [-1] \qquad [+6] \qquad [0] \qquad [-2] \qquad \text{[oxidation states]}$$

- Concentrated sulphuric acid cannot be used to prepare hydrogen iodide

Sulphuric acid as a sulphonating agent

An important industrial use of sulphuric acid is the sulphonation of straight chain alkylbenzenes in the production of detergents:

$$CH_3(CH_2)_{10}CH_2C_6H_5(l) + H_2SO_4(l) \rightarrow CH_3(CH_2)_{10}CH_2C_6H_5SO_3H(l) + H_2O(l)$$

GROUP VII – THE HALOGENS

> Halogens and halides are very popular with examiners.

The similarities between the elements and their compounds are greater in this group than in any other p-block group. So you must study the group properties in some detail. Remember that fluorine, as the first element in the group, will show some atypical properties. You will need also to study the group trends from fluorine to iodine. These elements and their compounds have a lot in common because the outer p-subshell of their atoms is just one electron short of the configuration of the neighbouring noble gas atoms.

ELECTRONIC CONFIGURATION	m.pt. /°C	b.pt. /°C	N_p	$E^{\ominus}$/V	HYDRIDES	OXIDES
F [He]$2s^2 2p^5$	−220	−188	4.0	+2.87	H_2F_2 weak acid	OF_2
Cl [Ne]$3s^2 3p^5$	−101	−35	3.0	+1.36	HCl strong acid	Cl_2O; ClO_2; Cl_2O_7
Br [Ar]$3d^{10}4s^2 4p^5$	−7	59	2.8	+1.09	HBr strong acid	Br_2O; BrO_2
I [Kr]$4d^{10}5s^2 5p^5$	114	184	2.5	+0.54	HI strong acid	IO_2; I_2O_5

N_p = Pauling's electronegativity index

Fig. 12.31 Properties of the halogens

PHYSICAL PROPERTIES

- Halogens are non-metals that exist as diatomic covalent molecules

The increase in melting (and boiling) points and standard enthalpy changes of melting (and boiling) with atomic number down the group is related to an increase in van der Waals forces caused by more electrons in the atom. Chlorine, bromine and iodine dissolve slightly in water: $Cl_2(g) \rightarrow Cl_2(aq)$.

CHEMICAL PROPERTIES

The aqueous chlorine (as well as aqueous bromine and iodine) disproportionates in the water: e.g.

$$Cl_2(g) \rightarrow Cl_2(aq) + H_2O(l) = Cl^-(aq) + ClO^-(aq) + 2H^+(aq)$$

$$\uparrow \qquad\qquad\qquad \uparrow \qquad \uparrow$$

$$[0] \qquad\qquad\qquad [-1] \qquad [+1] \qquad \text{[oxidation states]}$$

Fluorine does not dissolve in water. Instead, water burns in fluorine to form hydrogen fluoride and oxygen!

$$F_2(g) + H_2O(l) \rightarrow 2HF(g) + \tfrac{1}{2}O_2(g); \quad \Delta H^\ominus = -257\,kJ\,mol^{-1}$$

- Halogens act as oxidising agents by gaining electrons from metals or by sharing electrons with less electronegative non-metals

The halogens can act as oxidising agents by completing their outer p-subshell of electrons to form stable anions with the same electronic configurations as their corresponding noble gases: e.g.

$$Cl(2.8.7) + e^- \rightarrow Cl^-(2.8.8) \quad \text{[Ar 2.8.8]}$$

Fluorine is the most electronegative of all the elements, so its oxidation is -1 in all its compounds. The other three halogens can form compounds in which their oxidation numbers can range from -1 to $+7$.

- The electronegativity and the reactivity of the halogens decreases with atomic number down the group from fluorine to iodine

Reaction with metals

Halogens combine with most metals to form halides which are often largely ionic in character. However, the size and the polarisability of the halide anion increases down the group. Consequently, iodides are more covalent in character than fluorides (you should compare the properties of aluminium fluoride and aluminium chloride). With d-block metals such as iron, fluorine and chlorine can oxidise the metal to a higher oxidation state than iodine can: e.g.

$$Fe(s) + 1\tfrac{1}{2}Cl_2(g) \xrightarrow{\text{heat}} FeCl_3(s); \qquad \Delta H^\ominus = -400\,kJ\,mol^{-1}$$
$$\text{[iron(III) chloride]}$$

$$Fe(s) + I_2(s) \xrightarrow{\text{heat}} FeI_2(s); \qquad \Delta H^\ominus = -113\,kJ\,mol^{-1}$$
$$\text{[iron(II) iodide]}$$

Reaction with non-metals

Halogens form covalent compounds with other non-metals including hydrogen, oxygen and even themselves: e.g.

$$I_2(s) + Cl_2(g) \rightarrow 2ICl(l) \qquad \text{[iodine monochloride]}$$
$$I_2(s) + 3Cl_2(g) \rightarrow 2ICl_3(s) \qquad \text{[iodine trichloride]}$$

HYDRIDES

The hydrogen halides are binary compounds that exist as covalently bonded diatomic molecules with the general formula HX. Hydrogen fluoride has an exceptionally high boiling point owing to intermolecular hydrogen bonding caused by the high electronegativity of fluorine: see Fig. 12.32. Since the HX bond length increases and the bond energy decreases with the increasing atomic number of the halogen, the stability of these hydrides decreases down the group.

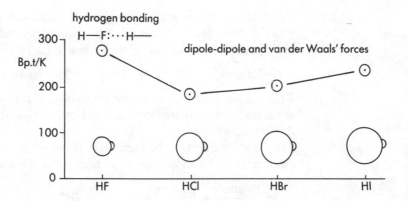

Fig. 12.32 Boiling points of the group 7 hydrides

Preparation

HF, HCl and HBr can all be made by direct synthesis in an exothermic reaction: e.g.

$$H_2(g) + Cl_2(g) \xrightarrow[\text{electrical spark}]{\text{flame, light or}} 2HCl(g); \quad \Delta H^\ominus = -184 \, \text{kJ mol}^{-1}$$

However, hydrogen iodide is an endothermic compound and cannot be made this way. On the contrary, if you plunge a red hot silica rod into a tube of the gas, iodine is produced as a purple vapour and a black deposit on the glass:

$$2HI(g) \xrightarrow{\text{heat}} H_2(g) + I_2(s); \quad \Delta H^\ominus = -53 \, \text{kJ mol}^{-1}$$

You could prepare HCl in the laboratory by adding concentrated sulphuric acid to sodium chloride:

$$H_2SO_4(l) + NaCl(s) \rightarrow NaHSO_4(s) + HCl(g)$$

The method is also used for HF but the product is regarded as too dangerous for practical work in schools and colleges. The principle of the method is the displacement of a more volatile acid (HX) from its salt by a less volatile acid (H_2SO_4). Sulphuric acid is not satisfactory for preparing HBr or HI because it also oxidises the halide to halogen:

$$H_2SO_4(l) + 2HBr(g) \rightarrow Br_2(l) + 2H_2O(l) + SO_2(g)$$
$$H_2SO_4(l) + 8HI(g) \rightarrow 4I_2(s) + 4H_2O(l) + H_2S(g)$$

Instead of sulphuric acid you could use phosphoric acid because it is fairly involatile and it is not an oxidising agent. It works well for hydrogen bromide but not for hydrogen iodide because the latter decomposes so readily on heating:

$$H_3PO_4(l) + NaBr(s) \xrightarrow{\text{heat}} NaH_2PO_4(s) + HBr(g)$$

Properties

Hydrogen halides (HX) and ammonia (NH_3) react to form a white smoke of ammonium halides (NH_4X);

$$X\text{—}H(g) + :NH_3(g) \xrightarrow[\text{needed as catalyst}]{\text{trace of moisture}} X^-[H:NH_3]^+(s)$$

Chlorine reacts with excess ammonia to form nitrogen and hydrogen chloride:

$$3Cl_2(g) + 2NH_3(g) \rightarrow N_2(g) + 6HCl(g)$$
$$\uparrow \qquad\quad \uparrow \qquad\quad \uparrow \qquad\quad \uparrow$$
$$[0] \qquad [-3] \qquad [0] \qquad [-1] \qquad \text{[oxidation states]}$$

The HCl(g) gives a white smoke of ammonium chloride with the excess ammonia. This is why a bottle of conc. aqueous ammonia is used to test for chlorine gas leaks. If excess ammonia is not used, the dangerously explosive liquid NCl_3 is formed!

Hydrogen halides give misty fumes in damp air and dissolve rapidly in water to give acidic solutions:

$$HCl(g) + H_2O(l) \rightarrow H_3O^+(aq) + Cl^-(aq)$$
hydrogen   hydrochloric acid
chloride

Hydrofluoric acid is atypical in being a weak acid ($pK_a = 3.3$) because the hydrogen fluoride dimerizes in water, due to hydrogen bonding. This decreases the number of protons donated to the water molecules: e.g.

$$H_2O(aq) + H\text{—}F \cdots H\text{—}F(aq) \rightarrow H_3O^+(aq) + [F \cdot\cdot H\text{—}F]^-(aq)$$

KHF_2 can be prepared as a crystalline solid! This is why you may see the formula H_2F_2 in some textbooks. Hydrofluoric acid also attacks glass! Hence its use in etching glass and in frosting light bulbs.

Hydrofluoric acid can be oxidised to fluorine but only by the technically difficult industrial electrolysis of an anhydrous liquid mixture of KF and HF at $100\,°C$:

$$2F^- \xrightarrow[\text{graphite}]{\text{anode}} F_2 + 2e^-; \qquad 2H^+ + 2e^- \xrightarrow[\text{steel}]{\text{cathode}} H_2$$

By contrast, hydroiodic acid, HI(aq), is rather unstable and turns brown because it is readily oxidised by oxygen in the air:

$$4H_3O^+(aq) + 4I^-(aq) + O_2(g) \rightarrow 6H_2O(l) + 2I_2(aq)$$

Chemical agents such as manganese(IV) oxide and acidified manganate(VII) ions will even oxidise hydrochloric acid to chlorine:

$$MnO_4^-(aq) + 8H^+(aq) + 5Cl^-(aq) \rightarrow Mn^{2+}(aq) + 4H_2O(l) + 2\tfrac{1}{2}Cl_2(g)$$

We often prepare chlorine in the laboratory by adding conc. hydrochloric acid to potassium manganate(VII) crystals. On the industrial scale chlorine (along with sodium or sodium hydroxide and hydrogen) is manufactured by electrolysis of sodium chloride:

$$2Cl^- \xrightarrow[\text{graphite}]{\text{anode}} Cl_2 + 2e^-; \qquad 2Na^+ + 2e^- \xrightarrow[\text{mercury}]{\text{cathode}} 2Na$$

All four 'hydrohalidic' acids (aqueous hydrogen halides) have the typical properties of aqueous acids: e.g. they react with alkalis, basic oxides, carbonates and hydrogen-carbonates to produce the halide salts (the word halogen means 'salt former');

$$HCl(aq) + KOH(aq) \rightarrow KCl(aq) + H_2O(l)$$
$$2HBr(aq) + MgO(s) \rightarrow MgBr_2(aq) + H_2O(l)$$
$$2HI(aq) + Na_2CO_3(s) \rightarrow 2NaI(aq) + H_2O(l) + CO_2(g)$$

The alkali metal halides are ionic and soluble in water. The silver halides become increasingly covalent in character and less soluble in water down the group from AgF to AgI owing to the increasing size and polarisability of the halide anion. Silver fluoride is soluble in water and forms hydrates! You may use aqueous silver nitrate, together with nitric acid and ammonia, in your practical work, when testing for halides: see Fig. 12.33.

You should be able to use a table of standard electrode potentials to predict that chlorine could displace bromine and iodine from bromides and iodides, and that bromine could displace iodine from iodides: e.g.

$$Cl_2(aq) + 2I^-(aq) \rightarrow 2Cl^-(aq) + I_2(aq)$$

$$\text{reducing agents (I$^-$ strongest to Cl$^-$ weakest)}$$
$$\downarrow$$

$I_2(aq),\ 2I^-(aq)\|Pt$	$E^\ominus = +0.54\,V$
$Br_2(aq), 2Br^-(aq)\|Pt$	$E^\ominus = +1.09\,V$
$Cl_2(aq), 2Cl^-(aq)\|Pt$	$E^\ominus = +1.36\,V$

$$\uparrow$$

oxidising agents (Cl$_2$ strongest to I$_2$ weakest)

When you add drops of aqueous chlorine to aqueous iodide ions, the solution may not turn very brown. And if you then add 1,1,1-trochloroethane, you may not see this organic liquid turn purple. One possible explanation is that the displaced iodine reacts with excess chlorine to form iodine monochloride (ICl) or iodine trichloride (ICl_3) which will dissolve in the organic liquid but will not produce a purple colour.

OXIDES

The halogens form various typical non-metal oxides. They are covalent and acidic. They are mostly unstable and often dangerously explosive. Iodine(V) oxide is exceptional in

	Cl^-(aq)	Br^-(aq)	I^-(aq)
add HNO_3(aq) and $AgNO_3$(aq)	white ppt. AgCl darkens in sunlight	off-white ppt. AgBr darkens in sunlight	yellow ppt. AgI stays yellow
add NH_3(aq) to ppt.	ppt. dissolves completely	ppt. partially dissolves	ppt. unaffected
add Cl_2(aq) and CCl_3CH_3(l)	aq. soln. colourless org. liq. colourless	aq. soln. brown org. liq. red-brown	aq. soln. brown org. liq. purple
add Br_2(aq) and CCl_3CH_3(l)	aq. soln. brown org. liq. red-brown	aq. soln. brown org. liq. red-brown	aq. soln. brown org. liq. purple

HNO_3(aq) allows only silver halides to precipitate

NH_3(aq) forms $[Ag(NH_3)_2]^+$(aq) and lowers the Ag^+(aq) concentration enough to prevent the precipitation of any AgCl and to allow the precipitation of some of the AgBr and almost all of the AgI.

Fig. 12.33 Testing for halides

being a stable, white crystalline solid. It can be prepared by oxidising iodine with hot conc. nitric acid and then heating the resulting iodic(V) acid to dehydrate it:

$$10HNO_3(aq) + I_2(s) \xrightarrow{heat} 2HIO_3(s) + 10NO_2(g) + 4H_2O(l)$$

$$2HIO_3(s) \xrightarrow{heat} I_2O_5(s) + H_2O(g)$$

Fluorine reacts atypically with aqueous sodium hydroxide to form fluorine oxide (F_2O) (or more accurately called oxygen difluoride OF_2) in which oxygen has its highest oxidation number of +2:

$$F_2(g) + 2OH^-(aq) \rightarrow OF_2(g) + H_2O(l)$$

OXOACIDS AND OXOSALTS

Fluorine does not form oxoacids. The oxoacids of the other halogens are mostly unstable (but see iodic(V) acid above). They are found only in aqueous solution and usually formed by the disproportionation of a halogen: e.g.

$$Cl_2(g) + 3H_2O(l) = 2H_3O^+(aq) + Cl^-(aq) + ClO^-(aq)$$

[0]   [−1]   [+1]

Oxosalts are formed when halogens react by disproportionation with alkalis: e.g.

$$Cl_2(g) + 2NaOH(aq) \xrightarrow[alkali]{cold\ dilute} H_2O(l) + NaCl(aq) + NaClO(aq)$$

$$3Cl_2(g) + 6NaOH(aq) \xrightarrow[alkali]{hot\ conc.} 3H_2O(l) + 5NaCl(aq) + NaClO_3(aq)$$

These equations may be written ionically without the spectator ions (Na^+);

$$Cl_2(g) + 2OH^-(aq) \rightarrow H_2O(l) + Cl^-(aq) + ClO^-(aq)$$
                                                              [chlorate(I)]

$$3Cl_2(g) + 6OH^-(aq) \rightarrow 3H_2O(l) + 5Cl^-(aq) + ClO_3^-(aq)$$
                                                              [chlorate(V)]

Chlorate(V) forms in hot alkali because the chlorate(I) disproportionates:

$$3ClO^-(aq) \rightarrow 2Cl^-(aq) + ClO_3^-(aq)$$

[+1]   [−1]   [+5]   [oxidation states]

If potassium chlorate(V) is gently heated just above its melting point (356 °C), the liquid gradually becomes pasty as the oxosalt disproportionates into potassium chloride (m.pt. 770 °C) and potassium chlorate(VII) (m.pt. 610 °C);

$$6KClO_3(s) \xrightarrow{heat} 5KCl(s) + KClO_4(s)$$

[+5]   [−1]   [+7]   [oxidation states]

On stronger heating and/or with manganese(IV) oxide as catalyst, the potassium chlorate(VII) decomposes into potassium chloride and oxygen:

$$KClO_4(s) \longrightarrow KCl(s) + 2O_2(g)$$

- Oxoacids and oxosalts may disproportionate if the oxidation number of the halogen is positive but below the maximum possible value

You should be able to predict the redox reactions of the oxoacids and their salts using a table of standard electrode potentials. For example, aqueous iodate(V) ions will oxidise aqueous iodide ions to iodine:

$$IO_3^-(aq) + 6H^+(aq) + 5I^-(aq) \rightarrow 3I_2(aq) + 3H_2O(l)$$

reducing agent
↓

$$I_2(aq), 2I^-(aq)|Pt \qquad E^\ominus = +0.54\,V$$
$$[2IO_3^-(aq) + 12H^+(aq)],[I_2(aq) + 6H_2O(l)]|Pt \qquad E^\ominus = +1.19\,V$$

↑
oxidising agent

- Oxoacids and oxosalts of halogens are generally good oxidising agents

GROUP 0 – NOBLE GASES

This is a group of monatomic gases with very similar physical properties. For example, they all have very low boiling points. The elements show the usual trends with increasing atomic number down the group: e.g. the boiling point increases, the atomic radius increases and the first ionisation energy decreases.

They are generally unreactive. Helium, neon and argon do not seem to form any compounds at all. Xenon will combine with fluorine: e.g.

$$Xe(g) + 2F_2(g) \xrightarrow{400\,°C\ 6\,atm} XeF_4(s)$$

You should be able to predict the shape of a molecule of xenon(IV) fluoride (see Fig. 12.34) by applying the Valence Shell Electron Pair Repulsion (VSEPR) rules: see Chapter 4. The lack of reactivity is usually related to the electronic configurations of the noble gases. All the electron shells of all the atoms are filled.

The uses of the noble gases mainly depend upon their unreactive nature: e.g. helium instead of flammable hydrogen for airships, neon for filling fluorescent tubes and argon for an inert atmosphere in the high-temperature reduction of titanium(IV) chloride to titanium.

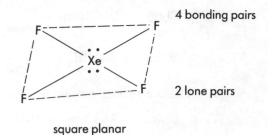

Fig. 12.34 The shape of the xenon(IV) fluoride molecule

EXAMINATION QUESTIONS AND OUTLINE ANSWERS

Q1 a) Give an account of the *redox* chemistry of the anions of nitrogen and sulphur. Your answer should include reference to the chemistry of NO_2^-, NO_3^-, S^{2-}, SO_3^{2-}, $S_2O_3^{2-}$ and $S_2O_8^{2-}$. You should consider the relative stabilities of the various oxidation states exhibited by nitrogen and sulphur, and include a range of reactions to illustrate your account. Marks will be gained for properly balanced equations. *(18)*

b) When a heavy metal nitrate is heated to produce the metal oxide, nitrogen dioxide and oxygen, the nitrogen dioxide and oxygen are always produced in the ratio $4:1$. By considering the oxidation state changes of the nitrogen and the oxygen show why this should be so, and hence write an equation for the thermal decomposition of aluminium nitrate.   (7)

(ULSEB 1988)

Outline answers

Q1

(a) *Write your answer in the form of short paragraphs and make it look attractive by writing equations on separate lines so they stand out.*

(b) $NO_3^- \longrightarrow NO_2 + O_2$

   +5 -2   +4   0

ox. no. of N falls from $+5 \rightarrow +4$ (1 unit)

ox. no. of O rises from $-2 \rightarrow 0$ (2 units)

$\Rightarrow$ There must be two N to one O to balance the oxidation number changes

$\Rightarrow$ $2N : \frac{1}{2}O_2$

or 4 mole NO_2 produced for 1 mole O_2

EXAMINATION QUESTIONS AND TUTOR'S ANSWERS

Q2 a) Give the oxidation number of chlorine in each of the species in the table.   (5)

	Cl_2	ClO_3^-	ClO^-	CCl_4	ClO_4^-
Oxidation number	0	$+5$	$+1$	-1	$+7$

b) The equation for the reaction between chlorine and hot concentrated sodium hydroxide solution is given below.

$$3Cl_2 + 6NaOH \rightarrow 5NaCl + NaClO_3 + 3H_2O$$

Write two balanced ion-electron equations (half-equations), one for the reduction process and one for the oxidation process involved in this reaction.   (3)

reduction process $Cl_2 + 2e^- \rightarrow 2Cl^-$

oxidation process $\frac{1}{2}Cl_2 + 6OH^- \rightarrow ClO_3^- + 3H_2O + 5e^-$

c) In a reaction between an **alkaline** solution of $KMnO_4$ (concentration $0.106\,mol\,dm^{-3}$) and a solution of KI ($0.0600\,mol\,dm^{-3}$), it was found that $25.0\,cm^3$ of the KI solution reacted completely with $28.3\,cm^3$ of the $KMnO_4$ solution. Calculate the ratio of the amounts of $KMnO_4$ and KI which react. Given that the manganese precipitates as MnO_2, deduce the final oxidation state of the iodine.   (5)

amount of KI ∝ 25.0 × 0.0600

amount of KMnO₄ ∝ 28.3 × 0.106

⇒ ratio of reacting amounts of KMnO₄ : KI is

(28.3 × 0.106) : (25.0 × 0.0600) = 2:1

KMnO₄ → MnO₂

+7   +4   falls by 3 units of ox. no.

⇒ Iodine must rise by 2×3 units to balance

⇒ final oxidation state is +5

(ODLE 1988)

Q3 Aluminium chloride may be prepared in the laboratory using the apparatus below.

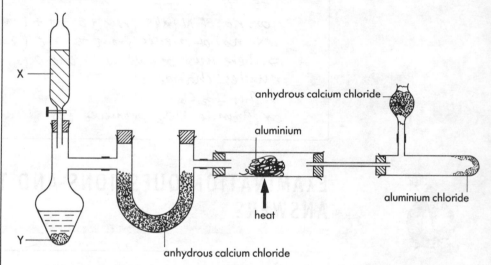

a) i) Name the substances X and Y which may be used to generate chlorine.

X   *concentrated Hydrochloric acid*

Y   *potassium manganate (VII)*

ii) Why is the preparation carried out in a fume cupboard?

chlorine is toxic

hydrogen chloride fumes are corrosive

iii) What is the purpose of the tubes containing anhydrous calcium chloride?

Keep moisture out of the reaction mixture and the product   (4)

b) A sample of aluminium chloride weighing 0.1 g was vaporised at 350 °C and one atmosphere pressure to produce 19.2 cm³ of vapour.

i) Calculate a value for the relative molecular mass of aluminium chloride under the conditions of the experiment ($PV = nRT$ where $R = 0.082\,\text{atm}\,\text{dm}^3\,\text{K}^{-1}\,\text{mol}^{-1}$)

$PV = nRT$ ⇒ $1 \times \dfrac{19.2}{1000} = n \times 0.082 \times (350 + 273)$

⇒ $n = 3.75 \times 10^{-4}$ mol

⇒ $\dfrac{0.1}{m} = 3.75 \times 10^{-4}$

⇒ $M = 267$

Relative molecular mass = ___267___

ii) Suggest a molecular formula for aluminium chloride in the gaseous state at 350 °C using the data calculated above. (Relative atomic masses; Al = 27; Cl = 35.5)

$$M_r (AlCl_3) = 133.5 = \frac{1}{2} \times 267$$

$$Al_2Cl_6$$

Formula _____ (4)

c) Aluminium chloride is white when pure, though it often appears yellow because it contains *traces* of iron(III) chloride.

i) Describe how you would show the presence of iron(III) ions in a sample of aluminium chloride.

dissolve sample in distilled water and add aqueous sodium thiocyanate to see a red colour develop

ii) Write a balanced equation for the reaction involved in part i).

$$Fe^{3+}(aq) + SCN^-(aq) \rightarrow Fe(SCN)^{2+}(aq)$$ (3)

d) Aluminium chloride fumes in air by hydrolysis setting free hydrogen chloride.

i) What is meant by hydrolysis?

decomposition by water

ii) Write a balanced equation for the reaction taking place.

$$AlCl_3(s) + H_2O(l) \rightarrow Al(OH)_3(s) + HCl(g)$$

iii) How would you test for the hydrogen chloride given off in the reaction?

use ammonia gas (from conc. aqueous ammonia) and observe a white smoke (3)

e) Aluminium chloride is used as a catalyst in the Friedel-Crafts reaction. An example of this reaction is that of benzene with chloromethane to form methylbenzene (toluene).

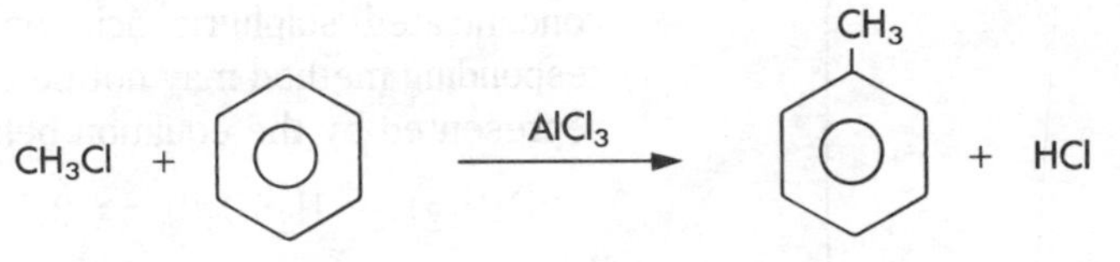

With an excess of chloromethane three structural isomers of dimethylbenzene are possible.

i) Draw the structures of the three isomers.

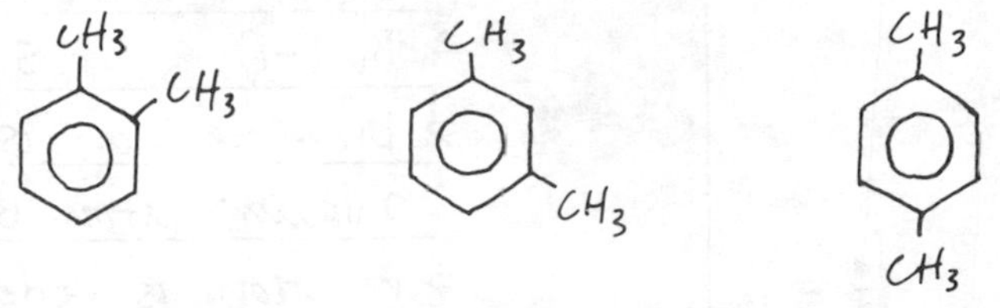

ii) Name a modern technique by which you would separate these three isomers.

chromatography

(4)
(NISEC 1988)

STUDENT'S ANSWER WITH EXAMINER'S COMMENTS

Q4 This question concerns the halogen elements and some of their compounds.

a) The electron configuration of fluorine is $1s^2 2s^2 2p^5$. Give the electron configurations of the

i) chlorine **atom**

" Correct "

$$1s^2 \; 2s^2 \; 2p^6 \; 3s^2 \; 3p^5 \qquad \checkmark$$

ii) bromide **ion**

" $3d^{10}$ missing. "

$$1s^2 \; 2s^2 \; 2p^6 \; 3s^2 \; 3p^6 \; 4s^2 \; 4p^6 \qquad \times$$

(2)

b) The electron affinities of chlorine and bromine are -364 and $342 \, \text{kJ mol}^{-1}$ respectively. Explain what is meant by electron affinity.

" Good answer "

Molar enthalpy of electron gain for the process
$$Cl_{(g)} + e^- \rightarrow Cl^-_{(g)}$$

(2)

c) The radius of a fluorine atom is $0.072 \, \text{nm}$, while that of a fluoride ion is $0.136 \, \text{nm}$. Explain why the ionic radius is larger than the atomic radius.

The F atom gains an electron to form F^-
This fills the outer shell. The electrons 'spread out' to make room. $\checkmark$

(1)

d) The boiling point of hydrogen fluoride is $293 \, \text{K}$ whereas those of hydrogen chloride and hydrogen bromide are $188 \, \text{K}$ and $206 \, \text{K}$ respectively. Account for the anomalously high value of hydrogen fluoride.

" This holds the molecules together more strongly, so a higher temperature is needed to make them boil. "

Hydrogen bonding
H-F --- H-F

(2)

e) Hydrogen chloride may be prepared in the laboratory by the action of concentrated sulphuric acid on potassium chloride crystals but a corresponding method may not be used for hydrogen bromide, as the reaction represented by the equation below occurs.

$$2HBr(g) + H_2SO_4(l) \rightleftharpoons 2H_2O(l) + SO_2(g) + Br_2(g)$$

i) Deduce the oxidation states of all the atoms in this equation and explain why the reaction may be regarded as redox.

HBr	H_2SO_4	SO_2	Br_2
$(+1)+(-1)=0$	$2\times(+1)+x+4\times(-2)=0$	$x+2\times(-2)=0$	0
$\Rightarrow Br(-1)$	$S(+6)$	$S(+4)$	$Br(0)$

$\Rightarrow$ bromide ion is oxidised from -1 to 0
sulphur atom is reduced from $+6$ to $+4$
$\Rightarrow$ reaction is redox $\checkmark$

ii) Which acid could be used to react with potassium bromide to prepare pure hydrogen bromide?

Orthophosphoric (v) acid $\checkmark$

iii) Give the equation for the reaction between potassium chloride and concentrated sulphuric acid.

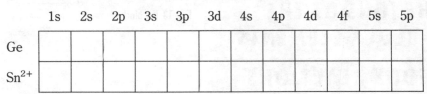

$$KCl_{(s)} + H_2SO_4 _{(l)} \rightarrow KHSO_4 _{(s)} + HCl_{(g)} \checkmark$$

(6)
(Total 13 marks)
(AEB 1988)

Good. K_2SO_4 here would be wrong – a common mistake.

PRACTICE QUESTIONS

Study Chapter 12 thoroughly to prepare answers to this kind of question. It would help the examiner if you presented your comparisons in a two-column table. Remember to include equations for chemical properties.

Q5 a) Illustrate the unique features of carbon in Group IV by
   i) comparing and accounting for the different physical states of CO_2 and SiO_2 at room temperature and pressure,
   ii) comparing and accounting for the differing behaviour towards water of CCl_4 and $SnCl_4$,
   iii) describing the structure of the graphite form of carbon and accounting for its physical properties. *(18)*

 b) Illustrate the increase in metallic character as Group IV is descended by referring to **two** differences in physical properties between silicon and tin, and to the reaction of these elements with concentrated hydrochloric acid. *(5)*

 c) Illustrate the difference in stability between the +2 and +4 oxidation states of tin and lead by
   i) comparing the action of heat on tin(IV) oxide and lead(IV) oxide,
   ii) comparing the behaviour of aqueous solutions of tin(II) nitrate and lead(II) nitrate towards aqueous iron(III) nitrate. *(7)*
(JMB 1988)

Q6 This question concerns the elements carbon, silicon, germanium, tin and lead in Group 4 of the Periodic Table.
 a) Give the electronic configuration of Ge and Sn^{2+}. The atomic numbers of germanium and tin are 32 and 50 respectively.

	1s	2s	2p	3s	3p	3d	4s	4p	4d	4f	5s	5p
Ge												
Sn^{2+}												

How do you account for the fact that tin forms the 2+ ion but germanium does not? *(3)*

 b) The melting point of carbon is almost 4000 °C whereas that of tin is 232 °C. How do you account for this difference? *(3)*

 c) Explain why PbO_2 liberates chlorine from concentrated hydrochloric acid but SnO_2 does not. *(3)*

 d) Tetrachloromethane and tetrachlorosilane are both thermodynamically (energetically) unstable with respect to reaction with water, but only tetrachloromethane is kinetically stable.
   i) Explain what is meant by

      • thermodynamic stability   • kinetic stability

   ii) How do you account for this difference in kinetic stability? *(4)*
(ULSEB 1988)

GETTING STARTED

The d-block elements are the metals in the horizontal rows crossing the periodic table between group 2 and group 3. They form alloys, and frequently have coloured compounds and complex ions in various oxidation states. The metals can act as heterogeneous catalysts and their ions as homogeneous catalysts. The emphasis is on the first row from scandium to zinc. The physical properties and chemical behaviour of particular elements are considered in order to illustrate the general features arising from the incomplete d-subshells of the first transition metal series from titanium to copper: see Fig. 13.1.

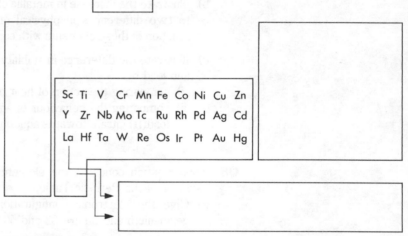

Fig. 13.1 d-block elements

ESSENTIAL PRINCIPLES

THE d-BLOCK METALS

In the first and second rows across the periodic table, the s- and p-block elements, Li-Ne and Na-Ar, change dramatically in their behaviour from reactive alkali metals on the left to reactive non-metal halogens and unreactive noble gases on the right. We explain this by saying that each increase in nuclear charge has a large effect on the outer shell electrons because (i) the inner shell shielding does not change and (ii) the electrons in the same outer shell do little to shield each other from the attractive force of the nucleus: see Fig. 13.2. By contrast, in the first and second series crossing and the periodic table from group 2 to group 3, the d-block elements, Sc-Zn and Y-Cd, show relatively little change in their behaviour. We explain this by saying that each increase in nuclear charge has only a small effect on the outer 4s or 5s subshell because the nucleus is shielded by more electrons in the inner 3d or 4d subshell: see Fig. 13.2.

ELEMENTS FROM LEFT TO RIGHT ACROSS THE SECOND PERIOD

inner shells shielding outer shell
↓   ↓

Na $[1s^2 2s^2 2p^6]$ $3s^1$ Si $[1s^2 2s^2 2p^5]$ $3s^2 3p^2$ Cl $[1s^2 2s^2 2p^6]$ $3s^2 3p^5$
11+|| <·······>   14+|| <··········>   17+|| <··········>

↑
nuclear charge attracts outer shell

nuclear attraction (<· · ·>) for outer shell electrons increases considerably
whilst inner shell shielding (||) remains unchanged
explains changes in atomic radius and first ionisation energy being large

r/nm	0.191	0.118	0.099
E_{m1}/kJ mol^{-1}	496	789	1251

ELEMENTS FROM LEFT TO RIGHT ACROSS THE FIRST TRANSITION SERIES

inner shells shielding outer shell
↓   ↓   ↓

Ti [Ar] $[3d^2]$ $4s^2$ Mn [Ar] $[3d^5]$ $4s^2$ Ni [Ar] $[3d^8]$ $4s^2$
22+ <· ·>||||<· ·>   25+ <· · ·>||||||<· · ·>   28+ <· · · ·>||||||||<· · · ·>

↑
nuclear charge attracts outer shell

increased nuclear attraction (<· · ·>) for outer shell electrons
diminished by greater inner shell shielding (||)
explains changes in atomic radius and first ionisation energy being small

r/nm	0.132	0.139	0.134
E_{m1}/kJ mol^{-1}	658	717	737

Fig. 13.2 Nuclear charge and inner electron shell shielding

PHYSICAL PROPERTIES

All the first series d-block elements are regarded as typical metals. You should therefore take their remarkably similar physical properties to be characteristic of a metal: see Fig. 13.3. The hardness, high melting points and high enthalpy changes upon melting indicate very strong metallic bonding compared to that in the alkali metals. You could say that the bonding involves delocalisation of some of the inner d-subshell electrons as well as the outer s-subshell electrons.

- The characteristic physical properties of the transition metals (Ti-Cu) may be related to the incomplete 3d-subshell

METALLIC PROPERTY	ASSOCIATED USES
ductile, malleable, dense, high m.pt., very high b.pt., hard and high tensile strength	vast range of equipment and machinery structures, structural materials and tools requiring good mechanical characteristics
electrically and thermally conducting	copper utensils, copper wire, nichrome heating elements
lustrous and sonorous	coins, chromium plating, musical instruments

Fig. 13.3 Physical properties and uses of transition metals

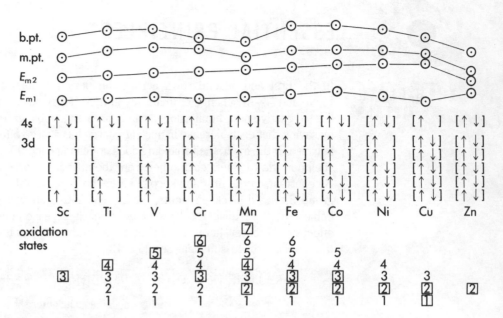

Fig. 13.4 Electronic configuration and physical properties

Compare graphs of m.pt./°C, b.pt./°C, E_{m1}/kJ mol^{-1} and E_{m2}/kJ mol^{-1} against atomic number for the series Sc-Zn: see Fig. 13.4. Notice that manganese has a relatively low melting point and boiling point. And notice that chromium and copper have relatively low first ionisation energies but relatively high second ionisation energies. You could explain these features by the extra stability of having 5 or 10 electrons in the d-subshell. In manganese the electrons in the half-filled d-subshell participate less readily in the delocalised metallic bonding. In chromium and copper the more stable half-filled and filled d-subshell means only 1 electron in the 4s-subshell and great shielding for it.

- A half-filled 3d-subshell (5 electrons) or a completely filled 3d-subshell (10 electrons) means extra stability

Alloys

An alloy may form when a molten mixture of two or more metals is cooled. The solid alloy may be (i) a simple mixture of tiny crystals of the separate metals, (ii) a solid solution of the atoms of the metals randomly occupying positions in each others lattices, (iii) an intermetallic compound of the metals, or (iv) something in between these three classes.

- Transition metals form alloys which are often solid solutions because their different atoms are very similar in size

Alloying produces greater hardness and tensile strength, usually at the expense of lesser ductility and malleability. You could explain this by saying that the different atoms of one component make it more difficult for the planes of the same atoms of the other component to slip over one another.

- Steels are extremely important alloys of iron with other transition metals such as chromium, manganese, cobalt and carbon

Paramagnetism

The extent to which a material is attracted by a magnetic field is related to the number of unpaired electrons in the atoms or ions of the material. The paramagnetism of the d-block metal cations is zero for Sc^{3+} and Zn^{2+}. For the cations of the metals from Ti (Ti^{3+}) to Cu (Cu^{2+}), the paramagnetism increases and then decreases with atomic number, reaching a maximum for Fe (Fe^{3+}). The ferromagnetism shown by metallic iron, cobalt and nickel is an extreme case of paramagnetism.

HETEROGENEOUS CATALYSIS

- Many of the d-block transition metals are industrially important heterogeneous catalysts

Two well-known examples are iron in the Haber process (synthesis of NH_3 from N_2 and $3H_2$) and nickel in margarine production (hydrogenation of oils to fats). Chromium, cobalt, copper, iron and nickel are used to catalyse a wide range of reactions in the petrochemical industry. In the main, these are gas-phase reactions involving reactants in the gaseous state.

- The solid catalyst uses its 3d and 4s subshells to adsorb reactants and lower activation energies by weakening bonds in their molecules

CHEMICAL PROPERTIES

The p-block and s-block metals show little or no variety in oxidation number. In sharp contrast, the d-block transition metals display a wide range of oxidation numbers: see Fig. 13.4. The numbers in circles are the most important.

> Make sure you understand oxidation numbers. They are always tested.

- The variable oxidation numbers arise because the 3d electrons are comparable in energy to the 4s electrons and also involved in bonding

OXIDATION NUMBERS

The oxidation number is a number assigned to an element according to the following set of rules applied in a priority order:

Rule		ox.no.
1	Oxidation number of an uncombined element	0
2	Sum of oxidation numbers of elements in uncharged formula	0
3	Sum of oxidation numbers of elements in charged formula	charge
4	Oxidation number of fluorine in any formula	-1
5	Oxidation number of an alkali metal in any formula	$+1$
6	Oxidation number of an alkaline earth metal in any formula	$+2$
7	Oxidation number of oxygen (except in peroxides $= -1$)	-2
8	Oxidation number of halogen in metal halides	-1
9	Oxidation number of hydrogen (except in metal hydrides $= -1$)	$+1$

The underlying principle behind these rules is to imagine electrons transfer completely to the more electronegative atom when two atoms combine and then to define the oxidation number for each element as the number of electrons to be added to each combined atom to make it uncharged. Since fluorine is the most electronegative of all the elements and it has 7 valence electrons, you would always imagine a combined F atom to have one negative charge. So you imagine subtracting one electron to make a combined F atom uncharged. Hence the oxidation number of combined fluorine is always -1.

INORGANIC NOMENCLATURE

- The name of a compound should specify unambiguously its composition

You should already be familiar with some of the conventions for naming compounds and writing their formulae. For example, the ending -ide usually means a binary (two element) compound e.g. magnesium oxide, MgO, hydrogen chloride, HCl, with the symbol of the more electronegative element on the right of the formula. You should also be aware of breaks from these conventions. For example, we do not call water dihydrogen oxide, the compounds sodium cyanide (NaCN) and sodium hydroxide (NaOH) contain three elements not two and we write NaOH not NaHO!

Oxidation numbers are used in the Stock system of naming inorganic compounds of elements with variable valence in an attempt to specify unambiguously the composition of the compounds. If there is no variable valence then you will not need to use oxidation numbers in a name.

- The positive oxidation number of an element is written as a Roman numeral in brackets immediately after the relevant part of the name

You should use Roman numerals in the names but not the formulae of the compounds of some p-block elements and most d-block transition metals: see Fig. 13.5. Notice how the name of the element depends upon the part played by the element in the compound.

ammonium vanadate(V)	NH_4VO_3
copper(II) dichromate(VI)	$CuCr_2O_7$
iron(III) hexacyanoferrate(II)	$Fe_4[Fe(CN)_6]$
potassium chromate(VI)	K_2CrO_4
tetracarbonylnickel(O)	$Ni(CO)_4$

Fig. 13.5 Names of some inorganic compounds

- When the transition metal is part of the cation, the name is the same as the element
- When the d-block or p-block element is part of the anion, the name (derived from the English or Latin) ends in -ate

BALANCING REDOX EQUATIONS

- A redox reaction involves an increase in oxidation number of one element and a simultaneous decrease in oxidation number of another
- Disproportionation is a redox involving a simultaneous increase and decrease in oxidation number of the same element

You can use the changes in oxidation numbers to work out balanced equations for redox reactions: see Fig. 13.6.

Here is a summary of the steps you take:

1 Write down correct formulae of reactants and products
2 Assign oxidation numbers using rules in priority order
3 Find the change in oxidation number of reductant and oxidant
4 Balance these two changes in oxidation numbers
5 Check and balance if necessary the remaining atoms
6 Check charge balance if there are ions in the equation

> This is important. Get plenty of practice balancing equations.

If at step 5 you find the equation needs oxygen atoms, you can write in $H_2O(l)$ or $OH^-(aq)$: e.g.

$$SO_2(aq) + Br_2(aq) \rightarrow SO_4^{2-}(aq) + 2Br^-(aq)$$

↑ two oxygens   ↑ four oxygens $\Rightarrow$ add $2H_2O(l)$

$$SO_2(aq) + 2H_2O(l) + Br_2(aq) \rightarrow SO_4^{2-}(aq) + 2Br^-(aq)$$

step 1   write down correct formulae of all reactants and products:
$$Cr_2O_7^{2-}(aq) + H^+(aq) + SO_3^{2-}(aq) \rightarrow 2Cr^{3+}(aq) + SO_4^{2-}(aq)$$

step 2   apply priority rules to assign oxidation numbers:
$$Cr_2O_7^{2-}(aq) + H^+(aq) + SO_3^{2-}(aq) \rightarrow 2Cr^{3+}(aq) + SO_4^{2-}(aq)$$
↑ +6   ↑ +4   ↑ +3   ↑ +6

step 3   calculate rise and fall of oxidation numbers:
(a rise of 6 − 4 = 2)   +4 ↓   +6 ↓
$$Cr_2O_7^{2-}(aq) + H^+(aq) + SO_3^{2-}(aq) \rightarrow 2Cr^{3+}(aq) + SO_4^{2-}(aq)$$
↑ $2 \times (+6) = 12$ ↑ $2 \times (+3) = 6$ (a fall of 12−6 = 6)

step 4   balance the rise and fall of oxidation numbers:
$$Cr_2O_7^{2-}(aq) + H^+(aq) + 3SO_3^{2-}(aq) \rightarrow 2Cr^{3+}(aq) + 3SO_4^{2-}(aq)$$
↑ total fall of 6 will equal 3x a rise of 2 to balance Cr and S atoms

step 5   balance other atoms without changing balance of redox atoms:
$$Cr_2O_7^{2-}(aq) + H^+(aq) + 3SO_3^{2-}(aq) \rightarrow 2Cr^{3+}(aq) + 3SO_4^{2-}(aq)$$
↑ seven oxygens   ↑ nine oxygens   ↑ twelve oxygens
$\Rightarrow$ 7 + 9 − 12 = four oxygens needed on the right hand side; i.e. $4H_2O(l)$
$$Cr_2O_7^{2-}(aq) + H^+(aq) + 3SO_3^{2-}(aq) \rightarrow 2Cr^{3+}(aq) + 3SO_4^{2-}(aq) + 4H_2O(l)$$
↑ one hydrogen   ↑ eight hydrogens
$\Rightarrow$ multiply by eight to give fully balanced equation
↓
$$Cr_2O_7^{2-}(aq) + 8H^+(aq) + 3SO_3^{2-}(aq) \rightarrow 2Cr^{3+}(aq) + 3SO_4^{2-}(aq) + 4H_2O(l)$$

step 6   check the charge balance:
$$Cr_2O_7^{2-}(aq) + 8H^+(aq) + 3SO_3^{2-}(aq) \rightarrow 2Cr^{3+}(aq) + 3SO_4^{2-}(aq) + 4H_2O(l)$$
↑ (2−)   ↑ $8 \times (1+)$ ↑ $3 \times (2-)$ ↑ $2 \times (3+)$ ↑ $3 \times (2-)$
$\Rightarrow$ (2−)   +   (8+)   +   (6−)   =   (6+)   +   (6−)   ✓ balanced

Fig. 13.6 Balancing equations for redox reactions

If the equation needs hydrogen atoms, you can add H^+: e.g.

$$SO_2(aq) + 2H_2O(l) + Br_2(aq) \rightarrow SO_4^{2-}(aq) + 2Br^-(aq) + 4H^+(aq)$$

Notice that writing $H_2O(l)$ on the left and $2H^+(aq)$ on the right of an equation has the same effect as writing $2OH^-(aq)$ on the left and $H_2O(l)$ on the right. This is due to $H^+(aq) + OH^-(aq) \rightarrow H_2O(l)$, an acid-base (**not** redox) reaction. Consequently, for example, the following equations are equivalent:

$$Cl_2(aq) + 2OH^-(aq) \rightarrow Cl^-(aq) + ClO^-(aq) + H_2O(l)$$
$$Cl_2(aq) + H_2O(l) \rightarrow Cl^-(aq) + ClO^-(aq) + 2H^+(aq)$$

If at step 6 you find the charges are not balanced, go through steps 1 to 5 again looking carefully for errors.

REACTIONS OF METALS

REACTION OF METALS WITH ACIDS

- When a metal reacts with aqueous hydrogen ions to form hydrogen, the reaction is a redox not an acid-base reaction

The standard electrode potential, $E^\ominus$, gives a quantitative measure of the energetic feasibility of a metal to reduce $H^+(aq)$ to $H_2(g)$: see Chapter 8. For the first transition metal series, the variation in the values of $E^\ominus$ is rather wide and irregular because the property depends on several factors: i.e. atomisation energy and ionisation energies of the metal and hydration energies of the cations. Under standard conditions, reaction with $H^+(aq)$ to form $H_2(g)$ is energetically feasible for the following metals, starting with the most likely reductant: titanium, manganese, vanadium, chromium, iron, cobalt, nickel. They are all more electropositive than hydrogen. And the more electropositive the metal, the more negative is its $E^\ominus$ value relative to hydrogen. However, many are kinetically rather stable and quite resistant to corrosion.

- The standard electrode potential of copper is $+0.34$ volt so copper does not react with acids to form hydrogen

Copper does react with aqueous nitric acid to form oxides of nitrogen. The precise nature of this redox reaction depends upon the concentration of the acid and the temperature: e.g.

$$Cu(s) + 4HNO_3(aq) \xrightarrow[\text{acid}]{\text{concentrated}} Cu(NO_3)_2(aq) + 2H_2O(l) + 2NO_2(g)$$

$$3Cu(s) + 8HNO_3(aq) \xrightarrow[\text{acid}]{\text{mod. conc}} 3Cu(NO_3)_2(aq) + 4H_2O(l) + 2NO(g)$$

The metal reduces the (nitrogen in) NO_3^- ion instead of the hydrogen ion.

Concentrated nitric acid makes some metals e.g. iron and chromium, passive by converting their surface into an extremely tough unreactive oxide.

REACTION OF METALS WITH WATER

The metals show no simple pattern in their behaviour towards water. For example, chromium is resistant to attack but iron rusts when exposed to water and oxygen. The rusting process involves a complicated set of electrochemical reactions which include the oxidation of the iron and the reduction of the oxygen:

$Fe(s) \rightarrow Fe^{2+}(aq) + 2e^-$ oxidation number increases from 0 to $+2$
$Fe^{2+}(aq) \rightarrow Fe^{3+}(aq) + e^-$ oxidation number increases from $+2$ to $+3$
$O_2(g) + 2H_2O(l) + 4e^- \rightarrow 4OH^-(aq)$ oxidation number decreases from 0 to -2

The very costly world-wide task of rust prevention is tackled by (i) trying to keep water and/or oxygen out of contact with the iron surface (by oiling, painting, plastic-coating, tin-plating, etc) or (ii) providing the electrons for the reduction of the oxygen from a source other than the iron, e.g. a more reactive 'sacrificial' metal such as magnesium or zinc:

[sacrificial metal] $Zn(s) \rightarrow Zn^{2+}(aq) + 2e^-$ [electrons for oxygen]

Red-hot iron reduces steam to hydrogen; a reaction not welcomed by firemen spraying water into a burning steel and concrete building:

$$3Fe(s) + 4H_2O(l) \xrightarrow{\text{red heat}} Fe_3O_4(s) + 4H_2(g)$$

REACTION OF METALS WITH NON-METALS

The metals form a wide variety of halides, oxides and sulphides but not necessarily by direct reaction with halogens, oxygen or sulphur. The oxidation number of the metal and the stability of the compound will depend upon such factors as the oxidising power of the non-metal, the relative sizes of the atoms and the structure of the compound.

Halides

Fluorine attacks all metals and forces them into their highest oxidation states: e.g. vanadium(V) fluoride and chromium(VI) fluoride. Otherwise the highest oxidation number in the transition metal halides is +4 in chlorides such as $TiCl_4$. The lowest oxidation number is +1 in copper(I) chloride and copper(I) iodide.

- The covalent character of the halides increases with increasing atomic number of the halogen and increasing oxidation number of the metal

You could make iron(III) chloride by heating iron wool in chlorine:

$$Fe(s) \ + \ 1\tfrac{1}{2}Cl_2(g) \ \xrightarrow{\text{heat}} \ FeCl_3(s); \ \Delta H^{\ominus} \ = \ -400 \, kJ \, mol^{-1}$$

and iron(II) chloride by heating iron wool in hydrogen chloride:

$$Fe(s) \ + \ 2HCl(g) \ \xrightarrow{\text{heat}} \ FeCl_2(s); \ \Delta H^{\ominus} \ = \ -342 \, kJ \, mol^{-1}$$

Copper(I) chloride can be obtained by reducing copper(II) chloride with copper in the presence of conc. hydrochloric acid. The resulting dark brown solution contains copper(I) ions stabilised as a complex anion. When this is poured into distilled water (oxygen-free or containing some sulphur dioxide as an anti-oxidant) a white precipitate of copper(I) chloride is formed:

$$Cu(s) \ + \ CuCl_2(aq) \ \xrightarrow[\text{then into cold water}]{\text{hot conc. HCl(aq)}} \ 2CuCl(s)$$

Copper(I) iodide is formed quantitatively when aqueous iodide is added to aqueous copper(II) ions:

$$2Cu^{2+}(aq) \ + \ 4I^-(aq) \rightarrow 2CuI(s) \ + \ I_2(aq)$$

You can use this reaction to determine the concentration of the aqueous copper(II) ions because you can titrate the iodine with aqueous sodium thiosulphate.

Oxides and hydroxides

The d-block metals form a variety of oxides: see Fig. 13.7. The highest oxides are covalent and acidic: e.g. manganese(VII) oxide is a green oil with a simple molecular structure. The

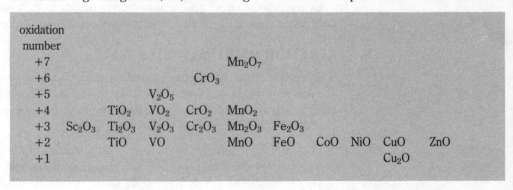

oxidation number									
+7					Mn_2O_7				
+6				CrO_3					
+5		V_2O_5							
+4	TiO_2	VO_2	CrO_2	MnO_2					
+3	Sc_2O_3	Ti_2O_3	V_2O_3	Cr_2O_3	Mn_2O_3	Fe_2O_3			
+2		TiO	VO		MnO	FeO	CoO	NiO	CuO	ZnO
+1									Cu_2O	

Fig. 13.7 Some d-block oxides

rest of the oxides are largely ionic and neutral, amphoteric or basic depending upon the oxidation number of the metal. Those with an oxidation number of +2 are mostly basic: e.g. copper(II) oxide reacts with acids to form copper(II) salts:

$$CuO(s) \ + \ 2H^+(aq) \rightarrow Cu^{2+}(aq) \ + \ H_2O(l)$$

If you add aqueous sodium hydroxide to aqueous metal cations, you should see precipitates appear. These precipitates may be gelatinous or granular and are often coloured: see Fig. 13.8. In some cases, the precipitate will disappear when you add excess alkali because the hydroxide is amphoteric: e.g.

$$Zn(OH)_2(s) \ + \ 2OH^-(aq) \rightarrow Zn(OH)_4^{2-}(aq) \ [\text{tetrahydroxyzincate}]$$

AQUEOUS CATION	COLOUR OF PRECIPITATE
chromium(III)	bluish-green
manganese(II)	white (rapidly turns brown)
iron(II)	green (white when pure)
iron(III)	reddish brown
cobalt(II)	blue (gradually turns pink)
nickel(II)	green
copper(II)	pale blue
zinc	white

Fig. 13.8 Coloured precipitates with aqueous alkali

In other cases, the precipitate may disappear because you are using aqueous ammonia as the alkali and it forms a complex amine cation: e.g.

$$Cu(OH)_2(s) + 4NH_3(aq) \rightarrow [Cu(NH_3)_4]^{2+}(aq) + 2OH^-(aq)$$

You may regard these precipitates as insoluble hydroxides but their composition is often uncertain. In some cases they are more appropriately called hydrated oxides because they dehydrate on heating to form the oxide: e.g. you should see the blue gelatinous precipitate of copper(II) hydroxide turn to a black granular precipitate if you gently boil the suspension:

$$Cu(OH)_2(s) \xrightarrow{\text{heat}} CuO(s) + H_2O(l)$$

In your practical work you may have done these reactions in test-tubes and deduced from your observations the identity of the transition metal cation.

The oxidising power of the oxides increase with increasing oxidation number of the metal. You may know that in 1774 the Swedish chemist Karl Wilhelm Scheele discovered chlorine by heating hydrochloric acid with manganese(IV) oxide:

$$MnO_2(s) + 4HCl(aq) \xrightarrow{\text{heat}} MnCl_2(aq) + 2H_2O(l) + Cl_2(g)$$

The highest oxides react with alkali to form oxoanions that retain the highest oxidation state: e.g.

$$Mn_2O_7(l) + 2OH^-(aq) \rightarrow 2MnO_4^-(aq) + H_2O(l)$$
$$\quad\quad \uparrow \quad\quad\quad\quad\quad\quad\quad\quad \uparrow$$
$$\quad [+7] \quad\quad\quad\quad\quad\quad\quad [+7] \quad\quad \text{[oxidation state unchanged]}$$

We may prepare chlorine by adding drops of conc. hydrochloric acid to potassium manganate(VIII) crystals:

$$5Cl^-(aq) + MnO_4^-(aq) + 8H^+(aq) \rightarrow Mn^2(aq) + 4H_2O(l) + 2\tfrac{1}{2}Cl_2(g)$$
$$\quad\quad\quad\quad\quad\quad\quad \uparrow \quad\quad\quad\quad\quad\quad\quad\quad \uparrow$$
$$\text{reduction from } [+7] \quad\quad\quad\quad\quad \text{to } [+2] \quad\quad \text{[fall in oxidation number]}$$

And in your practical work you may titrate aqueous reducing agents (e.g. iron(II) sulphate) with potassium manganate(VII) solution, acidified with sulphuric acid but not hydrochloric acid:

$$5Fe^{2+}(aq) + MnO_4^-(aq) + 8H^+(aq) \rightarrow Mn^{2+}(aq) + 4H_2O(l) + 5Fe^{3+}(aq)$$
$$\quad\quad\quad\quad \uparrow \quad\quad\quad\quad\quad\quad\quad\quad \uparrow$$
$$\text{reduction from } [+7] \quad\quad\quad\quad\quad \text{to } [+2] \quad\quad \text{[fall in oxidation number]}$$

THE IONS OF THE d-BLOCK ELEMENTS

The electrons in the ions of the s- and p-block metals are in complete shells so they can be excited by absorbing energy in the ultraviolet region of the spectrum but not in the visible region. Consequently the s- and p-block compounds are usually colourless or white. Scandium, zinc and copper(I) compounds are colourless for the same reason. On the other hand, the electrons in the incomplete 3d-subshell of the transition metal ions can be excited by absorbing energy in the visible region of the spectrum. Consequently the transition metal compounds and their aqueous solutions are frequently coloured: see Figs. 13.8 and 13.9.

Unpaired d-electrons cause many of the transition metal compounds to be paramagnetic. If you hang a sample of an iron(III) compound in a glass tube from a chemical balance, you may detect as much as a 0.1 g weight increase when you hold a strong magnet underneath the sample.

COLOUR	FORMULA	NAME
crystalline solids		
red	CrO_3	chromium(VI) oxide
orange	$Na_2Cr_2O_7$	sodium dichromate(VI)
yellow	Na_2CrO_4	potassium chromate(VI)
green	$FeSO_4.4H_2O$	iron(II) sulphate-4-water
blue	$CuSO_4.5H_2O$	copper(II) sulphate-5-water
indigo	$[Cu(NH_3)_4]SO_4.H_2O$	tetraamminecopper(II) sulphate-1-water
violet	$NH_4Fe(SO_4)_2.12H_2O$	ammonium iron(III) sulphate-12-water
aqueous solutions		
purple	$MnO_4^-(aq)$	manganate(VII) anion
green	$MnO_4^{2-}(aq)$	manganate(VI) anion
yellow	$VO_4^{3-}(aq)$	vanadate(V) anion
blue	$VO^{2+}(aq)$	oxovanadium(IV) cation
green	$[V(H_2O)_6]^{3+}(aq)$	hexaaquavanadium(III) cation
violet	$[V(H_2O)_6]^{2+}(aq)$	hexaaquavanadium(II) cation
yellow	$CrO_4^{2-}(aq)$	chromate(VI) anion
orange	$Cr_2O_7^{2-}(aq)$	dichromate(VI) anion

Fig. 13.9 Transition metals form coloured compounds

AQUEOUS CATIONS AND OXOANIONS

The most important characteristics of the d-block ions are the variety and relative stability of their oxidation states, the formation and relative stability of various complexes and the ability to act as homogeneous catalysts. You will find the emphasis is upon the chemistry of the d-block ions in aqueous solution.

REDOX REACTIONS

In the series from scandium to zinc, the range of available oxidation states reaches a maximum of seven with manganese: see Fig. 13.4. For the elements from Sc-Mn inclusive you can relate the maximum oxidation number with the number of 3d and 4s electrons in the atom. The decrease in the range of oxidation states from manganese to zinc could be related to the decrease in the number of unpaired 3d electrons in the atom.

- The stability of the higher oxidation states decreases with atomic number from left to right across the series Sc-Zn
- The stability of the +2 oxidation state relative to the +3 state increases with atomic number from left to right across the series
- Compounds with elements in the intermediate oxidation states tend to disproportionate
- The greater stability of Mn^{2+} (w.r.t. Mn^{3+}) and Fe^{3+} (w.r.t. Fe^{2+}) may be related to the stability of the half-filled 3d-subshell
- Transition metal ions tend to exist as hydrated cations in low oxidation states and oxoanions at high oxidation states
- The relative stability of the oxidation states is extremely important and usually discussed in terms of standard electrode potentials, $E^\ominus$

PREDICTING REDOX REACTIONS

> You really must understand electrochemical cells. Study Chapter 8.

Make sure you understand electrochemical cells and standard electrode potentials before continuing with this section: see Chapter 8.

Questions that ask you to use $E^\ominus$ values are very popular in examinations. You will have to predict the energetic but not kinetic feasibility of a redox reaction and to work out a balanced equation for the feasible reaction. The examiners will certainly not expect you to remember $E^\ominus$ values and they will usually give you details of the electrode systems either as you would find them in your data book: e.g.

$$[MnO_4^-(aq) + 8H^+(aq)],[Mn^{2+}(aq) + 4H_2O(l)]|Pt \quad E^\ominus = +1.51\,V$$

or in the form of a half-cell reduction reaction: e.g.

$$MnO_4^-(aq) + 8H^+(aq) + 5e^- \rightarrow Mn^{2+}(aq) + 4H_2O(l)$$

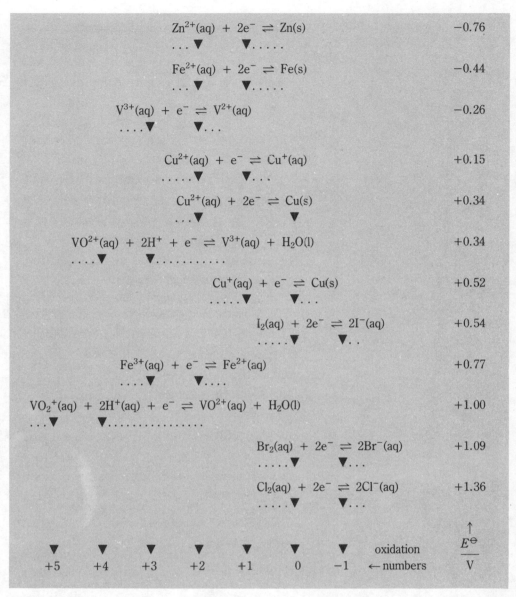

$$Zn^{2+}(aq) + 2e^- \rightleftharpoons Zn(s) \qquad -0.76$$

$$Fe^{2+}(aq) + 2e^- \rightleftharpoons Fe(s) \qquad -0.44$$

$$V^{3+}(aq) + e^- \rightleftharpoons V^{2+}(aq) \qquad -0.26$$

$$Cu^{2+}(aq) + e^- \rightleftharpoons Cu^+(aq) \qquad +0.15$$

$$Cu^{2+}(aq) + 2e^- \rightleftharpoons Cu(s) \qquad +0.34$$

$$VO^{2+}(aq) + 2H^+ + e^- \rightleftharpoons V^{3+}(aq) + H_2O(l) \qquad +0.34$$

$$Cu^+(aq) + e^- \rightleftharpoons Cu(s) \qquad +0.52$$

$$I_2(aq) + 2e^- \rightleftharpoons 2I^-(aq) \qquad +0.54$$

$$Fe^{3+}(aq) + e^- \rightleftharpoons Fe^{2+}(aq) \qquad +0.77$$

$$VO_2^+(aq) + 2H^+(aq) + e^- \rightleftharpoons VO^{2+}(aq) + H_2O(l) \qquad +1.00$$

$$Br_2(aq) + 2e^- \rightleftharpoons 2Br^-(aq) \qquad +1.09$$

$$Cl_2(aq) + 2e^- \rightleftharpoons 2Cl^-(aq) \qquad +1.36$$

							oxidation	$\dfrac{E^{\ominus}}{V}$
+5	+4	+3	+2	+1	0	−1	← numbers	

Fig. 13.10 Chart of $E^{\ominus}$ values

Quite often these ion-electron half-equations are printed on charts: see Fig. 13.10. The scale on the vertical axis gives values of $E^{\ominus}/V$ but notice that the 'negative' values are at the top and the positive values at the bottom! The horizontal axis is marked off to show the oxidation states with the 'positive' values on the left and the negative values on the right!

When you are using a chart or table of ion-electron half-equations to work out a balanced equation for an energetically feasible reaction there are some important points to remember:

1   Before you attempt to combine two ion-electron half-equations you must make sure they each involve the same number of electrons
2   For the greatest energetic feasibility the two ion-electron half-equations should be as far apart as possible
3   You combine the *oxidant* (the left-hand side of an ion-electron half-equation) *bottom left* in the chart and the *reductant* (the right-hand side of a half-equation) *top right* in the chart. This gives you the reactants for the left-hand side of your balanced equation
4   You combine the right-hand side of the lower half-equation and the left-hand side of the upper half-equation to obtain the products for the right-hand side of your balanced equation
5   You should 'cancel down' any formulae (often H^+, OH^- or H_2O) on both sides of your equation and check that your final equation is balanced and does not contain any 'spare' electrons.

Study carefully the two examples in Fig. 13.11. Remember that your prediction is based on the combination of ion-electron half-reactions that occur under standard conditions at electrodes. The reaction may well not take place when you put together the oxidant and reductant under actual experimental conditions away from their electrodes. Metals, for example, are often protected by a tough oxide film.

Q   Could iron(III) ions react with metallic copper?

A   Yes $Fe^{3+}(aq)$ is the oxidant down on the left (at $E^\ominus = +0.77\,V$)
     $Cu(s)$ is the reductant up on the right (at $E^\ominus = +0.34\,V$)

Double the ion-electron half-equation for the reduction of the iron(III) to balance the electrons in the other half-equation:

$$\Rightarrow \quad 2Fe^{3+}(aq) + 2e^- \rightarrow 2Fe^{2+}(aq)$$

Add the 'lower' LHS and the 'upper' RHS of the half-equations:

reactants $2Fe^{3+}(aq) + Cu(s) \rightarrow$

Add the 'lower' RHS and the 'upper' LHS of the half-equations:

$$\rightarrow 2Fe^{2+}(aq) + Cu^{2+}(aq) \text{ products}$$

Combine the two results to give a complete balanced equation:

$$2Fe^{3+}(aq) + Cu(s) \rightarrow 2Fe^{2+}(aq) + Cu^{2+}(aq)$$

[This reaction is used for etching printed circuit boards]

Q   Could aqueous copper(I) ions disproportionate?

A   Yes $Cu^+(aq)$ is the oxidant down on the left (at $E^\ominus = +0.15\,V$)
     $Cu^+(aq)$ is the reducant up on the right (at $E^\ominus = +0.52\,V$)

The electrons are already balanced (1 in each half-equation).

Add the 'lower' LHS and the 'upper' RHS of the half-equations:

reactants $Cu^+(aq) + Cu^+(aq) \rightarrow$

Add the 'lower' RHS and the 'upper' LHS of the half-equations:

$$\rightarrow Cu^{2+}(aq) + Cu(s) \text{ products}$$

Combine the two results to give a complete balanced equation:

$$2Cu^+(aq) \rightarrow Cu(s) + Cu^{2+}(aq)$$

Q   Could zinc reduce aqueous vanadate(V) ions?

A   Yes $VO_2^+(aq)$, $VO^{2+}(aq)$ and $V^{3+}(aq)$ would all be oxidants down on the left
     (at $E^\ominus = +1.00\,V$, $+0.34\,V$ and $-0.26\,V$)
     $Zn(s)$ is the reductant up on the right (at $E^\ominus = -0.76\,V$)

reactants $2VO_2^+(aq) + 4H^+(aq) + Zn(s) \rightarrow$
$$\rightarrow 2VO^{2+}(aq) + 2H_2O(l) + Zn^{2+}(aq) \text{ products}$$
$$\Rightarrow 2VO_2^+(aq) + 4H^+(aq) + Zn(s) \rightarrow 2VO^{2+}(aq) + 2H_2O(l) + Zn^{2+}(aq)$$
reactants $2VO^{2+}(aq) + 4H^+(aq) + Zn(s) \rightarrow$
$$\rightarrow 2V^{3+}(aq) + H_2O(l) + Zn^{2+}(aq) \text{ products}$$
$$\Rightarrow 2VO^{2+}(aq) + 4H^+(aq) + Zn(s) \rightarrow 2V^{3+}(aq) + 2H_2O(l) + Zn^{2+}(aq)$$
reactants $2V^{3+}(aq) + Zn(s) \rightarrow$
$$\rightarrow 2V^{2+}(aq) + Zn^{2+}(aq) \text{ products}$$
$$\Rightarrow 2V^{3+}(aq) + Zn(s) \rightarrow 2V^{2+}(aq) + Zn^{2+}(aq)$$

[If you add zinc to acidified aqueous ammonium vanadate(V) you should see the colour change from yellow to blue to green to violet: see Fig. 13.9.]

Fig. 13.11 Predicting redox reactions from $E^\ominus$ values

CATALYTIC ACTIVITY

If you add aqueous potassium peroxodisulphate(VI) (persulphate) to a colourless aqueous solution containing potassium iodide, sodium thiosulphate and starch, some time elapses before the solution turns blue:

$$S_2O_8^{2-}(aq) + 2I^-(aq) \rightarrow 2SO_4^{2-}(aq) + I_2(aq)$$

You might expect this reaction to be slow because both reactants are anions and ions with the same charge will repel one another. The blue colour caused by iodine molecules combining with starch will not appear until all the sodium thiosulphate (which reduces iodine molecules back to iodide ions) is used up. If you repeat this experiment but with a few drops of aqueous iron(II) *or* iron(III) sulphate in the reaction mixture, the blue colour appears in far less time. You could explain the catalytic effect of the iron cations as follows:

$$I_2(aq) + 2e^- \rightarrow 2I^-(aq)$$
$$2Fe^{3+}(aq) + 2e^- \rightleftharpoons 2Fe^{2+}(aq)$$
$$S_2O_8^{2-}(aq) + 2e^- \rightarrow 2SO_4^{2-}(aq)$$

Peroxodisulphate(VI) anions are rapidly reduced to sulphate ions by any iron(II) cations that are oxidised to iron(III) ions:

$$S_2O_8^{2-}(aq) + 2Fe^{2+}(aq) \rightarrow 2SO_4^{2-}(aq) + 2Fe^{3+}(aq)$$

These iron(III) cations are then rapidly reduced back to iron(II) ions by any iodide anions that are oxidised to iodine molecules:

$$2Fe^{3+}(aq) + 2I^-(aq) \rightarrow 2Fe^{2+}(aq) + I_2(aq)$$

If you monitor the purple colour of acidified aqueous potassium manganate(VII) during the oxidation of aqueous ethanedioic (oxalic) acid, you may find that the reaction begins very slowly, gradually speeds up and then eventually slows down again. The reaction accelerates because manganese(II) cations (one of the products) catalyses the redox reaction between the manganate(VII) ions and the ethanedioic acid molecules.

■ A reaction is autocatalytic if it is catalyzed by one of its products

If you add pink aqueous cobalt(II) chloride in slightly alkaline aqueous rochelle salt (sodium potassium tartrate or 2,3-dihydroxybutanedioate) to warm aqueous hydrogen peroxide, the mixture quickly turns green and gives off oxygen gas. As the effervescence subsides the green solution returns to its original pink colour. You could explain this catalysis along similar lines to the explanation for the oxidation of iodide by persulphate:

$$\tfrac{1}{2}O_2(g) + H_2O(l) + 2e^- \rightarrow 2OH^-(aq)$$
$$[\text{green solution}]\ 2Co^{3+}(aq) + 2e^- \rightleftharpoons 2Co^{2+}(aq)\ [\text{pink solution}]$$
$$H_2O_2(aq) + 2e^- \rightarrow 2OH^-(aq)$$

The hydrogen peroxide oxidises the cobalt(II) to cobalt(III). The cobalt(III) is reduced back to cobalt(II) as it oxidises the hydroxide ions to oxygen gas and water. The net reaction is:

$$H_2O_2(aq) \rightarrow \tfrac{1}{2}O_2(g) + H_2O(l)$$

The pink and green solutions probably contain cobalt(II) and cobalt(III) complex ions involving 2,3-dihydroxybutanedioate ions, hydroxide ions and water molecules as ligands.

■ Many aqueous transition metal ions can act as homogeneous catalysts for redox reactions occurring in aqueous solution

COMPLEX FORMATION

AQUA IONS

In aqueous solution the s-block cations are hydrated because their positive charges attract the polar water molecules whereas the smaller, more polarising transition metal cations exist as aqua ions because they can use their vacant orbitals to accept electron pairs and form coordinate (dative) bonds with the oxygen atom in water molecules. The number of molecules bonded to the central ion is called the coordination number and may be 2, 4 (fairly common) or 6 (most common). The shape of the aqua ion depends upon the coordination number but cannot be predicted by the valence shell electron pair repulsion theory: see Fig. 13.12.

Fig. 13.12 Shapes of complex ions

linear [CuCl₂]⁻

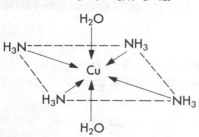

square planar [Ni(CN)₄]²⁻

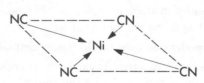

tetrahedral [CoCl₄]²⁻

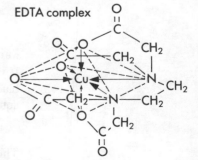

octahedral [Cu(NH₃)₄(H₂O)₂]²⁺

octahedral

EDTA complex

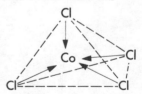

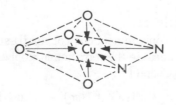

Acidity

The central ion polarises the bonded water molecules sufficiently to allow them to lose protons and cause the aqua ion to act as an acid: e.g.

$$[Fe(H_2O)_6]^{3+}(aq) \rightleftharpoons [Fe(H_2O)_5OH]^{2+}(aq) + H^+(aq)$$

Since the polarising power of the central ion increases with increasing oxidation number, aqua cations with a triple charge are much more acidic than those with a double charge. This is how you may explain the fact that aqueous iron(III) salts but not iron(II) salts react with carbonates to produce carbon dioxide:

$$2H^+(aq) + CO_3^{2-}(aq) \rightarrow H_2O(l) + CO_2(g)$$
$$\uparrow$$

protons from the hexaaquairon(III) cation

The polarising power of the iron(III) cation may explain why iron(III) carbonate does not exist. CO_3^{2-} is a large, very polarisable anion.

Hydrolysis

'Iron(III) alum' (ammonium iron(III) sulphate-12-water) crystals contain the hexa-aquairon(III) cation and are pale violet in colour. When you dissolve the crystals in water you get a yellow-brown solution that gradually produces a brown precipitate:

$$[Fe(H_2O)_5OH]^{2+}(aq) \rightleftharpoons [Fe(H_2O)_4(OH)_2]^+(aq) + H^+(aq)$$
$$[Fe(H_2O)_4(OH)_2]^+(aq) \rightleftharpoons [Fe(H_2O)_3(OH)_3](s) + H^+(aq)$$

When making solutions, we usually add acid to aqueous iron salts to minimise this hydrolysis. According to the law of chemical equilibrium, if we increase the concentration of one product (the $H^+(aq)$ ion) then the concentration of the other products must decrease and that of the reactants must increase: see Chapter 7.

When analysing solutions we may add alkali to aqueous metal salts to maximise this hydrolysis and produce coloured precipitates that identify the cation: e.g.

$$[Fe(H_2O)_6]^{2+}(aq) + 2OH^-(aq) \rightarrow [Fe(H_2O)_4(OH)_2](s) \quad \text{dark green ppt.}$$

You can usually simplify these equations as follows:

$$Cu^{2+}(aq) + 2OH^-(aq) \rightarrow Cu(OH)_2(s) \quad \text{light blue ppt.}$$

The composition of many of these precipitates is much more complicated than that shown by a simple formula like $Ni(OH)_2$ but such complexities are beyond the requirements of A-level.

COMPLEX IONS

The $[Fe(H_2O)_6]^{3+}(aq)$ and $[Fe(H_2O)_5OH]^{2+}(aq)$ ions are called complex cations. The H_2O molecule and the OH^- ion are called ligands.

- A complex ion consists of a central metal cation with six (four or two) ligands (molecules or ions) datively bonded to it
- A ligand is an atom or small group of atoms datively bonded by its lone electron pair to a central transition metal cation
- A monodentate (one tooth) ligand forms one dative bond with the central ion
- A bidentate (two teeth) ligand consists of several atoms, two of which are each able to form one dative bond with the central ion
- A hexadentate ligand consists of many atoms, six of which each form one dative bond with the central ion: see Fig. 13.13

Relative stabilities

If you add ammonia to aqueous copper(II) sulphate the pale blue solution becomes deep blue in colour because the tetraamminecopper(II) cation is more stable than the copper(II) aqua ion: see Chapter 7.

$$4NH_3(aq) + [Cu(H_2O)_6]^{2+}(aq) \rightarrow [Cu(NH_3)_4(H_2O)_2]^{2+}(aq) + 4H_2O(l)$$

This is the principle underlying the test for copper(II) ions or for ammonia molecules.

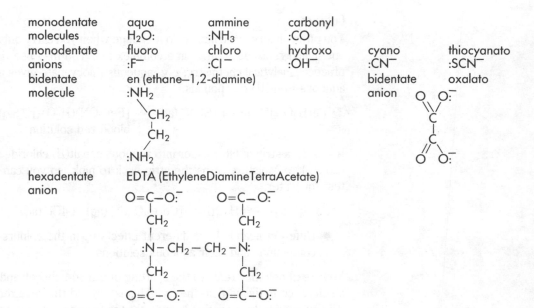

monodentate molecules	aqua H_2O:	ammine $:NH_3$	carbonyl $:CO$		
monodentate anions	fluoro $:F^-$	chloro $:Cl^-$	hydroxo $:OH^-$	cyano $:CN^-$	thiocyanato $:SCN^-$
bidentate molecule	en (ethane–1,2-diamine)			bidentate anion	oxalato
hexadentate anion	EDTA (EthyleneDiamineTetraAcetate)				

Fig. 13.13 Some common ligands

If you add concentrated hydrochloric acid to aqueous copper(II) sulphate, chloride ions replace the water molecules as ligands. The solution becomes deep green. This colour appears because an aqueous solution containing tetrachlorocuprate(II) anions is yellow and an aqueous solution containing copper(II) aqua ions is blue. The yellow and blue colours together appear deep green. You may write the equation as follows:

$$4Cl^-(aq) + Cu^{2+}(aq) \rightleftharpoons [CuCl_4]^{2-}(aq)$$

Notice that the Cl^- ligands are covalently bound to the central copper ion by dative bonds. Therefore they do not form a precipitate of silver chloride with aqueous silver nitrate.

■ Anionic ligands dative-covalently bound to the central ion are shown inside the [ ] of the formula and are not free to act as anions

If you add aqueous 'EDTA' (ethylenediaminetetraacetic acid) to the dark blue solution and to the deep green solution, the ammonia molecules and the chloride ions are replaced by the hexadentate ligand to give a pale blue solution:

$$EDTA^{4-}(aq) + [Cu(NH_3)_4]^{2+}(aq) \rightarrow [Cu(EDTA)]^{2-}(aq) + 4NH_3(aq)$$
$$EDTA^{4-}(aq) + CuCl_4^{2-}(aq) \rightarrow [Cu(EDTA)]^{2-}(aq) + 4Cl^-(aq)$$

The increase in entropy accompanying the replacement of six monodentate ligands by one hexadentate ligand largely accounts for the high stability of EDTA complexes and for the use of EDTA in complexometric analysis.

Notice that the net charge on a complex depends upon the number of ligands, their charges and the oxidation state of the central ion to which the ligands are bonded. You should be able to work out the net charge on a complex from its composition and the oxidation state of the central ion. And you should also be able to work out the oxidation number of the central ion from the formula and charge of the complex: see Fig. 13.14.

1 What is the charge on the tetraamminedichlorochromium(III) complex ion?

the formula has two Cl^- atoms and four NH_3 molecules so the ligands contribute 2 negative charges

the central chromium cation has an oxidation number of $+3$ so it contributes 3 positive charges

hence the net charge on the complex ion is $1+$ and it is a cation:

$[CrCl_2(NH_3)_4]^+$

2 What is the oxidation number of copper in the $[Cu(CH)_4]^{3-}$ complex ion?

the complex ion is an anion with a net charge of $3-$

the formula has four CN^- anions as ligands that contribute 4 negative charges

hence the central copper ion must contribute 1 positive charge so its oxidation number is $+1$

$[Cu(CN)_4]^{3-}$ would be called the tetracyanocuprate(I) anion

Fig. 13.14 Two simple problems on complex ions

Colours

You often observe a striking colour change when you carry out a test tube reaction in which one ligand replaces another in a complex ion. You may use these colour changes in your practical analytical chemistry: e.g. aqueous thiocyanate gives a vivid blood-red colour with aqueous iron(III) compounds:

$$[Fe(H_2O)_6]^{3+}(aq) + SCN^-(aq) \rightarrow [Fe(SCN)(H_2O)_5]^{2+}(aq) + H_2O(l)$$
$$\text{blood-red solution}$$

If you dip a strip of filter paper into aqueous cobalt(II) chloride then gently heat it, the strip turns blue. Moisture turns the strip back to pink, so you can use this colour change as a test for water:

$$CoCl_4^{2-}(aq) + 6H_2O(l) \rightarrow [Co(H_2O)_6]^{2+}(aq) + 4Cl^-(aq)$$

■ Different ligands have different effects upon the colours of transition metal compounds and their aqueous solutions

Your textbook may refer to the splitting of the 3d-subshell and you may even come across a reference to ligand-field theory. This is beyond the requirements of A-level chemistry. All you need to know is that ligands change the amount of energy (still in the visible region of the spectrum) needed to excite the 3d-subshell electrons of the central transition metal ion.

Stabilisation of oxidation states

If you put into a test tube colourless aqueous potassium iodide and starch you can turn it blue-black by adding aqueous iron(III) chloride but not by adding aqueous potassium hexacyanoferrate(III). In both cases the oxidation state of the iron is +3 but it is more stable and therefore a weaker oxidising agent complexed with CN^- instead of H_2O ligands:

$$E^{\ominus}/V$$

$$[Fe(CN)_6]^{3-}(aq) + e^- \rightleftharpoons [Fe(CN)_6]^{2-}(aq) \qquad +0.36$$
$$\tfrac{1}{2}I_2(aq) + e^- \rightleftharpoons I^-(aq) \qquad +0.54$$
$$[Fe(H_2O)_6]^{3+}(aq) + e^- \rightleftharpoons [Fe(H_2O)_6]^{2+}(aq) \qquad +0.77$$

■ Different ligands have different effects upon the stability of the oxidation state of a transition metal

Shapes of complex ions

Complex ions may be linear, tetrahedral, square planar or octahedral: see Fig. 13.12. Octahedral is the most common and is adopted even by the hexadentate EDTA complex ions: see Fig. 13.13. Octahedral complexes can exhibit *cis-trans* (geometrical) isomerism and optical isomerism: see Fig. 13.15.

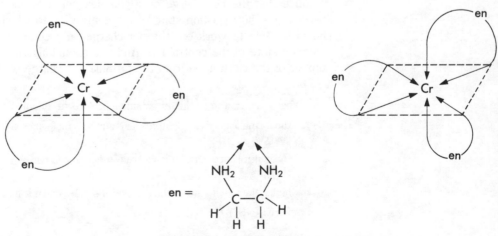

Fig. 13.15 Stereoisomerism in complex ions

en =

1.2-diaminoethane [bidentate ligand]

Nomenclature

You name the complex by starting with the names of the ligands in alphabetical order (ignoring numerical prefixes such as di, tetra, hexa) and ending with the name and oxidation number of the central ion: e.g.

$[Cu(NH_3)_4(H_2O)_2]^{2+}$(aq)	tetraamminediaquacopper(II) ion
$[Fe(CN)_6]^{3-}$(aq)	hexacyanoferrate(III) ion
$[Fe(CN)_6]^{4-}$(aq)	hexacyanoferrate(II) ion
$[Fe(SCN)(H_2O)_5]^{2+}$(aq)	pentaaquathiocyanatoiron(III) ion

Notice that the ending 'ate' shows the complex ion to be an anion.

Carbonyl compounds

Transition metals react with carbon monoxide to give neutral complexes called carbonyls. The oxidation state of the transition metal remains at 0 and its electron configuration is increased to that of a noble gas by the dative bonding of the :CO ligands. For example, cobalt (atomic no. 27) which requires 9 electrons to achieve the configuration of krypton (atomic no. 36) forms a carbonyl with the following formula:

$$(CO)_4Co\text{-}Co(CO)_4$$

The formation and decomposition of tetracarbonylnickel(0) plays a vital part in the industrial recovery and purification of metallic nickel:

$$Ni(s) \ + \ 4CO(g) \ \underset{200\,°C}{\overset{60\,°C}{\rightleftharpoons}} \ Ni(CO)_4(g)$$

EXAMINATION QUESTION AND OUTLINE ANSWER

Q1 a) List *five* typical properties of the transition elements, illustrating each property with a suitable example. *(10)*

  b) Give a detailed explanation of the following changes, writing balanced equations wherever possible.

  Granulated zinc was added to a solution of potassium dichromate(VI) in dilute sulphuric acid and the orange solution turned green. When the reaction was repeated in the absence of air the green solution finally turned to a sky-blue colour, but the solution became green again when shaken with air. *(9)*

  c) Cobalt(III) chloride dissolves in water to give a pink solution. When concentrated hydrochloric acid is added the solution becomes deep blue. Draw diagrams of the complex cobalt ions present before and after the addition of concentrated hydrochloric acid. Explain why the colour change occurs and why it can be reversed on addition of water to the blue solution. *(6)*

  (WESSEX (specimen))

Outline answer

Q1 (a) see chapter 13

  (b) Zinc reduces the chromium to +3 in air and +2 in absence of air
  $Zn(s) \rightarrow Zn^{2+}(aq) + 2e^-$
  in air $Cr_2O_7^{2-}(aq) + 14H^+(aq) + 6e^- \rightarrow 2Cr^{3+}(aq) + 7H_2O(l)$
  not in air $Cr^{3+}(aq) + e^- \rightarrow Cr^{2+}(aq)$

  (c) $[Co(H_2O)_6]^{2+}(aq)$ gives pink solution

    Cl^- ligands give $CoCl_4^{2-}(aq)$ — blue solution

EXAMINATION QUESTION AND TUTOR'S ANSWER

Q2 Potassium dichromate(VI) is a useful reagent in the laboratory. It may be used in the preparation of other chromium compounds or as an oxidising agent.

a) Chrome alum may be prepared by the reduction of potassium dichromate(VI).

i) State the formula and colour of chrome alum.

$KCr(SO_4)_2 \cdot 12H_2O$ *violet*

ii) Name a reducing agent that is commonly used in the preparation of chrome alum.

sulphur dioxide

iii) Describe the colour change that takes place during the reduction.

Orange to deep green
(5)

b) Potassium dichromate(VI) is a powerful oxidizing agent and will oxidize tin(II) to tin(IV). The half equations are:

$$Cr_2O_7^{2-} + 14H^+ + 6e^- \rightarrow 2Cr^{3+} + 7H_2O$$
$$Sn^{2+} \rightarrow Sn^{4+} + 2e^-$$

i) Write a balanced equation for the reaction of tin(II) ions with dichromate(VI) ions.

$$3Sn^{2+} + Cr_2O_7^{2-} + 14H^+ \rightarrow 3Sn^{4+} + 2Cr^{3+} + 7H_2O$$

ii) $20\,cm^3$ of a solution of tin(II) ions required $18.4\,cm^3$ of a $0.1\,mol\,dm^{-3}$ potassium dichromate(VI) solution for complete oxidation.
Calculate the mass of tin ions contained in $1\,dm^3$ of the solution.

$(Sn = 119)$

3 mol Sn^{2+} oxidised by 1 mol $Cr_2O_7^{2-}$

$\Rightarrow$ Concentration of the Sn^{2+} is

$$0.1 \times \frac{18.4}{20} \times 3 = 0.276 \; mol\,dm^{-3}$$

$\Rightarrow$ mass of tin ions is 119×0.276 g dm^{-3}
$$= 32.8g$$

Answer $\underline{32.8\,g}$
(6)

c) Potassium dichromate(VI) is used in the oxidation of alcohols. The product of the oxidation depends on the structure of the alcohol. Draw the structures of the oxidation products, if any, when the following alcohols are gently heated, with acidified potassium dichromate(VI) solution. If you consider that the alcohol does not undergo oxidation write 'no reaction'.

$CH_3CH_2CH_2OH$ $\xrightarrow{\;O\;}$ $CH_3-CH_2-C{\overset{\displaystyle O}{\underset{\displaystyle H}{\lesssim}}}$

$CH_3CH_2CH(OH)CH_3$ $\xrightarrow{\;O\;}$ $CH_3-CH_2-\overset{\displaystyle O}{\underset{\displaystyle \|}{C}}-CH_3$

$(CH_3)_3COH$ $\xrightarrow{\;O\;}$ *no reaction*
(4)

d) In solution the dichromate(VI) ion is in equilibrium with the chromate(VI) ion.

$$2CrO_4^{2-} + 2H^+ \rightleftharpoons Cr_2O_7^{2-} + H_2O$$

i) Name a reagent that you would add to the equilibrium mixture in order to move the equilibrium

to the left *aqueous sodium hydroxide*

to the right *hydrochloric acid*

ii) Describe the expected colour change when the equilibrium moves to the left.

from orange to yellow

(3)
(NISEC 1988)

STUDENT'S ANSWER WITH EXAMINER'S COMMENTS

Q3 This question concerns the chemistry of iron, Fe, a transition element.

a) i) Insert arrows in the boxes below to show the electronic configuration of the ground state of iron.

ii) Write down the oxidation state(s) commonly exhibited by iron.

2 and 3

> **Include the sign: e.g. +2.**

b) i) Solid iron(III) nitrate, $Fe(NO_3)_3.9H_2O$, is pale violet in colour. On dissolving in water the resulting solution is yellow-brown but turns almost colourless (*very* pale violet) on adding nitric or sulphuric acid. State, giving a formula, the species which has

the very pale violet colour, $[Fe(H_2O)_6]^{3+}$ ✓

the yellow-brown colour. $[Fe(H_2O)_5OH]^{2+}$ ✓

ii) A solution of iron(III) chloride in water is yellow-brown. The colour is said to be due to the presence of the ion $[Fe(H_2O)_5Cl]^{2+}$. The colour is unchanged by the addition of either hydrochloric or nitric acid. Suggest an explanation for this.

> **Good effort.**

The complex ion is too stable to be attacked by the acids or
$[Fe(H_2O)_4Cl_2]^+$ is also yellow-brown ✓

iii) Explain why aqueous solutions of iron(III) salts have a pH less than 7.

> **Good.**

The hydrated ion can lose protons:
$[Fe(H_2O)_6]^{3+}_{(aq)} \rightarrow [Fe(H_2O)_5OH]^{2+}_{(aq)} + H^+_{(aq)}$ ✓

c) i) On treatment with sulphur dioxide, an aqueous solution of iron(III) chloride rapidly becomes pale green in colour. Explain, giving a balanced chemical equation.

> **Equation not balanced – check the changes: see Chapter 13.**

$Fe^{3+}_{(aq)} + SO_2(aq) + 2H_2O_{(l)} \rightarrow Fe^{2+}_{(aq)} + SO_4^{2-}(aq) + 4H^+_{(aq)}$

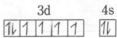

pale green
solution ✓

> **2 Fe³⁺(aq);**
>
> **2 Fe²⁺(aq)**

ii) After boiling out dissolved sulphur dioxide and cooling, the solution from c)i) above was treated with dilute sodium hydroxide solution. An almost white precipitate was initially produced which rapidly turned dark green and on standing, the precipitate at the top of the mixture became red-brown. Give an explanation, indicating the reason for the colour changes and the identities of the species produced.

$Fe^{2+}(aq) + 2OH^-(aq) \longrightarrow Fe(OH)_2 (s)$ white ✓

iron (II) hydroxide white when pure but oxidizes in air to $Fe(OH)_3 (s)$ – red-brown ✓

"Green colour may be partial formation of iron(III) compound."

d) A transition metal other than iron reacts with dilute sulphuric acid with effervescence giving a blue solution which turns dark green very rapidly on exposure to air. Give
   i) the name and symbol of this transitional metal,

   Chromium Cr ✓

   ii) the formula of the species in the blue solution,

   Cr^{2+}

   iii) the formula of the species in the dark green solution,

   Cr^{3+}

   iv) the oxidation states other than 2+ which this transition element may exhibit in its compounds.

   +6

"Complex needed for full marks e.g. $[Cr(H_2O)_6]^{2+}(aq)$"

"$[Cr(H_2O)_5OH]^{2+}(aq)$"

"and +3."

(10)

(WJEC 1988)

PRACTICE QUESTIONS

Q4 This is a fairly common question. You must know your facts and prepare thoroughly to give a good answer.

Describe the typical *chemical* properties of transition elements and their compounds. Illustrate your answer with examples selected from those transition elements that you have studied. Descriptions of physical properties are not required.

Show how a knowledge of the electronic structures of the transition elements can be used to account for some of the properties you have mentioned.

(NUFF 1988)

Q5　a)　Explain the terms
　　　i) ligand;
　　　ii) complex ion. (5)
　　b)　Write formulae to represent
　　　i) a cationic complex ion;
　　　ii) an anionic complex ion;
　　　iii) a neutral complex. (3)
　　c)　Give an example of a reaction involving a transitional metal compound in which a change in ligand is accompanied by a change in co-ordination number. (3)
　　d)　Draw structures to represent the two geometrical isomers of a transitional metal complex. Name the shape of the complex you have drawn. (3)

(AEB 1988)

INTRODUCTION TO ORGANIC THEORY

GETTING STARTED

Organic chemistry is the study of carbon compounds. Chemists have already identified and named more than 6 million organic compounds but the number in the catalogue is growing daily. The covalent molecular structure of these compounds is the key to their vast numbers and their division into classes. It is also the key to their synthesis, their chemical and physical properties and their uses. An understanding of structure is essential to our understanding of organic chemistry.

FORMULAE OF ORGANIC COMPOUNDS

FUNCTIONAL GROUPS

ISOMERISM

NOMENCLATURE

HOMOLOGOUS SERIES

REACTION MECHANISMS

TYPES OF REACTION

ESSENTIAL PRINCIPLES

FORMULAE OF ORGANIC COMPOUNDS

The five types of formula that you will encounter in organic chemistry are explained in Fig. 14.1. The most important are the formulae that show the sequence and arrangement of atoms and groups in a molecule: i.e. the structural, displayed (sometimes called graphic) and stereochemical formulae.

Empirical formula C_2H_4O	simplest possible formula
	Shows amounts of each element in a compound
	Deduced from composition by mass or from mass/charge ratio of parent molecule ion in mass spectrum
Molecular formula $C_4H_8O_2$	integral multiple of empirical formula
	Shows composition of molecule
	Deduced from empirical formula and relative molecular mass or from mass spectral analysis
Structural formula $(CH_3)_2CHCO_2H$	structure of a molecule
	Shows arrangement of atoms and groups in molecule
	Deduced from properties of a substance and from infra red absorption spectra and mass spectra
Displayed (graphic) formula	detailed structure of molecule
	Shows arrangement of all atoms and bonds in molecule as a projection of the structure onto a plane
	Deduced from structural formula by applying the rules of bonding to the atoms
Stereochemical formula	the spatial structure of a molecule
	Shows spatial arrangement of the bonds, atoms and groups in molecule as a 'perspective 3-D' formula
	Deduced from displayed formula by applying to the bonds, atoms and groups the principles of electron-pair repulsion and of electron delocalisation

Fig. 14.1 Types of formulae for organic compounds

FUNCTIONAL GROUPS

A functional group is an element e.g. Br, or combination of elements (e.g. OH) responsible for specific properties of an organic compound or class of compounds.

The functional groups you will probably meet are shown in Fig. 14.2. Most of these are for classes of compounds that you will deal with in your practical work. You will find more details of these various classes of organic compounds later.

CONVENTIONS FOR WRITING FORMULAE

You will soon realise that we follow certain rules when writing structural formulae. You may also discover that you can write a structural formula in more than one way. For example, here are four ways of writing the formula of hexanoic acid:

> You need to practise writing organic formulae.

$CH_3CH_2CH_2CH_2CH_2COOH$ $CH_3(CH_2)_4CO_2H$	These two are completely acceptable structural formulae because they show the molecular structure unambiguously
$C_5H_{11}COOH$ $C_5H_{11}CO_2H$	These two are not completely acceptable because the alkyl group could be isomers of hexyl e.g. 2- or 3-methylpentyl, etc.

When you write a structural formula, make sure it is conventionally correct and completely unambiguous. Students and examination candidates often confuse the alcohol and aldehyde functional groups and the benzene and cyclohexane rings: see Fig. 14.3. Don't lose marks by making these mistakes yourself!

FUNCTIONAL GROUP		PREFIX	SUFFIX	CLASS OF COMPOUNDS
>C:C<	C = C		ene	alkenes
C_6H_5—		phenyl	benzene	arenes
—OH	— O — H	hydroxy	ol	alcohols and phenols
—CHO	$-C{\overset{O}{\underset{H}{}}}$		al	aldehydes
>CO	> C = O	oxo	one	ketones and aldehydes
$-CO_2H$	$-C{\overset{O}{\underset{O-H}{}}}$		oic acid	carboxylic acids
$-CO_2-$	$-C{\overset{O}{\underset{O-}{}}}$		oate	esters and polyesters
—COCl	$-C{\overset{O}{\underset{Cl}{}}}$		oyl chloride	acyl chlorides
$(-CO)_2O$	anhydride structure		anhydride	acid anhydrides
$-NH_2$	$-N{\overset{H}{\underset{H}{}}}$	amino	amine	primary amines
$-CONH_2$	$-C{\overset{O}{\underset{N-H}{}}}$ with H below		amide	amides
$-NO_2$	$-N{\overset{O}{\underset{O}{}}}$	nitro		nitro compounds
—CN	— C ≡ N	cyano	nitrile	nitriles
—Hal	—Hal	halogeno		halogeno compounds

Fig. 14.2 Functional groups in organic compounds

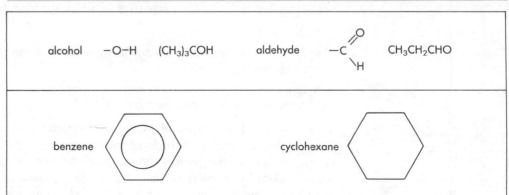

Fig. 14.3 Structural formulae frequently confused

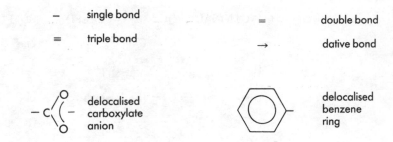

Fig. 14.4 Conventions for displayed formulae

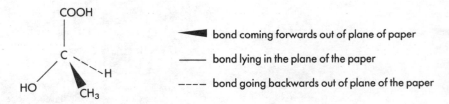

Fig. 14.5 Convention for a stereochemical formula

When you write a displayed (graphic) formula you may use the following conventions: see Fig. 14.4.

Many students find it difficult to draw 'perspective 3-D' formulae. You may find it helpful to use the conventions in Fig. 14.5 when you tackle a stereochemical formula. You need to be able to visualise an organic structure in three dimensions in order to draw a stereochemical formula. I believe models are essential. Your teacher or lecturer will probably provide molecular model kits for you to use at school or college. You can make your own very cheaply from coloured plasticine and cocktail sticks.

To deduce a stereochemical formula from a structural formula you need to know the bond angles in the molecule. You can work out approximate values for the angles by applying the simple rules based on the idea of repulsion between pairs of electrons: see Fig. 14.6. You should remember the number and type of electron pairs usually assigned to the following atoms in organic molecules – number in ( ) refers to ammonium or oxonium ions:

ELEMENT	BONDED PAIRS	LONE PAIRS
Carbon	4	0
Nitrogen	3 (4)	1 (0)
Oxygen	2 (3)	2 (1)
Sulphur	2	2
Halogen	1	3
Hydrogen	1	0

ISOMERISM

The ability of carbon atoms to bond together to form chains, rings and networks is the major reason for the existence of more than six million organic compounds. Many of these compounds may have the same molecular formula, yet each compound has a distinguishing set of physical and chemical properties by which we identify it.

Definitions:

■ Isomers are compounds having the same molecular formula but different physical and/or chemical properties

■ Isomerism is the existence of two or more compounds having the same molecular formula but different properties

There are two broad types of isomerism. Structural isomers have different groups of atoms in their molecules. Stereoisomers have the same groups of atoms in their molecules but the groups' relative positions in space are different. Figures 14.7 and 14.8 explain the various kinds of isomerism that you could encounter. You will find some typical examination questions on isomerism at the end of this unit.

1   valence shell electron pairs repel each other

2   mutual repulsion of four bonded pairs (BP) gives a tetrahedral shape:

$$(H-C-H \text{ bond angle} = 109.5°)$$

3   repulsion between a non-bonded lone pair (LP) and a bonded pair is greater than that between two bonded pairs:

$$(H-N-H \text{ bond angle} = 104.5°)$$

4   Lone Pairs repel each other even more strongly:

$$(H-O-H \text{ bond angle} = 104.5°)$$

5   Double bonded pairs are treated as though they are single bonded pairs but two atoms connected by a double bond cannot rotate about the bond as an axis:

$$(H-C-H \text{ bond angle} = 117.3°)$$

6   Triple bonded pairs are treated as though they are single bonded pairs:

$$(H-C-C \text{ bond angle} = 180.0°)$$

7   Delocalised bonds are treated as though they are single bonded pairs:

$$(C-C-C \text{ and } O-C-O \text{ bond angles} = 120.0°)$$

8   VSEPR theory gives only a qualitative prediction of bond angles and molecular shape:

(predict C–C–C bond angle in propanone to be 120°)

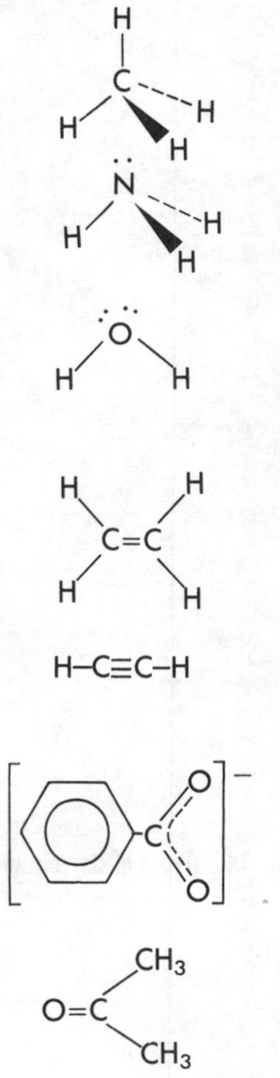

Fig. 14.6 Predicting molecular shapes by the VSEPR theory

chain isomers:

$CH_3-CH_2-CH_2-CH_2-CH_3$

pentane

$CH_3-CH-CH_2-CH_3$ (with CH_3 branch)

2-methylbutane

CH_3-C-CH_3 (with CH_3 above and CH_3 below)

2,2-dimethylpropane

position isomers:

$CH_3\,CH_2\,CH_2\,CH_2\,CH_2\,CH_2$ (Br on terminal)
1-bromohexane

$CH_3\,CH_2\,CH_2\,CH_2\,CH\,CH_3$ (Br on position 2)
2-bromohexane

$CH_3\,CH_2\,CH_2\,CH\,CH_2\,CH_3$ (Br on position 3)
3-bromohexane

functional group isomers:

$CH_3-CH_2-CH_2-CH_2-OH$ butan-1-ol

CH_3-CH_2-CHO propanal

$CH_3-CH_2-O-CH_2-CH_3$ ethoxyethane

$CH_3-CO-CH_3$ propanone

metamers:

Fig. 14.7 Types of structural isomerism

$CH_3-CH_2-CH_2-CO_2-CH_3$
methylbutanoate

$CH_3-CH_2-CO_2-CH_2-CH_3$
ethylpropanoate

$CH_3-CO_2-CH_2-CH_2-CH_3$
propylethanoate

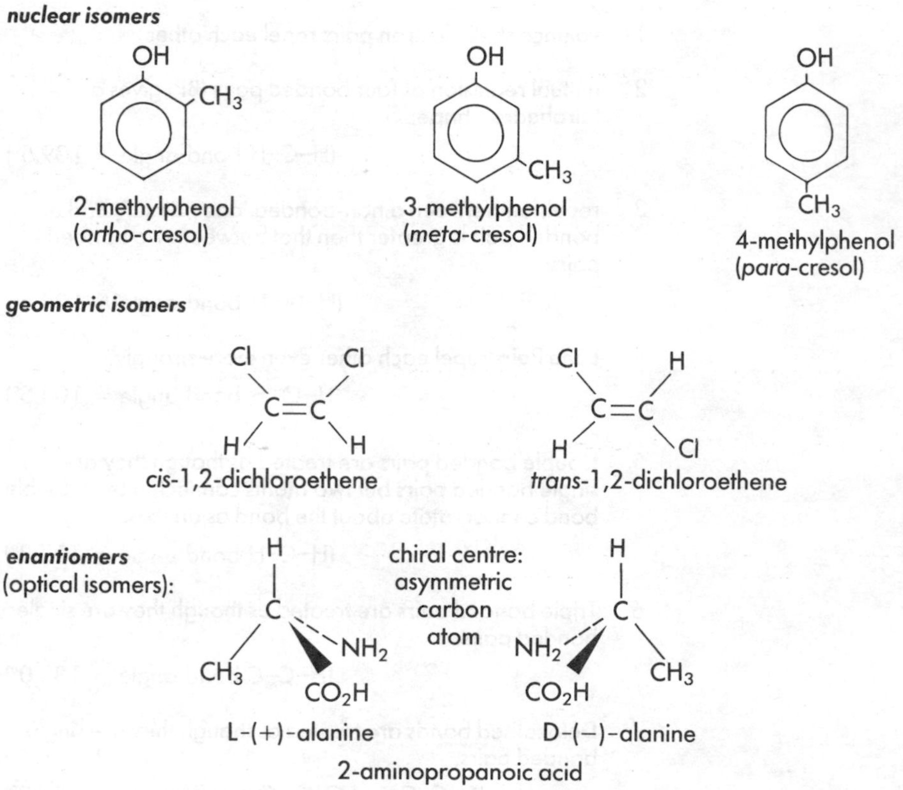
nuclear isomers
OH
CH3
2-methylphenol
(ortho-cresol)
OH
CH3
3-methylphenol
(meta-cresol)
OH
CH3
4-methylphenol
(para-cresol)
geometric isomers
Cl Cl
C=C
H H
cis-1,2-dichloroethene
Cl H
C=C
H Cl
trans-1,2-dichloroethene
enantiomers
(optical isomers):
H
C
NH2
CH3
CO2H
L-(+)-alanine
chiral centre:
asymmetric
carbon
atom
H
C
NH2
CH3
CO2H
D-(−)-alanine
2-aminopropanoic acid

Fig. 14.8 Types of stereoisomerism

DISCOVERING ISOMERS

You are very likely to be given a molecular formula and asked to work out the number and structure of the possible isomers. What do you do? You systematically apply to the atoms the following simple rules of bonding:

Element	C	H	O	O^+	N	N^+	Hal
No. bonds	4	1	2	3	3	4	1

And you start with the longest possible skeletal carbon chain.

What are the isomers with a molecular formula of C_5H_{12}?

The longest possible carbon chain is 5: C—C—C—C—C (1)

The end C-atoms take 3 H-atoms and the others take 2 H-atoms:

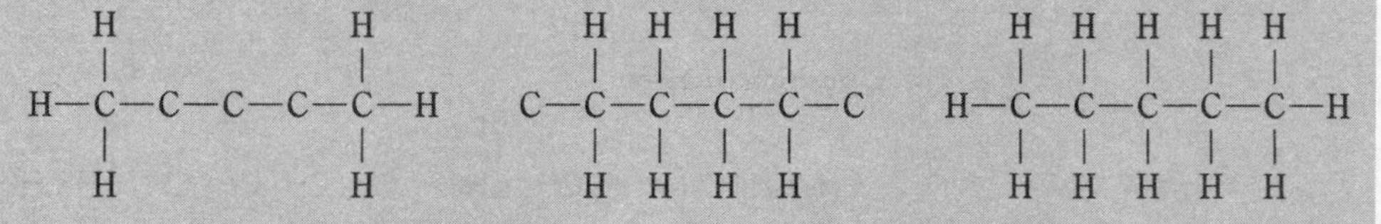

Now remove one C-atom from the chain and reattach it to any C-atom except an end carbon (because this usually restores the original chain length):

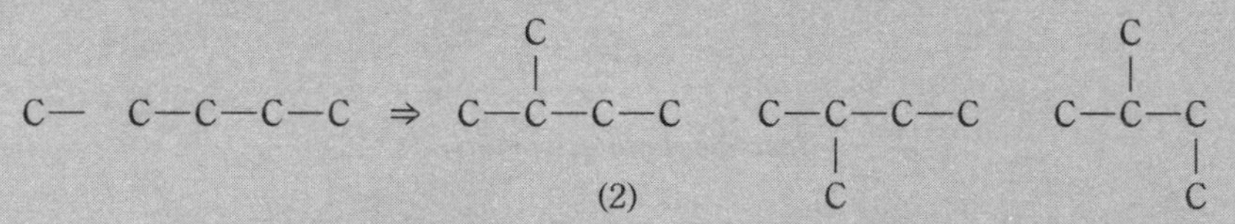

(2)

These three projections of skeletal carbon structures are all the same: a 'straight chain' of 4 C-atoms with a 'branch' of 1 C-atom on the second carbon atom from one end. These, and many more projections can be drawn because carbon atoms can rotate about the axis of single covalent bonds.

Now remove one C-atom from the longest chain of the 'branched' skeleton and reattach it to any C-atom except an end carbon:

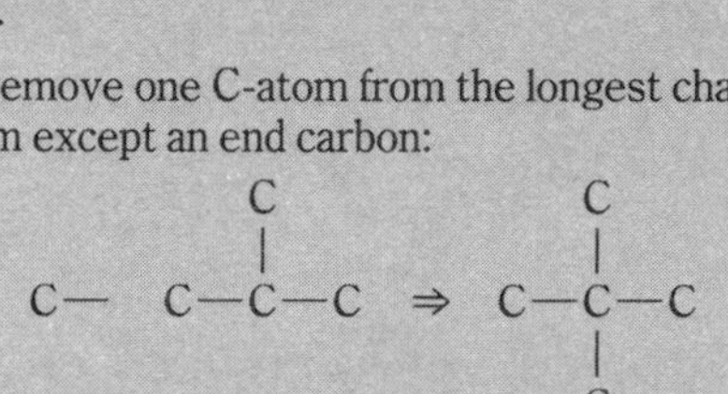

(3)

You have now found all the possible carbon skeletons because if you remove one C-atom from skeleton (3) it can be reattached to an end carbon only – this would produce skeleton (2) again! (Sometimes you can attach it to a carbon at the end of a branch but only if this does not produce the longer chain again!)

Finally, you must write down the complete formulae of the three isomers:

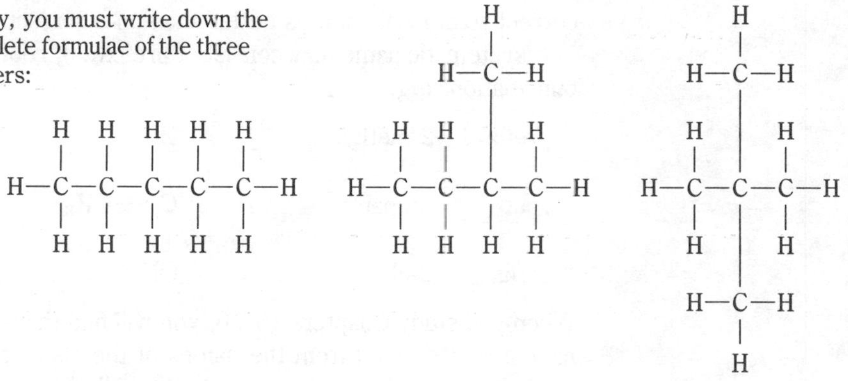

CH$_3$(CH$_2$)$_3$CH$_3$ CH$_3$CH$_2$CH(CH$_3$)$_2$ C(CH$_3$)$_4$

pentane   2-methylbutane   2,2-dimethylpropane

Only omit the hydrogen atoms if you are running out of time in an examination!

Fig. 14.9 Finding isomers: Case 1

What are the isomers with a molecular formula of C$_4$H$_{10}$O?

You start this problem in the same way as in case 1 by finding the possible carbon skeletons:

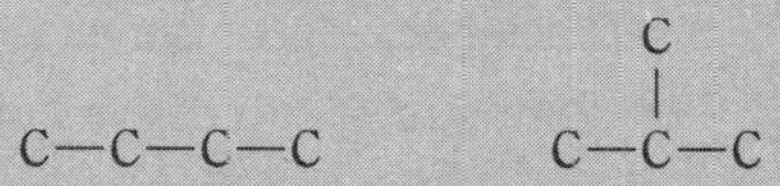

Now you consider what functional groups with an O-atom you can write using the given molecular formula:

C$_4$H$_9$OH $\Rightarrow$ C$_4$H$_9$—OH (hydroxyl) and C$_4$H$_{10}$—O— (ether)

The next step is systematically to fit the functional groups onto or into the possible carbon skeletons:

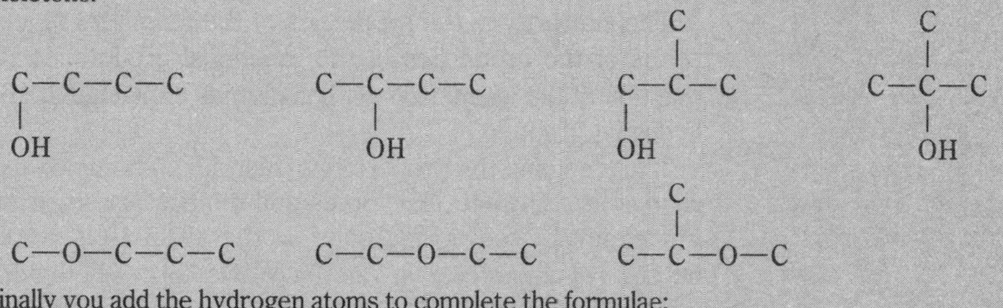

Finally you add the hydrogen atoms to complete the formulae:

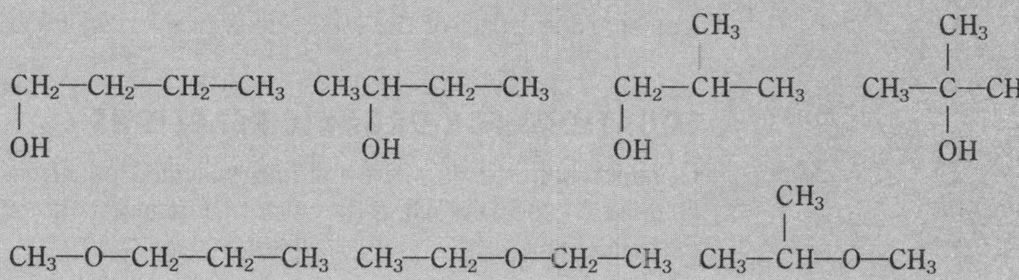

Fig. 14.10 Finding isomers: Case 2

NOMENCLATURE

Make sure you understand the rules for naming organic compounds.

In chemistry we give substances names (Latin: nomen – name; calare – call) that we can use as labels in place of long, detailed descriptions of the uniquely identifying properties of each substance. We put these labels on bottles, print them in catalogues and data books, and use them all the time when we talk and write to each other. Structure is so important in organic chemistry that we try to find names to tell us about the structure of the substance.

The International Union of Pure and Applied Chemists (IUPAC) has published 'Definitive Rules for the Nomenclature of Organic Chemistry'. You will find extracts (about 50 pages!) of the IUPAC rules in the 'rubber book'; the Handbook of Chemistry and Physics published by the Chemical Rubber Co. in the U.S.A. We try to follow these rules when naming organic compounds.

The Association for Science Education (ASE) has published a very much shorter set of rules (based on the IUPAC publications) for use in schools. Most examination boards have adopted these ASE rules for chemical nomenclature, symbols and terminology. The ASE publication includes a list of both recommended systematic and traditional names. Examiners will prefer that you use the recommended names for organic compounds. But a correct traditional name is better than an incorrect recommended name!

A systematic name may consist of prefix(es), root and suffix together with numbers and punctuation: e.g.

prefix	2-methyl	CH_3	
root	propan-	$CH_3 - C - CH_3$	2-methylpropan-2-ol
suffix	2-ol	OH	

When you study Chapters 16–20, you will find that the names of many organic compounds have a root derived from the names of the alkanes, that some names have no root e.g. methylamine and that there are both similarities and differences in the nomenclature of aromatic compounds.

HOMOLOGOUS SERIES

Compounds are divided into classes e.g. hydrocarbons, alcohols, carboxylic acids, on the basis of their composition and properties. Each class is subdivided into various homologous series (e.g. alkanes, primary monohydric alkanols, straight-chain fatty acids) on the basis of their composition and properties.

■ Each homologous series is distinguished by its general formula

■ Formulae of successive homologues differ by only one CH_2 group

■ Members of an homologous series always have the same functional group(s)

Note that two compounds with the same functional group need not be members of the same homologous series. For example, ethanol and cyclohexanol are not homologous. The homologous series of organic compounds important for AS- & A-level chemistry are dealt with in the following chapters.

The similarity in the properties of the members of an homologous series is explained in terms of the properties of the functional group(s) in their molecules. For example, the reaction of the alkan-1-ols with sodium is explained as the reduction of the hydrogen in the hydroxyl group.

The trends in the properties of a series of homologues may be explained in terms of the change in intermolecular forces and molecular complexity with increasing molecular size. For example, the boiling point of the alkan-1-ols increases with increasing molar mass because of an increase in van der Waals' forces between molecules. Their rate of reaction with sodium decreases with increasing molar mass because the hydroxyl group becomes a less significant part of the molecule with increasing carbon chain length.

EQUATIONS FOR ORGANIC REACTIONS

Organic chemists do write full and balanced equations, like those written by inorganic chemists, especially when they want to estimate the amounts of chemicals needed for an experiment:

$3C_2H_5OH + 2Na_2Cr_2O_7 + 8H_2SO_4 \rightarrow 3CH_3CO_2H + 2Na_2SO_4 + 2Cr_2(SO_4)_3 + 11H_2O$
3 mol   2 mol   8 mol   → 3 mol (if yield is 100%)

However, an 'equation' in organic chemistry is usually written to emphasise the structure(s) of the principal organic reactant(s) and product(s):

$$H-\underset{\displaystyle H}{\overset{\displaystyle H}{\underset{|}{\overset{|}{C}}}}-\underset{\displaystyle H}{\overset{\displaystyle H}{\underset{|}{\overset{|}{C}}}}-O-H \xrightarrow[\text{reflux}]{Cr_2O_7^{2-}(aq)/H^+(aq)} H-\underset{\displaystyle H}{\overset{\displaystyle H}{\underset{|}{\overset{|}{C}}}}-C \overset{\displaystyle O}{\underset{\displaystyle O-H}{}}$$

ethanol   ethanoic acid

You can see at a glance from such 'equations' that a reaction involves a change in the arrangement of atoms and groups in a molecule. One or more bonds must break and new bonds must form to produce a new molecule with a different structure.

REACTION MECHANISMS

" Mechanisms are often tested. "

Chemists often think of a chemical reaction as occurring in individual steps with each step involving two (one or three but rarely more) molecules or ions. We call this sequence of reaction steps (that we think may occur at the molecular level) a *reaction mechanism*. Kinetic and spectroscopic experiments often provide vital clues in our quest to understand and control organic reactions. You will find specific details of mechanisms proposed for certain reactions in the following chapters and in Chapter 9. What follows here in this unit is an overview of the general features of a reaction.

Carbon is tetracovalent and its atoms can link together to form chains, rings or networks. The bond energies are high so the links are strong. A carbon atom cannot extend its outer shell beyond eight electrons so one bond must break to allow a new one to form. The energy required leads to a high activation energy and makes the endothermic carbon-carbon bond breaking a slow process. The bonds that carbon forms with hydrogen, oxygen and other elements such as nitrogen, fluorine, chlorine and bromine are also energetically strong and kinetically stable.

TYPES OF REACTION

HOMOLYTIC AND HETEROLYTIC FISSION

Although hydrogen is slightly less electronegative than carbon, C—H bonds are treated like C—C bonds as being non-polar. Consequently in the reactions of alkanes, for example, a single covalent carbon-carbon bond is usually considered to break 'homolytically' (into two equal parts):

$$ -\overset{|}{\underset{|}{C}}-\overset{|}{\underset{|}{C}}- \quad \rightarrow \quad -\overset{|}{\underset{|}{C}}\cdot \quad \cdot\overset{|}{\underset{|}{C}}- $$

each carbon atom keeps one of the electrons from the bonding pair

An important feature of homolysis is that molecular fragments with unpaired electrons are formed. These fragments, often called free radicals, are very reactive and may take part in chain reactions: see Chapter 15.

■ A radical is a fragment with an unpaired electron

Chlorine on the other hand is more electronegative than carbon and the C—Cl bond is polar. Consequently, in the reactions of chloroalkanes for example, the carbon-chlorine bond is usually considered to break 'heterolytically' – (into two unequal parts):

$$ -\overset{|}{\underset{|}{C}}-Cl \quad \rightarrow \quad -\overset{|}{\underset{|}{C}}{}^+ \quad :Cl^- $$

chlorine atom departs with both electrons to form a stable anion

An important feature of heterolysis is that an atom or group departing with the bonding electron pair is leaving behind a carbon atom open to attack or being attacked by another atom with a lone pair. This attacking atom is part of an ion (e.g. OH^-, CN^-) or a molecule (e.g. H_2O, NH_3) and donates its lone pair to form a (coordinate or dative) covalent bond with the carbon atom being attacked.

In the above example, the attacking species donating an electron pair is called a nucleophile whilst the species being attacked and accepting an electron pair is called an electrophile. We describe this reaction step as a nucleophilic attack upon an electrophilic centre.

■ A nucleophile is an electron-pair donor (and a Lewis base)
■ An electrophile is an electron-pair acceptor (and a Lewis acid)

The proposed movement of electron-pairs in a step of a suggested reaction mechanism may be shown using 'curly arrows':

$$ H-O:^- \quad R-\overset{\overset{H}{|}}{\underset{\underset{H}{|}}{C}}-Cl \quad \rightarrow \quad H-O-\overset{\overset{H}{|}}{\underset{\underset{H}{|}}{C}}-R \quad + \quad :Cl^- $$

You must use these arrows correctly. A 'curly arrow' denotes the movement of an electron-pair. Do NOT use it to show the movement of atoms, ions or molecules! You must draw each 'curly arrow' very accurately. The 'tail' of the arrow must show exactly where the electron-pair is coming from. The 'head' of the arrow must show exactly where the electron-pair is going to.

The reactions of at least six million organic compounds seems mainly to depend upon two things: the movement of electron-pairs and the polar nature of bonds in the molecules. Consequently, only four fundamental types of reaction are needed to account for the great variety of chemical properties of all the various classes of organic compounds.

- Addition — a reaction in which an atom or group is added to a molecule
- Elimination — a reaction in which an atom or group is removed from a molecule
- Rearrangement — a reaction in which atoms or groups in a molecule change position
- Substitution — a reaction in which one atom or group in a molecule is replaced by another atom or group

A nucleophilic substitution is a reaction in which the arriving (and departing) atom (or group) is an electron-pair donor. When the attacking (and leaving) atom (or group) is an electron-pair acceptor, we describe the reaction as an electrophilic substitution: see Fig. 14.11.

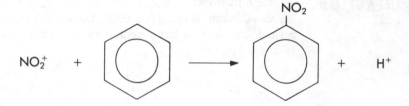

Fig. 14.11 Electrophilic substitution

- Condensation is NOT a fifth fundamental type of reaction since it may be regarded as an *addition* followed by an *elimination*

CLASSIFYING ORGANIC REACTIONS

The search for patterns in the behaviour of chemicals has led chemists to identify and name various classes of organic reactions: e.g.

acylation	esterification	hydrogenation	nitration
dehydration	halogenation	hydrolysis	polymerisation
diazotisation	hydration	isomerisation	sulphonation

You will probably meet most if not all of these classes of reaction in your organic chemistry course. In principle every class could be interpreted in terms of the four fundamental reaction types. You are not likely to have to do this but your syllabus may well require you to study reaction mechanisms for specific examples of one or two classes of reaction, e.g., chlorination of alkanes by a (free) radical chain mechanism and the hydrolysis of halogeno-alkanes by nucleophilic substitution. You will find more details and help in the following chapters of this book.

EXAMINATION QUESTION AND OUTLINE ANSWER

Q1 Isomerization in carbon compounds may be described as structural isomerism, cis-trans isomerism or chiral (optical) isomerism. Each of the following substances can exist in isomeric forms.

I 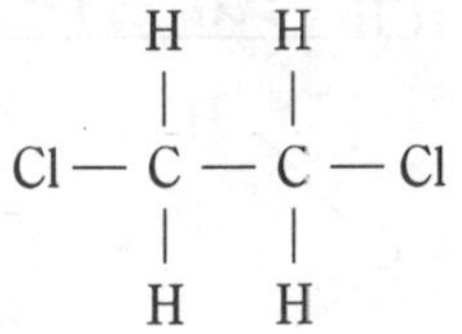

II 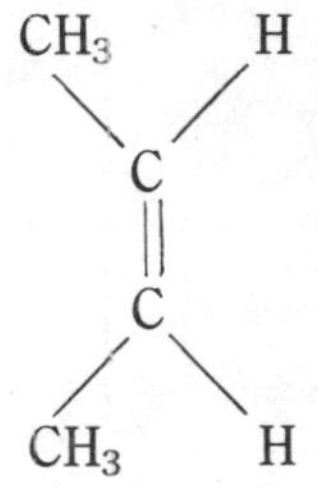

III 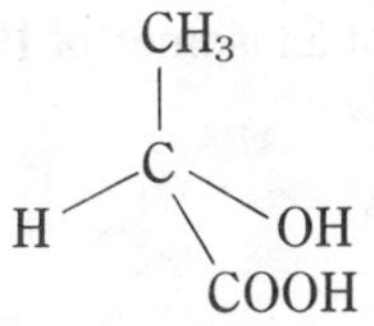

IV 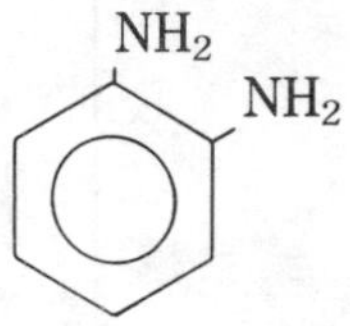

V CH_3COCH_3

Each of the substances I–V illustrates a particular type of isomerism. For each of the substances I–V:

a) draw the structure of ONE other isomer,
b) explain the type of isomerism taking place,
c) suggest how it might be possible to distinguish between the isomers using physical or chemical techniques where appropriate. *(28)*

(NISEC 1988)

Outline answer

Q1 (a) I 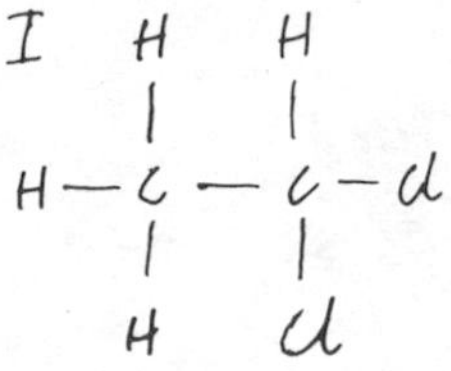

II 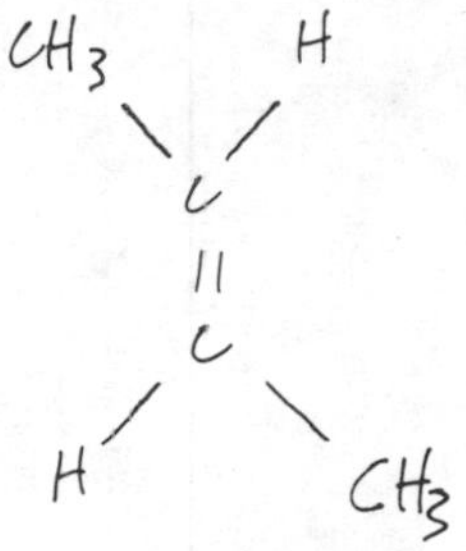

III 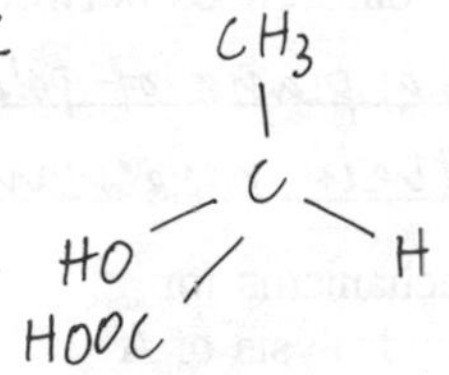

IV 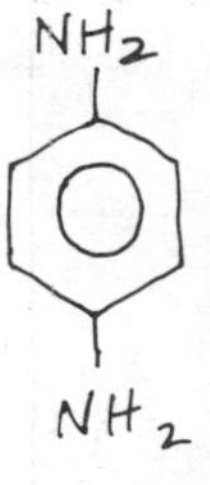

V $CH_3CH_2CO_2H$

(b) I structural II geometric
III optical IV nuclear
V functional group

(c) refer to your text book

EXAMINATION QUESTION AND TUTOR'S ANSWER

Q2 Give systematic names for each of the following compounds.

A $(CH_3)_3CBr$ _2 - Bromo - 2 - Methylpropane_

B $CH_3CH_2CH=CH$ _but - 1 - ene_

C

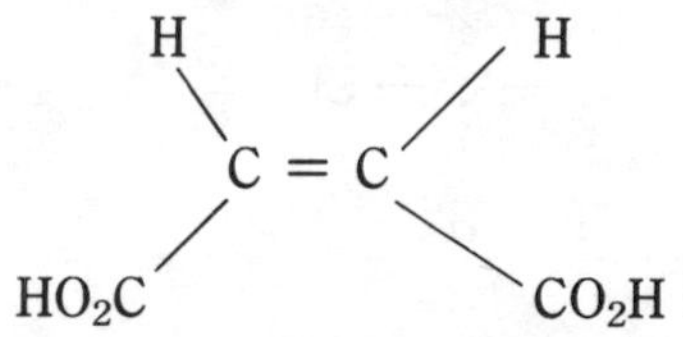

$$H \qquad\qquad H$$
$$\diagdown \qquad\qquad \diagup$$
$$C = C$$
$$\diagup \qquad\qquad \diagdown$$
$$HO_2C \qquad\qquad CO_2H$$ _cis -but -2- enedioic acid_

D $CH_3CH_2CH(CH_3)CH_2OH$ _2 - methylbutan - 1 - ol_ (5)

a) Give the structure of an isomer of **B** which exhibits geometric isomerism.

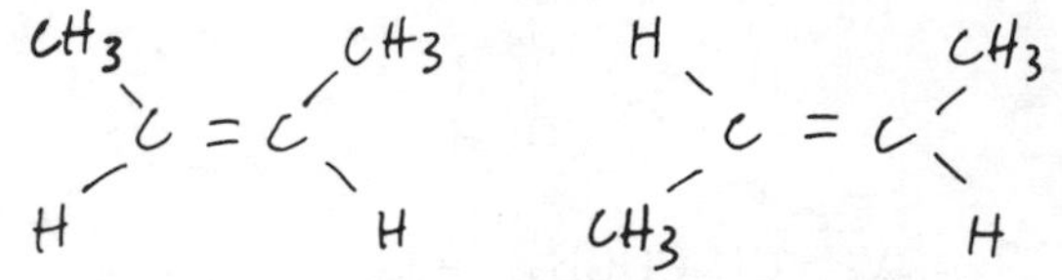

$$CH_3 \qquad CH_3 \qquad\qquad H \qquad CH_3$$
$$\diagdown C = C \diagup \qquad\qquad \diagdown C = C \diagup$$
$$\diagup \qquad \diagdown \qquad\qquad \diagup \qquad \diagdown$$
$$H \qquad\qquad H \qquad\qquad CH_3 \qquad\qquad H$$

How do you account for the separate existence of the geometric isomers?
rotation does not occur about a double bond
_____ (2)

b) On heating, **C** loses water. Suggest a structure for the resulting organic compound.

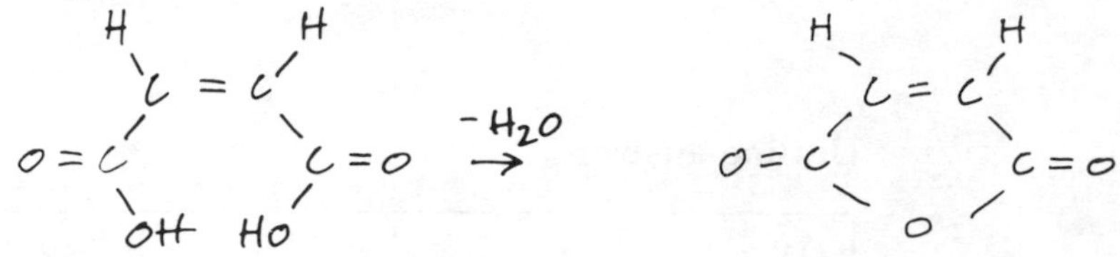

$$H \qquad\qquad H \qquad\qquad\qquad\qquad\qquad H \qquad\qquad H$$
$$\diagdown C = C \diagup \qquad\qquad\qquad\qquad\qquad \diagdown C = C \diagup$$
$$\diagup \qquad\qquad \diagdown \qquad\qquad\qquad\qquad \diagup \qquad\qquad \diagdown$$
$$O = C \qquad\qquad C = O \quad \xrightarrow{-H_2O} \quad O = C \qquad\qquad C = O$$
$$\diagdown \qquad\qquad \diagup \qquad\qquad\qquad\qquad\qquad \diagdown \qquad \diagup$$
$$OH \quad HO \qquad\qquad\qquad\qquad\qquad\qquad\qquad O$$

an internal
anhydride (2)

c) Which compound contains an asymmetric carbon atom?
_D_
_____

Give *two* differences between the resulting optical isomers.
they rotate the plane of polarized light in opposite
directions; their molecular structures are mirror images. (3)

d) Give mechanisms for
   i) the hydrolysis of **A**

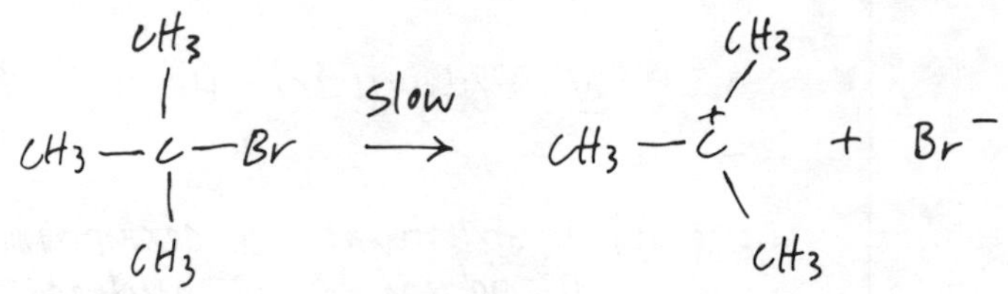

$$\qquad\qquad CH_3 \qquad\qquad\qquad\qquad\qquad CH_3$$
$$\qquad\qquad | \qquad\qquad\qquad slow \qquad\qquad \diagup$$
$$CH_3 - C - Br \quad \xrightarrow{\quad\quad} \quad CH_3 - C \quad + \quad Br^-$$
$$\qquad\qquad | \qquad\qquad\qquad\qquad\qquad\qquad \diagdown$$
$$\qquad\qquad CH_3 \qquad\qquad\qquad\qquad\qquad CH_3$$

$$(CH_3)_3 C^+ + :OH^- \xrightarrow{fast} (CH_3)_3 COH$$

ii) the addition of hydrogen bromide to **B**.

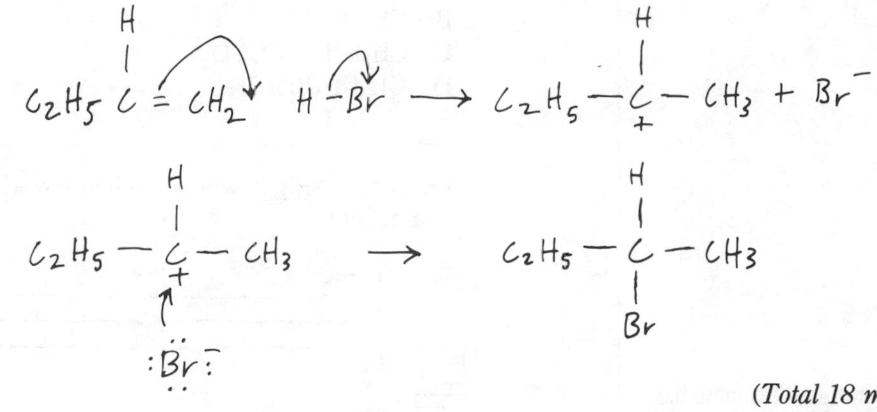

(6)
(*Total 18 marks*)
(ULSEB 1988)

STUDENT'S ANSWER WITH EXAMINER'S COMMENTS

Q3 a) The peak of highest mass/charge (*m/e*) ratio in the mass spectrum of a hydrocarbon has *m/e* of value 54.
(Take relative atomic masses: C = 12, H = 1.)

   i) Write down the molecular formula and the empirical formula of the hydrocarbon.   (2)

   molecular formula C_4H_6

   empirical formula C_2H_3

   ii) Draw **two** possible isomeric structural formulae **A** and **B** for the hydrocarbon.   (2)

A $CH_3 - C \equiv C - CH_3$ ✓ **B** $CH_3 CH_2 - C \equiv C - H$ ✓

> **These are alkynes.** Butadiene is also possible: $CH_2 = CH - CH = CH_2$

   b) Describe a test-tube experiment you could use to distinguish **X** and **Y**, giving the results for each compound.   (3)

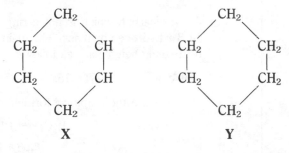

X   **Y**

> **Describe what you see!** X rapidly turns the orange-brown $Br_2(aq)$ colourless. Y only works slowly in sunlight.

shake each with aqueous bromine. X will give a positive result.

> **This is correct but not asked for.**

(ODLE 1988)

Q4 a) i) State which molecule(s) in the following list show(s) optical activity:

    A HOOCCH=CHCOOH

    B CFClBrH

    C $CH_3CH(OH)COOH$

    D $CH_3CH(Br)CH_3$

                  B and C ✓

ii) Describe briefly how a polarimeter works and how it is used to detect optical activity.

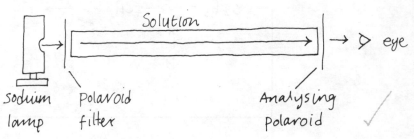

    sodium   Polaroid   Analysing

    lamp   filter   Polaroid ✓

Light is passed through a polarizing filter and the plane polarized light viewed at the other end.

b) i) Write down the systematic name of the following molecule:

$$CH_3$$
$$|$$
$$CH_3CH_2CH\,CH\,CH_2CH_2CH_3$$
$$|$$
$$CH_2CH_3$$

    3 – ethyl – 4 – methylheptane

ii) Draw a full structural formula for *cis*-but-2-ene.

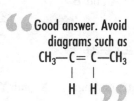

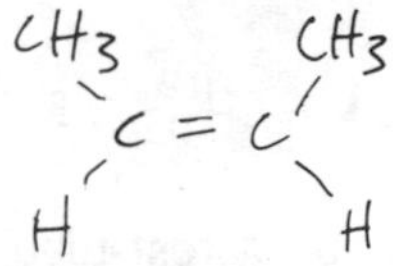

c) A hydrocarbon which is not a ring compound contains 85.7% carbon and 14.3% hydrogen by mass. The main peaks in its mass spectrum, in order of decreasing height, are as follows:

m/e 56; 29; 27; 15.

i) Calculate the empirical formula of the compound.

    $[A_r(H) = 1;\quad A_r(C) = 12.]$

% ÷ Aᵣ			Ratio of moles
C 85·7	12 =	7·14 →	1
H 14·3	1 =	14·3	2

    ⇒ empirical formula CH_2 ✓

ii) Calculate the molecular formula of the compound.

    $Mr (CH_2) = 14$ $Mr (compound) = 56$ ✓

    ⇒ molecular formula is C_4H_8 ✓

iii) Deduce the structure of the compound, giving your reasons, and
draw the structure below.

$m/e = 15$ corresponds to CH_3

27 could correspond to $CH\ CH_2$ i.e. $CH=CH_2$

29 " " to C_2H_5

$\Rightarrow$ structure could be $CH_3\ CH_2\ CH=CH_2$

66 **Good answer.** 99

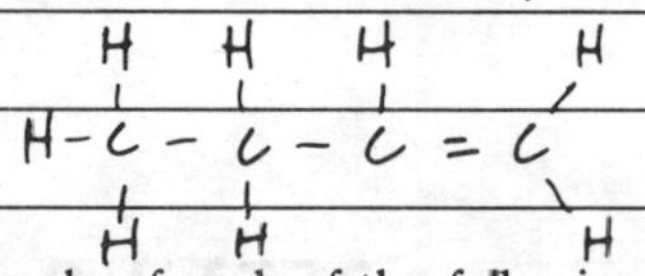

or $H-\overset{\overset{\displaystyle H}{|}}{\underset{\underset{\displaystyle H}{|}}{C}}-\overset{\overset{\displaystyle H}{|}}{\underset{\underset{\displaystyle H}{|}}{C}}-\overset{\overset{\displaystyle H}{|}}{C}=\overset{\overset{\displaystyle H}{|}}{\underset{\underset{\displaystyle H}{|}}{C}}$ but$-1-$ene

d) Give **one** example of each of the following:

  i) a free radical; _____ $Cl\cdot$ ✓

  ii) an electrophile; _____ NO_2^+ ✓

  iii) a nucleophile; _____ $\ddot{O}H^-$ ✓

  iv) heterolytic bond fission; _____ $\overset{.}{\underset{.}{C}}-Br \rightarrow \overset{.}{\underset{.}{C}}^+\quad :\overset{-}{\underset{..}{Br}}:$ ✓

  v) an elimination reaction. _____ $CH_3\ CH_2\ OH \rightarrow CH_2=CH_2 + H_2O$

(WJEC 1988)

CHAPTER

HYDRO-CARBONS

HYDROCARBONS AS FUELS

ALKANES

SUBSTITUTION REACTIONS OF METHANE

ALKENES

POLYMERISATION OF ALKENES

ARENES

HYDROGENATION OF BENZENE

GETTING STARTED

Hydrocarbons are compounds of carbon and hydrogen only. These compounds may be saturated or unsaturated and in the form of chains or rings. Hydrocarbons are mainly derived from coal, natural gas, petroleum and plants. Alkanes are saturated chain compounds and cycloalkanes are saturated ring compounds. Alkenes and alkynes are unsaturated chain compounds and cycloalkenes and cycloalkynes are unsaturated ring compounds. The aromatic hydrocarbons (arenes) form a major class of ring compounds consisting of the benzene, naphthalene and anthracene groups.

ESSENTIAL PRINCIPLES

HYDROCARBONS AS FUELS

All compounds of carbon and hydrogen will burn. The reaction is always exothermic and often explosive. The hydrogen will always burn to form water but carbon monoxide and carbon may be produced if combustion is incomplete because the supply of air is inadequate. Complete combustion, for example, in oxygen under pressure in a bomb calorimeter produces carbon dioxide and water:

$$C_xH_y + (x + y/4)O_2 \rightarrow xCO_2 + y/2H_2O$$

An explosive combustion is an extremely fast reaction involving a sudden large increase in the volume of gas. We may explain the high speed of the oxidation by a chain reaction mechanism involving oxygen and the hydrocarbon. We may explain the increase in gas volume by (a) the heat of combustion causing the carbon dioxide and water vapour to expand and (b) the formation of more gaseous products than gaseous reactants when $y > 4$ (because then the amount of gaseous products $[(x + y/2)]$ − the amount of gaseous reactants $[1 + (x + y/4)]$ will be positive $[y/4 − 1]$). We may explain the release of heat upon combustion as follows:

- When a bond breaks energy is taken in.
  Energy (ΔH_{bb}) is taken in when the carbon-carbon and carbon-hydrogen bonds in the hydrocarbon molecules and the oxygen-oxygen bonds in the oxygen molecules are broken. [$_{bb}$ means 'bond breaking']

- When a bond forms energy is given out.
  Energy (ΔH_{bf}) if given out when the carbon-oxygen bonds form to produce the carbon dioxide molecules and the hydrogen-oxygen bonds form to produce the water molecules. [$_{bf}$ means 'bond forming']

- In an exothermic reaction the heat given out by the formation of the bonds in the products is greater than the heat taken in when the bonds in the reactants are broken. ΔH ($= \Delta H_{bb} − \Delta H_{bf}$) is negative because $\Delta H_{bb} < \Delta H_{bf}$. The approximately constant difference (around 600 to 700 kJ mol^{-1}) in the heats of combustion of successive members of any homologous series represents the burning of CH_2 to form CO_2 and H_2O, or the net release of heat on breaking the CH and OO bonds and forming the CO and HO bonds.

In general, as x in the formula C_xH_y increases, there is an increase in the hydrocarbon's viscosity, melting point and boiling point. As the hydrocarbons become less volatile they become more difficult to ignite and less likely to burn completely in air.

Unsaturated hydrocarbons usually burn with a yellow smoky flame. The luminous flame you see when burning saturated hydrocarbons of high molar mass may be the burning of unsaturated hydrocarbons formed by the thermal cracking of the large molecules in the flame. In the oil refineries catalytic cracking of hydrocarbons is an essential process in the production of alkenes – the building bricks of the petrochemicals industry.

The saturated aliphatic hydrocarbons (alkanes) are generally less reactive than the unsaturated aliphatics and undergo predominantly *substitution* reactions. The unsaturated aliphatic hydrocarbons (alkenes) undergo a wider range of reactions that are predominantly *addition* reactions. Although they too are 'unsaturated', the aromatic hydrocarbons (arenes) undergo an even wider variety of reactions that are predominantly *substitution* reactions.

PHYSICAL PROPERTIES

Observation: Hydrocarbons are insoluble in water

Explanation: hydrocarbon   water
                      van der Waals' forces   van der Waals' forces
                                               dipole-dipole attractions
                                             hydrogen bonding
        Far more energy would have to be taken in when breaking the extra attractive forces between the water molecules than could be given out when forming only van der Waals' forces between the water and hydrocarbon molecules – hence oil and water do not mix!

Observation: Hydrocarbons float on water

Explanation:
structural unit	-CH_2-	H_2O
relative atomic mass	Carbon<Oxygen	
covalent radius	Carbon>Oxygen	

⇒ expect density of hydrocarbon<density of water

Observation: Boiling temperature increases with increasing molar mass: see Fig. 15.1

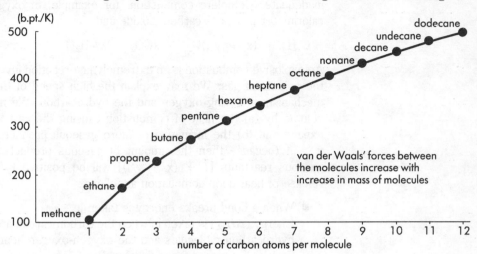

Fig. 15.1 Boiling temperatures of straight-chain alkanes

Explanation: van der Waals' forces of attraction increases as the number of atoms in the molecules increases

Observation: Boiling temperature of isomeric hydrocarbons generally decreases with increasing complexity of chain-branching: see Fig. 15.2

Explanation: van der Waals' forces are short range so some carbon atoms in one branched molecule may be too far away from similar carbon atoms in another molecule for them to attract one another.

hexane	68
3-methylpentane	63
2-methylpentane	60
2,3-dimethylbutane	58
2,2-dimethylbutane	50

Fig. 15.2 Boiling temperatures (°C) of isomers of C_6H_{14}

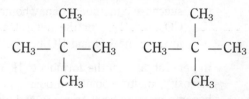

centre carbon atoms are too far apart
for van der Waals' attractive forces

ALKANES

The alkanes are a homologous series of saturated hydrocarbons with the general formula $C_nH_{(2n+2)}$.

NOMENCLATURE

❝ Learn the names of the first four alkanes by heart. ❞

Chemists still call the first four straight-chain homologues by their traditional names, make sure that you learn them: see Fig. 15.3.

We call the other members of the series: alkanes with five or more carbon atoms in a chain; names based on Greek or Latin words for the numbers. You probably already know some of these words: e.g. the Pentagon in Washington is a building with five corners (Greek; pente – five; gonia – angle) and the Decathlon is a two-day sporting competition with 5 events each day (Greek; deka – ten; athlon – prize in a contest).

You should know how to apply the following IUPAC (International Union of Pure and Applied Chemistry) rules for naming branched-chain isomers:

■ The root name is based on the longest continuous carbon chain and the name of the straight-chain alkane with the same number of carbons

■ The root is prefixed with names based on the shorter carbon branches and the names of the corresponding straight-chain alkanes

NUMBER OF CARBONS	MOLECULAR FORMULA	NAME	NUMBER OF CARBONS	MOLECULAR FORMULA	NAME
1	CH_4	methane	11	$C_{11}H_{24}$	undecane
2	C_2H_6	ethane	12	$C_{12}H_{26}$	dodecane
3	C_3H_8	propane	13	$C_{13}H_{28}$	tridecane
4	C_4H_{10}	butane	14	$C_{14}H_{30}$	tetradecane
5	C_5H_{12}	pentane	15	$C_{15}H_{32}$	pentadecane
6	C_6H_{14}	hexane	18	$C_{18}H_{38}$	octadecane
7	C_7H_{16}	heptane	20	$C_{20}H_{42}$	eicosane
8	C_8H_{18}	octane	21	$C_{21}H_{44}$	heneicosane
9	C_9H_{20}	nonane	30	$C_{30}H_{62}$	triacontane
10	$C_{10}H_{22}$	decane	100	$C_{100}H_{202}$	hectane

Fig. 15.3 Names of some straight-chain alkanes

- The number of identical branches: two, three, four, etc. is indicated by the appropriate adjunct: di-, tri-, tetra-, etc.

- The positions of the branches on the longest chain are numbered from the end giving the lower number for the initial branching point

- The prefixes are attached in alphabetical order of branch name and ignoring adjuncts such as di- and tri-: see Fig. 15.5

Note that -ane is the ending (suffix). We can name a whole new homologous series of organic compounds simply by replacing one letter in this ending: e.g. alkanes → alkenes → alkynes → alkanols → alkanals → alkanones → alkanoic acids. So pay very careful attention to the names of organic compounds. You cannot afford to make a single spelling mistake.

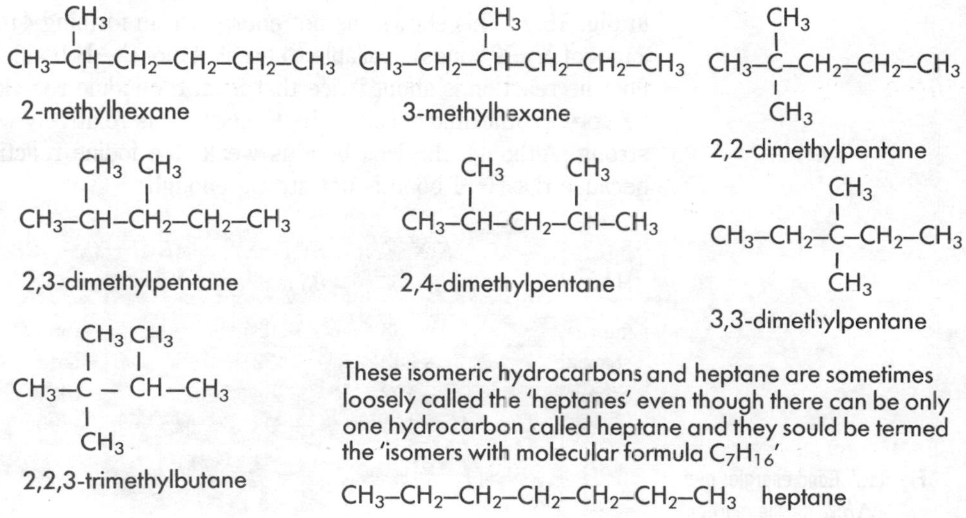

Fig. 15.4 Naming isomers of heptane

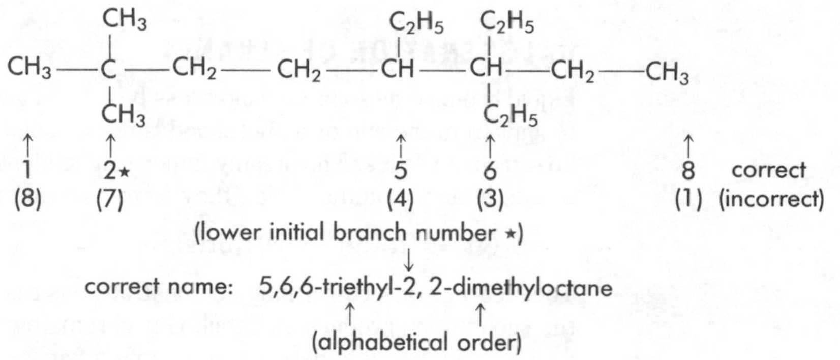

Fig. 15.5 Naming a hydrocarbon

COMBUSTION OF ALKANES

If there is plenty of air and not many carbon atoms in the alkane molecule, combustion will be complete:

$$C_nH_{(2n+2)} + \tfrac{1}{2}(3n+1)O_2 \rightarrow nCO_2 + (n+1)H_2O$$

NUMBER OF CARBON ATOMS IN MOLECULE	TYPES OF FUEL
1 – 2	natural gas
3 – 4	liquefied petroleum gas (LPG) – calor, camping gaz
5 – 10	petrol
11 – 15	aviation fuel, domestic heating oil
16 – 25	DERV (Diesel Engine Road Vehicle), industrial heating

Fig. 15.6 Alkanes as fuels

If there is insufficient air and many carbon atoms in the alkane molecule, some of the molecules will undergo incomplete combustion:

$$C_nH_{(2n+2)} + \tfrac{1}{2}(2n+1)O_2 \rightarrow nCO + (n+1)H_2O$$

Carbon monoxide (CO) is extremely poisonous. Alkanes are major constituents of our fuels: see Fig. 15.6. The formation of carbon monoxide by faulty or inefficient gas burners and combustion engines is another more serious and unwanted contribution to environmental pollution.

Halogen could replace oxygen as the oxidant in combustion. Fluorine usually reacts spontaneously and explosively but iodine is a very poor oxidant and rarely reacts directly with alkanes. Many alkane and chlorine (or bromine) mixtures (like most alkane and oxygen mixtures) will burn or explode. For example, a mixture of methane and chlorine burns when ignited by a flame and explodes when set off by light from a photographic flash. Hydrogen chloride, carbon and various chloromethanes could be formed according to the conditions of the reaction:

$$CH_4(g) + 2Cl_2(g) \rightarrow C(s) + 4HCl(g); \quad \Delta H^\ominus = -454 \, kJ \, mol^{-1}$$

The influence of the bond energies on the character of this reaction is illustrated in Fig. 15.7. This shows the net energy (from forming 4 mol of H—X bonds and breaking 2 mol of X—X bonds) available to break 4 mol C—H bonds. The energy available from the fluorine reaction is about twice that from the iodine reaction. The fluorine reaction would be very exothermic because the F—F bond is relatively weak and the H—F bond is very strong. Although the I—I bond is weak, the iodine reaction would be very endothermic because the H—I bond is not strong enough.

HALOGEN	$E(X-X)$ /kJ mol^{-1}	$E(H-X)$ /kJ mol^{-1}	$-4E(H-X) + 2E(X-X)$ /kJ mol^{-1}	$\Delta H^\ominus_{f,HX}$ /kJ mol^{-1}	$\Delta H^\ominus_{calc}$ /kJ mol^{-1}
fluorine	+158	+568	−1956	−271	−1009
chlorine	+243	+432	−1242	−92	−293
bromine	+192	+366	−1080	−36	−69
iodine	+151	+298	−890	+27	+183

$\Delta H^\ominus_{f,methane}$ taken as $-75 \, kJ \, mol^{-1}$

Fig. 15.7 Bond energies and $\Delta H^\ominus_{calc}$ for the reaction

HALOGENATION OF ALKANES

Liquid bromine dissolves in colourless hexane to give a red-brown solution. When exposed to light from the sun or a photoflood lamp the solution gives off steamy fumes and its red-brown colour fades. The steamy fumes can be identified as hydrogen bromide by the white smoke of ammonium bromide they form with ammonia gas:

$$NH_3(g) + HBr(g) \rightarrow NH_4Br(s)$$

Your teacher or lecturer may demonstrate this reaction but not expect you to know about the substitution products in detail. Gas-chromatographic analysis of the residual colourless liquid would probably detect and identify a variety of bromohexanes.

SUBSTITUTION REACTIONS OF METHANE

The gas phase reaction of bromine with methane is the simplest illustration of an organic substitution reaction involving a free radical chain mechanism. The proposed steps are as follows:

- Initiation $Br—Br \xrightarrow{\text{light}} 2Br\cdot$

  two free radicals formed by homolytic fission of the covalent bond when the bromine molecule absorbs light

- Propagation $Br\cdot + H—CH_3 \rightarrow H—Br + \cdot CH_3$

  bromine radical attacks methane molecule to produce a methyl radical and a hydrogen bromide molecule

  $Br—Br + \cdot CH_3 \rightarrow Br—CH_3 + \cdot Br$

  methyl radical attacks a bromine molecule to produce a bromine radical and a bromomethane molecule

  the net effect of the two propagation steps is:

  $Br_2 + CH_4 \rightarrow HBr + CH_3Br$

- Termination $\cdot CH_3 + \cdot CH_3 \rightarrow CH_3—CH_3$

  two methyl radicals form an ethane molecule

Detecting ethane in the mixture of bromomethanes provides supporting evidence for this proposed reaction mechanism.

The two propagation steps form a radical chain in which each step in the chain provides the radical for the next step in the chain. However, the bromination of methane actually produces four different substitution products in varying amounts depending upon the conditions:

$$
\begin{array}{cccc}
\text{Br} & \text{Br} & \text{Br} & \text{Br} \\
| & | & | & | \\
\text{H}-\text{C}-\text{H} & \text{H}-\text{C}-\text{Br} & \text{H}-\text{C}-\text{Br} & \text{Br}-\text{C}-\text{Br} \\
| & | & | & | \\
\text{H} & \text{H} & \text{Br} & \text{Br}
\end{array}
$$

We explain this by saying that the bromomethane molecule participates in the propagation steps:

$Br\cdot + H—CH_2Br \rightarrow H—Br + \cdot CH_2Br$

$Br—Br + \cdot CH_2Br \rightarrow CH_2Br_2 + H—Br$

In the same way the dibromomethane would take part to form tribromomethane which would participate in turn to produce tetrabromomethane.

- The ability of products (of a reaction) to become reactants (in the same or another reaction) often complicates our study of chemical changes

The chain mechanism is terminated by the removal of radicals. So the reverse of the initiation step may be regarded as a termination step:

$Br—Br \underset{\text{termination}}{\overset{\text{initiation}}{\rightleftharpoons}} Br\cdot + Br\cdot$

$Br\cdot + \cdot CH_3 \rightarrow CH_3Br$ is another possible termination step.

The substitution of hydrogen by chlorine in methane and in the methyl group (but not the benzene ring) of methylbenzene (toluene) is also regarded as an homolytic (free) radical mechanism. A mixture of products may again be formed:

| methane: | CH_4 | CH_3Cl | CH_2Cl_2 | $CHCl_3$ | CCl_4 |
| methylbenzene: | CH_3 | CH_2Cl | $CHCl_2$ | CCl_3 | |

Combustion of alkanes in air probably involves a more complicated free radical chain mechanism.

CRACKING OF ALKANES

Breaking down large hydrocarbon molecules into smaller ones is an operation of great importance to the petrochemical industry. The (catalytic and thermal) cracking processes are complicated molecular decompositions, rearrangements and recombinations involving free radical chain mechanisms. In the petroleum refinery, cracking helps to produce branched-chain alkanes and aromatic hydrocarbons for petrol with a high 'octane rating' – good 'anti-knock' characteristics. Cracking of petroleum not used for fuel produces the feedstock of alkenes required by the petrochemical industry.

ALKENES

The alkenes are a homologous series of unsaturated (acyclic) hydrocarbons with the general formula C_nH_{2n}. The double bond between two carbon atoms makes them more reactive than the alkanes.

NOMENCLATURE

The alkenes are named in a similar way to the alkanes but with the following important differences:

- The root name is based on the name of the straight-chain alkane with the same number of carbon atoms as the longest continuous carbon chain that contains the double bond
- The prefixes cis- and trans- are used to name simple geometric isomers with the same group on each of two doubly bonded carbon atoms

Fig. 15.8 illustrates the naming of some simple alkenes.

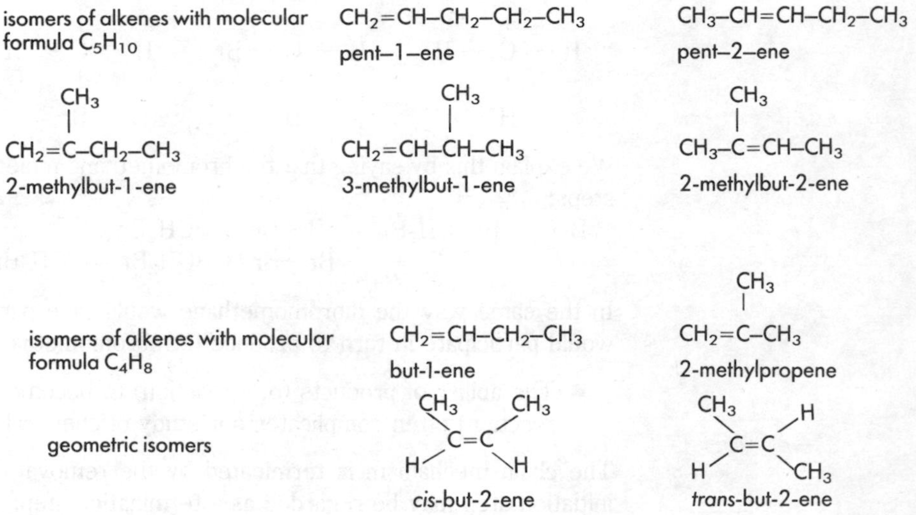

Fig. 15.8 Naming isomeric alkenes

LABORATORY PREPARATION

You can make and test impure samples of alkenes in your practical work by the dehydration of alcohols or the dehydrobromination of bromoalkanes. You can heat the alcohol with an excess of concentrated sulphuric acid or phosphoric acid or by passing its vapour over heated lumps of aluminium oxide, silicon oxide or pumice. If you use ethanol, the ethene will be contaminated with ethoxyethane:

$$CH_3{-}CH_2OH \xrightarrow[-H_2O]{\text{intramolecular dehydration}} CH_2{=}CH_2$$

$$2\,CH_3{-}CH_2OH \xrightarrow[-H_2O]{\text{intermolecular dehydration}} CH_3CH_2{-}O{-}CH_2CH_3$$

We usually find that this 'ethene' does not burn with a very smoky flame but it does rapidly decolourise aqueous bromine.

The ethene is less contaminated if you make it by warming bromoethane with potassium hydroxide in ethanol or passing its vapour over hot soda-lime:

$$CH_3{-}CH_2Br \xrightarrow{-HBr} CH_2{=}CH_2$$

INDUSTRIAL PRODUCTION

Alkenes are produced by the cracking (and dehydrogenation) of ethane and propane (from natural gas) and other alkanes are produced from light petroleum distillates in the naphtha range: e.g.

$$CH_3\!-\!CH_3 \xrightarrow{-H_2} CH_2\!=\!CH_2$$

$$C_3H_8 \longrightarrow CH_2\!=\!CH_2 + CH_4$$

OXIDATION OF ALKENES

We often test an organic substance to see how easily it ignites and how well it burns. Alkenes have nastier smells and burn with smokier flames than alkanes. The carbon-carbon double bond features in the molecular structure of insect attractants: e.g. pentacos-9-ene is one of the components in the sex attractant of the common house fly. And it probably influences the free radical chain mechanisms in the combustion reaction of the alkene with oxygen. But beware – nasty smells and smoky flames are not unique to alkenes! You will find them with saturated hydrocarbons of high molar mass, halogenoalkanes and other unsaturated compounds.

If you shake an alkene with acidified aqueous potassium manganate(VII) you will see the purple solution of MnO_4^- rapidly turn colourless (Mn^{2+}) but remain clear. If you use an alkaline solution, you will see the purple solution turn green (MnO_4^{2-}) then cloudy brown (MnO_2 ppt.):

$$>\!C\!=\!C\!< \xrightarrow{\text{alkaline manganate(VII)}} \underset{\underset{OH}{\displaystyle|}}{-C}-\underset{\underset{OH}{\displaystyle|}}{C}- \text{ (a diol)}$$

REDUCTION OF ALKENES

Alkenes react with hydrogen at the surface of a nickel catalyst. The hydrogen reduces the double bond to a single bond: see Fig. 15.9.

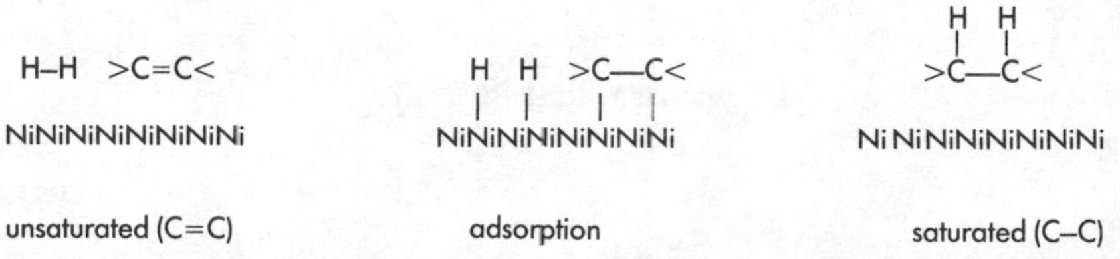

H–H >C=C<
NiNiNiNiNiNiNiNi
unsaturated (C=C)

H H >C—C<
NiNiNiNiNiNiNiNi
adsorption

H H
>C—C<
Ni NiNiNiNiNiNiNi
saturated (C–C)

Unsaturated edible oils are hydrogenated to saturated edible fats in the production of margarine: e.g.

CH₃(CH₂)₇CH=CH(CH₂)₇CO₂CH₂
CH₃(CH₂)₇CH=CH(CH₂)₇CO₂CH
CH₃(CH₂)₇CH=CH(CH₂)₇CO₂CH₂

$\xrightarrow[\text{Ni catalyst}]{3H_2(g)}$

CH₃(CH₂)₇CH₂–CH₂(CH₂)₇CO₂CH₂
CH₃(CH₂)₇CH₂–CH₂(CH₂)₇CO₂CH
CH₃(CH₂)₇CH₂–CH₂(CH₂)₇CO₂CH₂

olein (from olive oil) an ester of propane–1,2,3-triol and octadec-9-enoic (oleic) acid

stearin (in most fats) an ester of propane-1,2,3-triol and octadecanoic (stearic) acid

Fig. 15.9 Hydrogenation of alkenes

POLYMERISATION OF ALKENES

We can make the small molecules (monomers) of an alkene (unsaturated hydrocarbon) add together to form very large molecules (polymers) of an alkane (saturated hydrocarbon):

$$n >\!C\!=\!C\!< \longrightarrow \left(-\underset{|}{\overset{|}{C}}-\underset{|}{\overset{|}{C}}-\right)_n$$

This extremely important industrial process is generally performed in one of two ways:

■ High temperature and very high pressure (e.g. 200 °C and 1500 atm) chain reactions, initiated by radicals (e.g. trace of oxygen) that produce branched chain stereo-irregular (atactic) polymers with lower crystallinity, density and rigidity

Low density poly(ethene) is quite cheap to produce and is used to make 'plastic' film and bags for such things as food packaging.

■ Low temperature and low pressure (e.g. 60 °C and 2 atm) reactions in an inert solvent, using a Ziegler-Natta catalyst of titanium(IV) chloride with an organo-metallic aluminium compound (e.g. triethylaluminium – $Al(C_2H_5)_3$), that produce stereo-regular (isotactic and syndiotactic) polymers with higher crystallinity, density and rigidity

High density poly(ethene) is dearer to produce than low density poly(ethene) and is used to make rigid containers like milk crates. Some typical addition polymers are listed in Fig. 15.10. Addition polymers soften and melt when heated and become firm again when cooled. Consequently we call them thermoplastic polymers. In industry the alkenes are frequently known by their traditional names and the poly(alkenes) by their trade names: e.g.

systematic name:	ethene	poly(ethene)
traditional name:	ethylene	polyethylene
trade name:	–	polythene

MONOMER	FORMULA	POLYMER	STRUCTURE
ethene (ethylene)	$\begin{array}{c} H \quad\quad H \\ \diagdown\quad\diagup \\ C=C \\ \diagup\quad\diagdown \\ H \quad\quad H \end{array}$	poly(ethene) (polyethylene)	$\cdots-\overset{\displaystyle H}{\underset{\displaystyle H}{C}}-\overset{\displaystyle H}{\underset{\displaystyle H}{C}}-\cdots$
propene (propylene)	$\begin{array}{c} H \quad\quad CH_3 \\ \diagdown\quad\diagup \\ C=C \\ \diagup\quad\diagdown \\ H \quad\quad H \end{array}$	poly(ethene) (polypropylene)	$\cdots-\overset{\displaystyle H}{\underset{\displaystyle H}{C}}-\overset{\displaystyle CH_3}{\underset{\displaystyle H}{C}}-\cdots$
chloroethene (vinylchloride)	$\begin{array}{c} H \quad\quad Cl \\ \diagdown\quad\diagup \\ C=C \\ \diagup\quad\diagdown \\ H \quad\quad H \end{array}$	poly(chloroethene) (PolyVinylChloride)	$\cdots-\overset{\displaystyle H}{\underset{\displaystyle H}{C}}-\overset{\displaystyle Cl}{\underset{\displaystyle H}{C}}-\cdots$
tetrafluoroethene	$\begin{array}{c} F \quad\quad F \\ \diagdown\quad\diagup \\ C=C \\ \diagup\quad\diagdown \\ F \quad\quad F \end{array}$	poly(tetrafluoroethene) (PTFE)	$\cdots-\overset{\displaystyle F}{\underset{\displaystyle F}{C}}-\overset{\displaystyle F}{\underset{\displaystyle F}{C}}-\cdots$
phenylethene (styrene)	$\begin{array}{c} H \quad\quad C_6H_5 \\ \diagdown\quad\diagup \\ C=C \\ \diagup\quad\diagdown \\ H \quad\quad H \end{array}$	poly(phenylethene) (polystyrene)	$\cdots-\overset{\displaystyle H}{\underset{\displaystyle H}{C}}-\overset{\displaystyle C_6H_5}{\underset{\displaystyle H}{C}}-\cdots$
propenitrile (acrylonitrile)	$\begin{array}{c} H \quad\quad CN \\ \diagdown\quad\diagup \\ C=C \\ \diagup\quad\diagdown \\ H \quad\quad H \end{array}$	poly(propenitrile) (polyacrylonitrile)	$\cdots-\overset{\displaystyle H}{\underset{\displaystyle H}{C}}-\overset{\displaystyle CN}{\underset{\displaystyle H}{C}}-\cdots$
2-methylpropenoate (methyl methacrylate)	$\begin{array}{c} H \quad\quad CH_3 \\ \diagdown\quad\diagup \\ C=C \\ \diagup\quad\diagdown \\ H \quad\quad CO_2CH_3 \end{array}$	poly(2-methylpropenoate) (polymethylmethacrylate)	$\cdots-\overset{\displaystyle H}{\underset{\displaystyle H}{C}}-\overset{\displaystyle CH_3}{\underset{\displaystyle CO_2CH_3}{C}}-\cdots$

Fig. 15.10 Some addition polymers

ADDITION REACTIONS OF ALKENES

The high electron density of the C=C bond makes alkenes react as nucleophiles and undergo addition reactions with electrophiles. These additions follow the pattern shown by the following reaction mechanism:

> *Learn this mechanism. Use the curly arrows properly.*

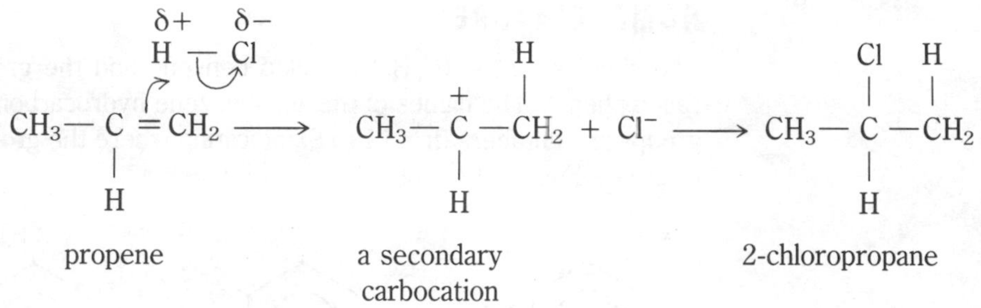

$$\overset{\delta+}{H}\underset{}{\overset{\delta-}{-}Cl}$$

propene   a secondary carbocation   2-chloropropane

Note that

- Addition occurs across the double bond to the two carbon atoms originally joined by it
- The addition product is 2-chloropropane NOT 1-chloropropane because the reaction follows Markovnikov's rule:
  When a molecule HZ adds to an unsymmetrical alkene, the hydrogen adds to the double bonded carbon with the greater number of H atoms.

We explain this by saying that a more stable secondary carbocation rather than a less stable primary carbocation is the intermediate formed in the first step of the mechanism.

We call these reactions electrophilic additions because the alkene behaves as a nucleophile in the first step. Why do we believe the first step produces a carbocation intermediate? One reason is that gaseous ethene reacts with bromine to form 1,2-dibromoethane only BUT reacts with bromine in the presence of lithium chloride (Li^+Cl^-) to form some $CH_2Cl—CH_2Br$ as well as $CH_2Br—CH_2Br$. So we suggest that in the second step the nucleophile Cl^- (from the lithium chloride) may also attack the electrophilic carbocation to complete the addition.

A halogen molecule like Br—Br has six lone pairs of electrons and is non-polar. So why should it behave as an electrophile? A completely dry gaseous mixture of bromine and ethene reacts in a glass flask but not if the inside of the flask is coated with parrafin wax! On this evidence we suggest that water molecules or ions in the glass surface may polarise some bromine molecules: e.g.

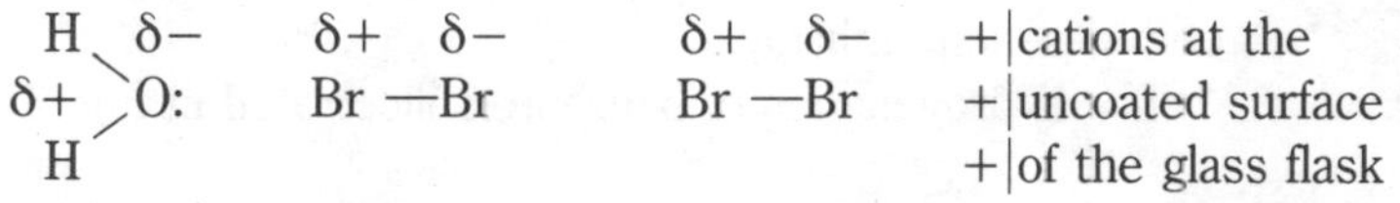

$$\overset{\delta+}{H}\overset{\delta-}{O:}\quad \overset{\delta+}{Br}—\overset{\delta-}{Br}\quad \overset{\delta+}{Br}—\overset{\delta-}{Br}\quad +|\text{cations at the}$$

The rapid decolorisation of aqueous bromine is used as a test for alkenes. This reaction may be explained as a fast addition reaction involving molecules of bromic(I) acid (hypobromous acid) present in the aqueous bromine equilibrium mixture:

$$Br_2(aq) + H_2O(l) \rightleftharpoons Br^-(aq) + H^+(aq) + \overset{\delta+}{Br}—\overset{\delta-}{OH}(aq)$$

2-Bromoethanol (Br—CH_2—CH_2—OH) is one of the products when ethene reacts with aqueous bromine and oxygen is more electronegative than bromine.

The addition reactions of the alkenes are typified by those of propene CH_3—CH=CH_2. They are summarised in Fig. 15.11.

MOLECULE	REAGENT AND CONDITIONS	MAIN PRODUCT(S)
Br—Br	bromine as a vapour or in an organic solvent	CH_3—CHBr—CH_2Br 1,2-dibromopropane
Br—OH	bromine in aqueous solution	CH_3—CHOH—CH_2Br 1-bromopropan-2-ol
H—Br	hydrogen bromide gas or conc. hydrobromic acid	CH_3—CHBr—CH_3 2-bromopropane
H_2O	conc. sulphuric acid followed by water	CH_3—CHOH—CH_3 propan-2-ol

Fig. 15.11 Electrophilic addition reactions of propene

ARENES   Arenes are hydrocarbons containing benzene rings. They were first prepared from compounds with fragrant aromas that were extracted from plant oils. And they were found to be more unsaturated than the alkenes.

NOMENCLATURE

The simplest arene (C_6H_6) is called benzene and the group (C_6H_5-) derived from it is called phenyl. The names of the alkylbenzene hydrocarbons include the names of the alkyl groups and numbers (from 1 to 6) indicating where the groups are attached to the benzene ring: see Fig. 15.12.

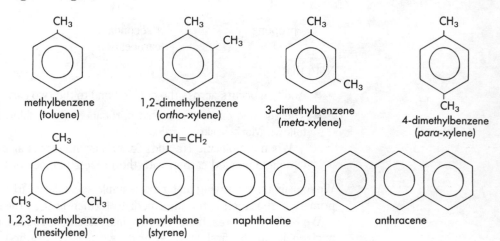

methylbenzene (toluene)

1,2-dimethylbenzene (*ortho*-xylene)

3-dimethylbenzene (*meta*-xylene)

4-dimethylbenzene (*para*-xylene)

1,2,3-trimethylbenzene (mesitylene)

phenylethene (styrene)

naphthalene

anthracene

Fig. 15.12 Naming some arenes

HYDROGENATION OF BENZENE

The standard molar enthalpy change for this reaction is about $-205\,kJ$. The reaction of 3 mol of H_2 with cyclohexene has a standard molar enthalpy change of about $-355\,kJ$. If the benzene ring contained three double bonds we should expect hydrogenation of those bonds to release 355 kJ and not 205 kJ of heat. Because the hydrogenation of benzene releases 150 kJ less heat than expected, it is as though the molecule has already released this heat when forming its less reactive ring structure which does NOT possess three double bonds. The 150 kJ is a measure of this extra stability of the benzene structure and is sometimes called the stabilisation energy. When benzene was said to have a resonance hybrid structure, we called the 150 kJ the resonance energy. Nowadays we call it the delocalisation enthalpy.

The hydrogenation of benzene is illustrated in Fig. 15.13.

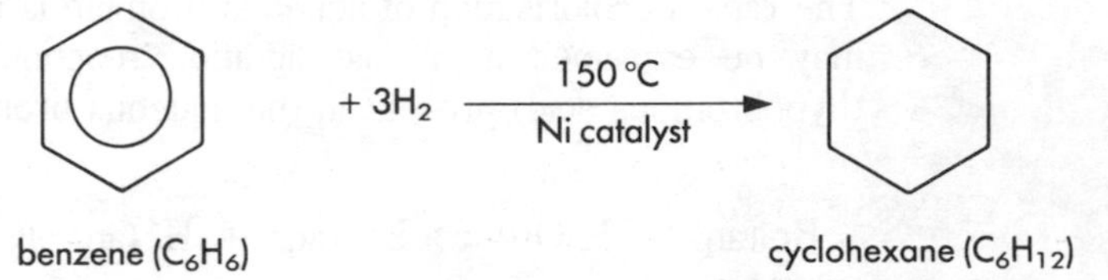

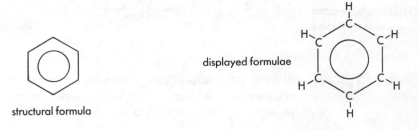

Fig. 15.13 Hydrogenation of benzene

benzene (C_6H_6) $+ 3H_2$ $\xrightarrow[\text{Ni catalyst}]{150\,°C}$ cyclohexane (C_6H_{12})

STRUCTURE OF BENZENE

X-ray diffraction analysis shows that benzene is a flat molecule: the six carbon and six hydrogen atoms are all in the same plane. The C—H bond length (0.108 nm) is normal for hydrocarbons but the distance between the centres of neighbouring carbon atoms (0.139 nm) is longer than a C=C double bond (0.134 nm) and shorter than a C—C single bond (0.154 nm). Benzene usually undergoes substitution reactions rather than addition reactions expected of highly unsaturated hydrocarbons. Although you will still come across the formula of benzene written as though the molecule had three double bonds, you should write the structure in the way shown in Fig. 15.14.

Fig. 15.14 Delocalised π-electrons in a benzene molecule

structural formula

displayed formulae

- The stability of the benzene ring structure is explained by the delocalization of π-electrons (represented by the circle)
- The delocalised electron cloud above and below the plane of the ring makes benzene nucleophilic
- The ring undergoes electrophilic substitutions rather than additions so the products keep the extra stability of π-electron delocalisation

TOXICITY OF BENZENE

Benzene is an extremely dangerous carcinogen. The liquid can be absorbed through the skin. It is volatile and flammable. The vapour forms explosive mixtures with air. Inhalation of even very low concentrations of the vapour can produce chronic effects.

- Practical work with benzene is banned in most schools and the less toxic methylbenzene (toluene) is used instead of benzene
- The chemical and physical properties of methylbenzene (toluene) are very similar to those of benzene but they are not identical

Methylbenzene may differ from benzene for two main reasons:

- The methyl group may take part in reactions not affecting the ring
- The methyl group may influence the way the benzene ring reacts.

The methyl group in methylbenzene is called a *substituent* because it has taken the place of, or substituted for, a hydrogen atom on the benzene ring. You should appreciate that substituents have an important influence on the properties of the benzene ring.

SUBSTITUTION REACTIONS OF ARENES

> **Benzene undergoes electrophilic substitution reactions.**

When benzene (colourless) is shaken with aqueous bromine (orange-red) then allowed to stand, the mixture separates into two layers. The upper layer is now a solution of bromine in benzene. It is red and it remains red. The lower aqueous layer is now a very pale orange and dilute solution of bromine. However, unlike the alkenes, benzene does NOT decolorise aqueous bromine by a rapid addition reaction!

- Arenes undergo electrophilic substitution reactions

Benzene reacts with bromine (or chlorine) in the presence of a catalyst (often called a halogen carrier) to form bromobenzene or chlorobenzene:

$$C_6H_6 + Br_2 \xrightarrow[\text{Fe(s) or FeBr}_3]{Br_2(l)} C_6H_5Br + HBr$$

We suppose that the catalyst turns the halogen into the reactive electrophile needed to attack the stable benzene ring: e.g.

$$Br\!-\!Br + FeBr_3 \rightarrow Br^+ + FeBr_4^-$$

The electrophile could use two of the six delocalised π-electrons to bond with a carbon atom in the ring: see Fig. 15.15. The remaining four electrons are delocalised over the other five carbon atoms which effectively share the positive charge. This 'delocalisation of the positive charge' stabilises the ring against nucleophilic attack by a bromide ion and prevents the addition reaction that occurs with alkenes.

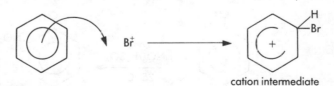

Fig. 15.15

cation intermediate

The cation intermediate loses a hydrogen ion (as a more stable cation and less reactive electrophile than Br^+) to restore the ring to its original stable arrangement of six delocalised π-electrons: see Fig. 15.16.

Figure 15.17 summarises the important electrophilic substitution reactions of benzene.

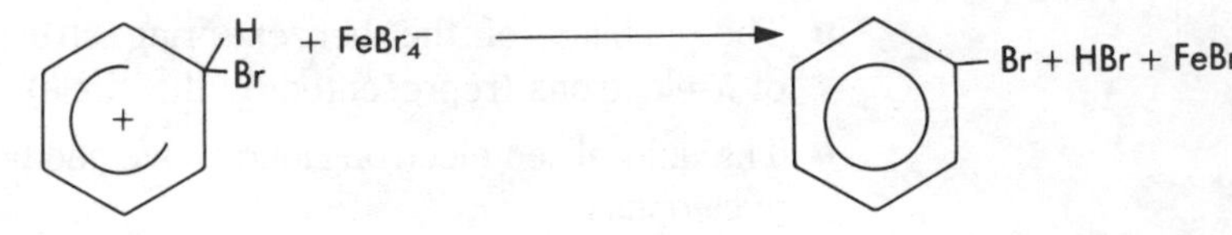

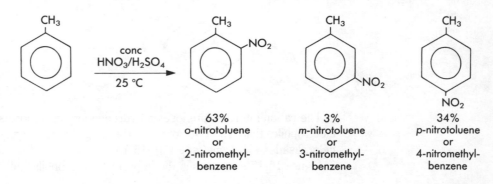

Fig. 15.16

ELECTROPHILE	REAGENTS AND CONDITIONS	SUBSTITUTION	PRODUCT
Cl^+	Cl_2 and anhydrous $AlCl_3$	Cl	chlorobenzene
CH_3^+	CH_3Cl and anhydrous $AlCl_3$	CH_3	methylbenzene
CH_3CO^+	CH_3COCl and anhydrous $AlCl_3$	$COCH_3$	phenylethanone
Br^+	Br_2 and Fe or $FeBr_3$	Br	bromobenzene
NO_2^+	conc. HNO_3 and H_2SO_4 at 45 °C	NO_2	nitrobenzene
SO_3	SO_3 in H_2SO_4 (fuming sulphuric) or conc. sulphuric under reflux	SO_3H	benzenesulphonic acid

Fig. 15.17 Electrophilic substitution reactions of benzene

The alkylation and acylation of benzene using anhydrous aluminium chloride as a catalyst is called the Friedel-Craft reaction.

The ring in methylbenzene is also attacked by electrophiles but three isomeric substitution products may be formed: see Fig. 15.18. Nitration of methylbenzene takes place faster and at a lower temperature than the nitration of benzene itself and the substitution occurs mainly at positions *ortho* and *para* to the methyl group. Theories to explain the speed and direction of this substitution usually assert that the methyl group is an *electron-donating* substituent and that the π-electron delocalisation is uneven. The methyl group makes the ring more nucleophilic by donating electronic charge to it. The uneven delocalisation of the electronic charge facilitates the attack of the electrophilic nitronium ion (NO_2^+) on the *2*- and *4*-carbon atoms in the ring. This theory leads to the following general principle:

- A substituted benzene compound forms mainly *1,2*- & *1,4*-substitution products and reacts faster than benzene if the group already attached to the ring is electron-donating: e.g. $-CH_3$, $-OH$, $-NH_2$
- A substituted benzene compound forms mainly *1,3*-substitution products and reacts slower than benzene if the group already attached to the ring is electron-attracting: e.g. $-NO_2$, $-CO_2H$, $-SO_3H$

Note that there are only single bonds in the electron-donating substituents but there is at least one double bond in the electron-attracting substituents. Whether a substituent is electron-donating or attracting depends in general not only upon the electronegativities of its atoms but also upon the interaction of their lone pair electrons with the delocalised

63%	3%	34%
o-nitrotoluene or 2-nitromethylbenzene	*m*-nitrotoluene or 3-nitromethylbenzene	*p*-nitrotoluene or 4-nitromethylbenzene

Fig. 15.18 Isomeric substitution products of methylbenzene

π-electrons of the benzene ring. When the substituent is just a halogen atom, the balance between these two factors leads to the following principle:

■ A halogenobenzene compound forms mainly *1,2-* & *1,4*-substitution products but reacts slower than benzene

OXIDATION OF ARENES

Benzene and toluene ignite easily and burn with very smoky flames. When any alkyl-benzene is refluxed with acidic or alkaline potassium manganate(VII) solution, the alkyl group is oxidised to a carboxyl group ($-CO_2H$) and the product is benzoic acid. BUT benzene and therefore the carbon ring resists oxidation: see Fig. 15.19.

Benzene undergoes radical chain addition reactions with chlorine in ultraviolet light (or bright sunshine) ultimately to form BHC (benzene hexachloride): see Fig. 15.20. BHC is actually a mixture of the nine isomers of benzene hexachloride with different arrangements of the H and Cl atoms above and below the ring! One of these isomers (about 15% of the mixture) called gamma-BHC, is an active insecticide sold commercially under the trade name 'Gammexane'.

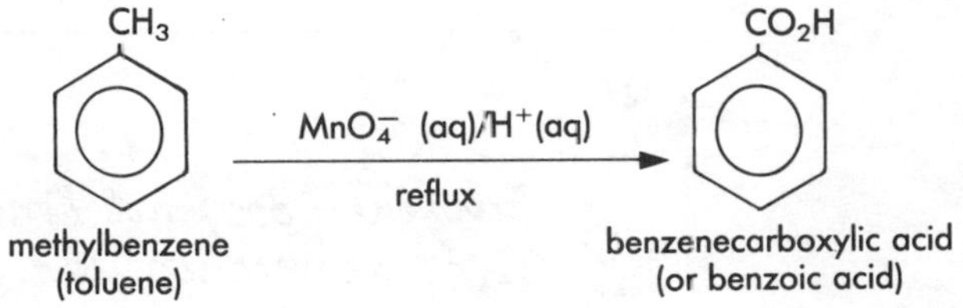

methylbenzene (toluene) → MnO_4^- (aq)/H^+(aq) reflux → benzenecarboxylic acid (or benzoic acid)

Fig. 15.19 Oxidation of side-chains

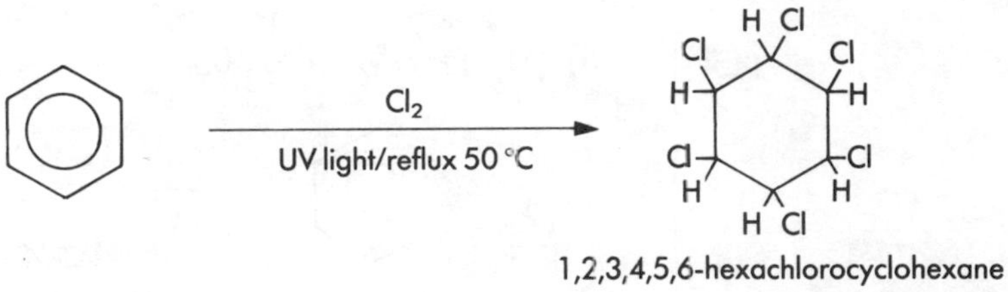

Cl_2 UV light/reflux 50 °C → 1,2,3,4,5,6-hexachlorocyclohexane

Fig. 15.20 Chlorination of benzene

Under similar conditions, alkylbenzenes undergo radical chain substitution reactions with chlorine replacing specifically the H-atoms on the carbon atom attached directly to the ring. We explain this by supposing that the intermediate radical is stabilised by delocalisation involving the unpaired electron and the π-electrons of the ring. For example, methyl-benzene will form chloromethylbenzene, dichloromethylbenzene and trichloromethyl-benzene. You can monitor the reaction by weighing the mixture and noting the increase in mass:

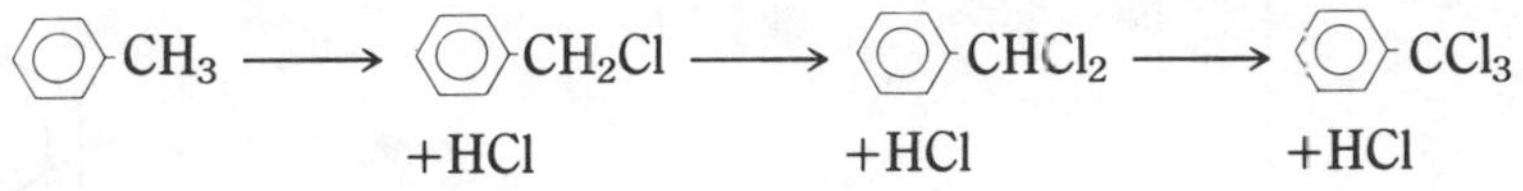

CH_3 ⟶ CH_2Cl ⟶ $CHCl_2$ ⟶ CCl_3
+HCl   +HCl   +HCl

COMPARISON OF AROMATIC AND ALIPHATIC COMPOUNDS

Just as a substituent group (e.g. 'OH') can influence the reactions of the benzene ring to which it is attached, so the benzene ring can influence the properties of the substituent group. Consequently phenol (C_6H_5OH) differs from hexanol ($C_6H_{13}OH$) in certain respects. This important point is considered in the chapters dealing with particular functional groups.

EXAMINATION QUESTION AND OUTLINE ANSWER

Q1 a) The enthalpies of hydrogenation of cyclohexene and benzene are $-120\,\text{kJ}\,\text{mol}^{-1}$ and $-208\,\text{kJ}\,\text{mol}^{-1}$ respectively.
Explain why the enthalpy of hydrogenation of benzene is not three times that of cyclohexene. *(6)*

b) i) By giving reagents and an equation, show how benzene may be converted into methylbenzene. *(3)*

ii) Write a mechanism for the reaction. *(4)*

c) By giving equations and essential conditions show how methylbenzene may be chlorinated

i) in the aromatic ring;

ii) in the side chain. *(6)*

(AEB)

Outline answer

Q1

(a) benzene does not contain 3 double bonds. $(3 \times 120 - 208)$ = 152 kJ represents the extra stability of the benzene ring due to delocalization

(b) (i) Friedel-Crafts

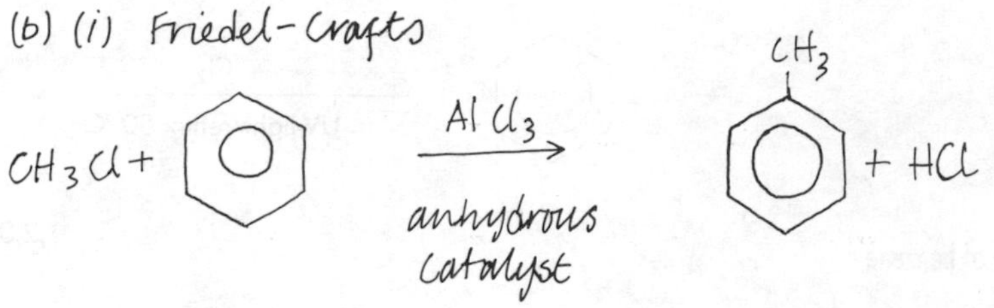

$$CH_3Cl + \text{(benzene)} \xrightarrow[\text{anhydrous catalyst}]{AlCl_3} \text{(methylbenzene, } CH_3\text{)} + HCl$$

(ii) $CH_3-Cl + AlCl_3 \longrightarrow CH_3^+ \ AlCl_4^-$

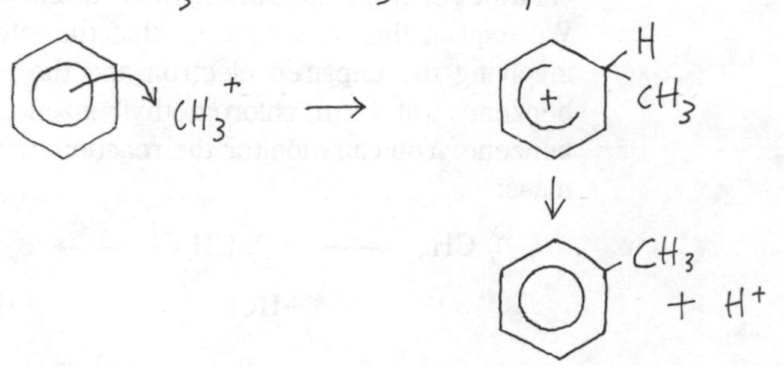

$$\text{(benzene)} \curvearrowright CH_3^+ \longrightarrow \text{(intermediate, } +, H, CH_3\text{)} \longrightarrow \text{(product, } CH_3\text{)} + H^+$$

(c) (i)

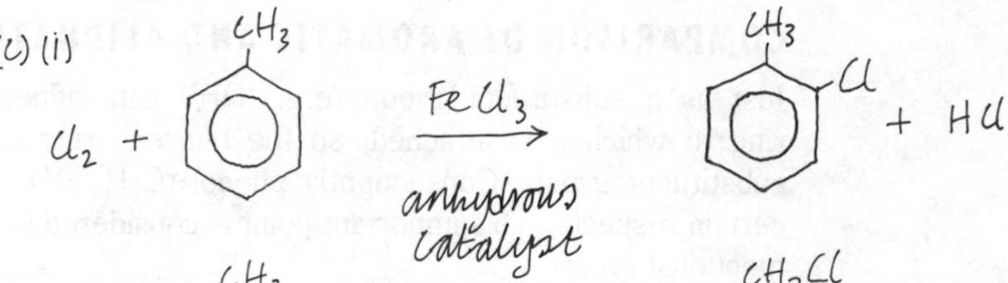

$$Cl_2 + \text{(methylbenzene, } CH_3\text{)} \xrightarrow[\text{anhydrous catalyst}]{FeCl_3} \text{(product, } CH_3, Cl\text{)} + HCl$$

(ii)

$$\text{(methylbenzene, } CH_3\text{)} \xrightarrow{UV \ light} \text{(product, } CH_2Cl\text{)} + HCl$$

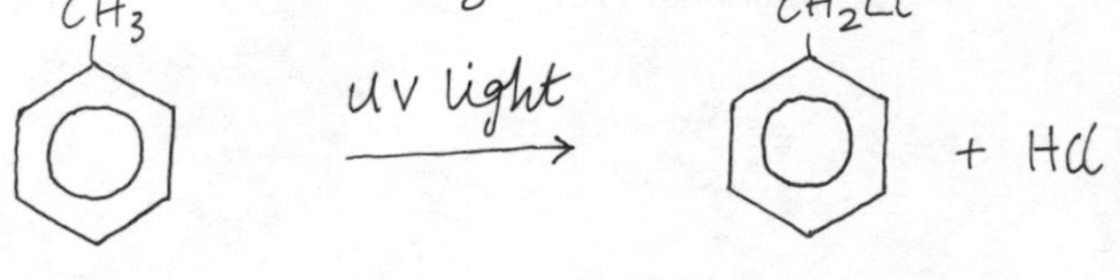

EXAMINATION QUESTION AND TUTOR'S ANSWER

Q2 The main source of hydrocarbons is now crude oil. Crude oil is used to produce hydrocarbons for two main purposes (i) energy and (ii) as feedstock for producing important useful chemical materials.

a)   i) Give the name of a hydrocarbon used commonly for energy production

Methane

   ii) Give the situation in which this hydrocarbon is used

in natural gas burners

   iii) Give the name of a hydrocarbon other than phenylethene

Propene

   iv) the name of an important chemical material obtained from the feedstock

Poly (ethene)   (4)

b) When phenylethene is treated with a peroxide which yields free radicals (R·) polymerization takes place.
One important step in the process may be represented:

$$R\cdot \ + \ CH_2{=}CH{-}C_6H_5 \ \rightarrow \ R{-}CH_2{-}CH{-}C_6H_5$$
phenylethene

   i) Show, by means of an equation, how this type of reaction leads to the formation of a polymer.

$R\,CH_2\ \overset{\bullet}{C}H\,C_6H_5 + CH_2 = CH\,C_6H_5 \rightarrow R\,CH_2\overset{C_6H_5}{\overset{|}{C}H}-CH_2-CH_2-\overset{\bullet}{C}H\,C_6H_5$

   ii) Explain why the polymer produced consists of chains having a variety of lengths

The termination step can happen at random

   iii) Explain how you think varying the amount of initiator might alter the average relative molecular mass of the product

more initiator will start more chains so there will be more short ones   (3)

c) The product in b) may be represented by

$$R{-}(CH_2{-}CH)_m{-}CH_2{-}\underset{\underset{C_6H_5}{|}}{C}{-}\underset{\underset{C_6H_5}{|}}{C}{-}CH_2{-}(CH{-}CH_2)_n{-}R$$

with C_6H_5 groups shown above the appropriate carbon atoms.

Explain why
   i) The chains are never branched *the free radical Ċ-atom is always at the same place next to the benzene ring.*

   ii) The —C_6H_5 groups appear on every second carbon atom except in one place where two occur together

Two benzene rings come together in the termination step

   iii) m and n are nearly always different

The chains grow separately from one another   (3)

d) Explain, with reasons, why you would expect the product in b) to be thermoplastic, rather than thermosetting.

The chains do not form cross-links with one another (2)

(*Total 12 marks*)
(ULSEB (AS specimen))

STUDENT'S ANSWER'S WITH EXAMINER'S COMMENTS

Q3   This question is concerned with some reactions of alkenes and their products.

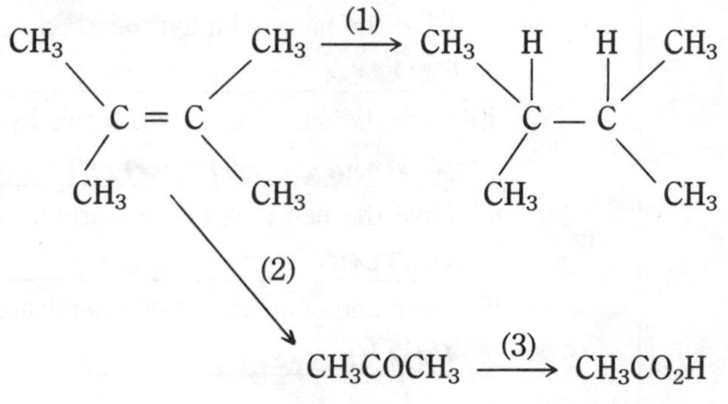

$$CH_3, CH_3 \xrightarrow{(1)} CH_3 \; H \;\; H \; CH_3$$
$$C=C$$
$$CH_3 \; CH_3 \qquad CH_3 \;\; CH_3$$

(2)

$$CH_3COCH_3 \xrightarrow{(3)} CH_3CO_2H$$

$$CH_3CH_2OH \xrightarrow{(4)} CH_3CHO$$

66 Give *all* the reagents. **99**

a)   What reagents would you use to carry out the above transformations?

(1)   *Hydrogen*   (2)

66 and nickel catalyst **99**

(2)   *Ozone*   (2)

66 then hot water **99**

(3)   *Bleaching powder*   (2)

66 as a suspension in water **99**

(4)   *Potassium dichromate (VI)*   (1)

66 and aqueous sulphuric acid **99**

b)   What procedure would you follow to obtain the best possible (maximum) yield in transformation (4)?

Procedure *continuous distillation (not reflux)* ✓ (2)
Explain why this procedure is used.

to prevent further oxidation of the aldehyde to the carboxylic acid ✓ (2)

c)   Draw the structural formula of the compound obtained when the alkene (A) reacts with H_2SO_4.

$$CH_3CH_2CH=CH_2 \qquad (A)$$

$$\overset{SO_4 \; H}{\underset{|}{CH_3 \; CH_2 \; CH-CH_3}} \; ✓ \qquad (2)$$

Give the structure of the product you would expect to be formed when alkene (A) reacts with iodine monochloride (I—Cl).

$$\underset{\underset{I}{|}}{CH_3 \; CH_2 \; CH-CH_2 \; Cl} \; ✓ \qquad (2)$$

(OCSEB 1988)

Q4   a)   For the photochemical chlorination of methane, write balanced equations to show:

i)   the initiation reaction: (1)

$$Cl-Cl_{(g)} \xrightarrow{light} 2Cl\cdot$$

ii)   **two** propagation reactions;   (2)

$$Cl\cdot + H-CH_3 \rightarrow H-Cl + \cdot CH_3$$
$$Cl-Cl + \cdot CH_3 \rightarrow Cl-CH_3 + Cl\cdot$$

iii)   a termination reaction;   (1)

$$2 \; \cdot CH_3 \rightarrow C_2H_6$$

iv) an *overall* equation showing CH_2Cl_2 as the organic product.   *(1)*

$$2 Cl_2(g) + CH_4(g) \rightarrow CH_2Cl_2(g) + 2HCl(g)$$

b) Methane can be successfully *brominated* using light of a lower frequency than that used for chlorination. Suggest an explanation for this.   *(3)*

The Br — Br bond is weaker than the Cl – Cl bond.

66 **Very good answers.** 99

(ODLE 1988)

PRACTICE QUESTION

Q5 a) Give a brief account of the use of petroleum as a source material for
   i) different types of fuels, and
   ii) simple hydrocarbons which are used for preparing organic compounds.   *(7)*
b) Using as examples the reactions of
   i) methane with chlorine, and
   ii) ethene with bromine,
   compare the reactions of alkanes and alkenes with the halogens.   *(5)*
c) Compare the chemical reactivities of benzene and ethene and account for the differences between them.   *(8)*

(WJEC 1988)

C H A P T E R

ORGANIC HALOGEN COMPOUNDS

HALOGENOALKANES

REACTION MECHANISMS

HALOGENOALKENES AND HALOGENOARENES

ACYL CHLORIDES

REACTION MECHANISMS

GETTING STARTED

The emphasis is upon comparing the effect of the organic molecular structure and of the four different halogens upon the reactivity of the halogen compound. Halogenoalkanes, halogenoalkenes and halogeno-arenes are three important classes of organic compounds containing carbon, hydrogen and halogen only. Of these, the halogenoalkanes are the most reactive. The acyl chlorides are an even more reactive class of organic halogen compound but they also contain oxygen. We shall also deal with these here, although they are sometimes separately considered under carboxylic acid derivatives.

ESSENTIAL PRINCIPLES

**HALOGENO-
ALKANES**

These are compounds formed when hydrogen atoms in an alkane are replaced by halogen atoms. They may be grouped into a variety of homologous series based on the number and type of halogen and on the nature of the carbon skeleton: e.g. the 1-chloroalkanes with no branching in the carbon chain would be one of the simplest homologous series.

NOMENCLATURE

The systematic method of naming halogenoalkanes is similar to the method of naming branched-chain alkanes.

- The root name is based on the longest continuous carbon chain and the name of the straight-chain alkane with the same number of carbons
- The root is prefixed with the numbers; di-, tri-, tetra-, etc. and the names of the halogens; fluoro-, chloro- bromo-, iodo-
- The positions of the halogens are given the lowest possible numbers and the prefixes are attached in alphabetical order of halogen:

$$F - \underset{\underset{F}{|}}{\overset{\overset{F}{|}}{C}} - \underset{\underset{H}{|}}{\overset{\overset{Cl}{|}}{C}} - Br \qquad \text{2-bromo-2-chloro-1,1,1-trifluoroethane}$$

('Halothane' – widely used anaesthetic)

You may find some simple compounds still given traditional names and referred to as primary, secondary or tertiary halogenoalkanes: e.g.

$$H - \underset{\underset{H}{|}}{\overset{\overset{H}{|}}{C}} - Cl \qquad CH_3 - \underset{\underset{H}{|}}{\overset{\overset{CH_3}{|}}{C}} - Br \qquad CH_3 - \underset{\underset{CH_3}{|}}{\overset{\overset{CH_3}{|}}{C}} - I$$

traditional:	methyl chloride	isopropyl bromide	tertiary butyl iodide
	(primary)	(secondary)	(tertiary)
systematic:	chloromethane	2-bromopropane	2-iodo-2-methylpropane

Play safe and use systematic nomenclature even for these!

PHYSICAL PROPERTIES

Halogenoalkanes have higher densities, melting points and boiling points than the corresponding alkanes. And these values increase with increasing atomic number of the halogen (from the fluoro- to the iodo-compounds) and with increasing halogenation of the alkane: see Fig. 16.1.

NAME	FORMULA	$T_b/°C$	NAME	FORMULA	$T_b/°C$
chloromethane	CH_3Cl	-24	fluoroethane	C_2H_5F	-38
dichloromethane	CH_2Cl_2	40	chloroethane	C_2H_5Cl	12
trichloromethane	$CHCl_3$	62	bromoethane	C_2H_5Br	38
tetrachloromethane	CCl_4	77	iodoethane	C_2H_5I	72

Fig. 16.1 Boiling temperatures of some halogenoalkanes

CHEMICAL PROPERTIES

- The halogenoalkanes are attacked by nucleophiles to undergo substitution and/or elimination reactions

The nature of the reaction depends upon the strength and polarity of the carbon-halogen bond, the structure of the alkyl group and the nucleophile's strength as a base (proton acceptor). It is also influenced by the temperature of the reaction and by the solvent's ionising properties (dipole moment and relative permittivity) but this is usually outside the scope of A-level chemistry.

- The reactivity of the halogenoalkanes increases as the strength of the carbon-halogen bond decreases from C—F to C—I

Many fluorocarbons, e.g. Poly(TetraFluoroEthene) PTFE, and chlorofluorocarbons, e.g. FREONS, are widely used because they are chemically inert and thermally stable. Chloro- and bromo-alkanes are widely used in organic syntheses because they are sufficiently stable to be prepared and purified but sufficiently reactive to be converted into alcohols, ethers, esters, nitriles, amines and Grignard reagents. In contrast the iodoalkanes are rather unstable, gradually decomposing and turning yellow even when stored in brown bottles to keep out light.

- The carbon atom becomes more electrophilic and heterolysis of the C—Hal bond more likely as the halogen becomes more electronegative
- Primary halogenoalkanes favour substitution reactions and tertiary halogenoalkanes favour elimination reactions
- Strongly basic nucleophiles favour elimination reactions and weakly basic nucleophiles favour substitution reactions

Hydrolysis of halogenoalkanes

As part of your practical work you may be asked to investigate the factors that affect the hydrolysis of halogenoalkanes. The *substitution* reaction may be represented by the following general equation:

$$H_2O + C_nH_{(2n+1)}X \rightarrow C_nH_{(2n+1)}OH + H^+ + X^-$$

Silver nitrate is used in qualitative analysis because it forms white (AgCl), off-white (AgBr) and pale-yellow (AgI) precipitates with aqueous halide ions: see Chapters 7 and 12. But note that it does not form these precipitates with halogen atoms *covalently* bonded to carbon in halogenoalkanes. So you could use silver nitrate to monitor the progress of the hydrolysis reaction.

A popular test-tube experiment uses the 1-halogenobutanes and the isomers of C_4H_9Br in aqueous ethanol (a mixed solvent to dissolve the ionic silver nitrate and the covalent halogenoalkane). 1-Iodobutane gives a yellow ppt. almost as soon as you add the silver nitrate. The 1-chlorobutane may take up to 15 minutes to produce a white precipitate. Of the three isomers of C_4H_9Br, the 2-bromo-2-methylpropane gives an off-white precipitate almost immediately, 2-bromobutane gives a precipitate sooner than the 1-bromobutane.

These observations are consistent with the above generalisations about the reactions of the halogenoalkanes. Other experiments show that these hydrolysis reactions are faster if you use aqueous alkali instead of water.

Dehydrohalogenation of halogenoalkanes

As part of your practical work you may be asked to prepare samples of an alkene from a halogenoalkane. This may be done by an *elimination* reaction which may be represented by the following general equation:

$$-\underset{\underset{|}{|}}{\overset{\overset{H}{|}}{C}} - \underset{\underset{|}{|}}{\overset{\overset{X}{|}}{C}} - + OH^- \rightarrow \ \ \diagdown_{\diagup}C=C_{\diagdown}^{\diagup} + X^- + H_2O$$

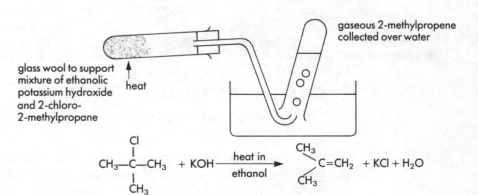

Fig. 16.2 Laboratory preparation of 2-methylpropene

glass wool to support mixture of ethanolic potassium hydroxide and 2-chloro-2-methylpropane

heat

gaseous 2-methylpropene collected over water

$$CH_3-\underset{\underset{CH_3}{|}}{\overset{\overset{Cl}{|}}{C}}-CH_3 + KOH \xrightarrow[\text{ethanol}]{\text{heat in}} \underset{CH_3}{\overset{CH_3}{\diagdown}}C=CH_2 + KCl + H_2O$$

You can prepare samples of 2-methylpropene in the laboratory by heating 2-chloro-2-methylpropane with a concentrated solution of potassium hydroxide in ethanol: see Fig. 16.2.

The alcoholic alkali reagent is particularly dangerous and eye protection is absolutely essential!

REACTION MECHANISMS

" Study these mechanisms thoroughly. "

■ The halogeno-compound loses the halogen as a halide ion (X^-) in both substitution and elimination reactions

In the substitution reaction the halogen is replaced by the more powerful nucleophile (e.g. OH^-) which attacks the electrophilic carbon atom. Primary halogenoalkanes favour the S_N2 mechanism in which substitution by the nucleophile takes place in one slow step involving 2 species: see Fig. 16.3.

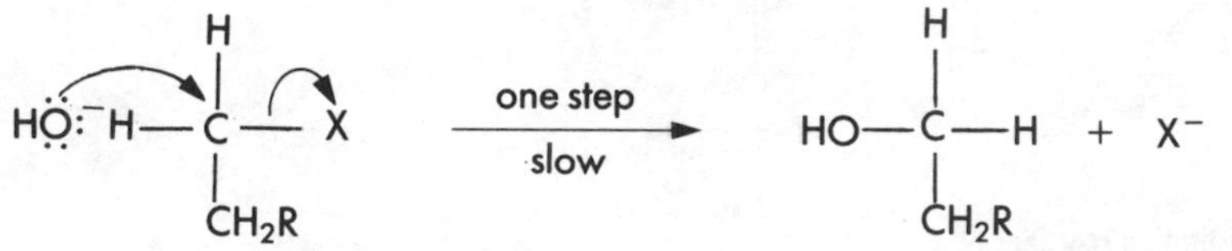

Fig. 16.3 S_N2 reaction mechanism

Tertiary halogenoalkanes favour the S_N1 mechanism in which substitution by the nucleophile takes place in one slow step involving 1 species: see Fig. 16.4. Secondary halogenoalkane substitution reactions may proceed by either or both mechanisms.

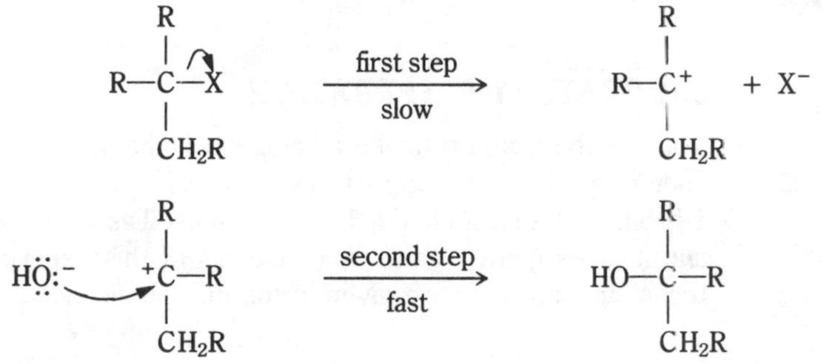

Fig. 16.4 S_N1 reaction mechanism

In elimination reactions the strongly (Lewis) basic nucleophile attacks a hydrogen atom on a (beta) carbon atom adjacent to the C—X carbon atom. Primary halogenoalkanes favour E_N2 reactions: see Fig. 16.5. And tertiary halogenoalkanes favour E_N1 reactions: see Fig. 16.6.

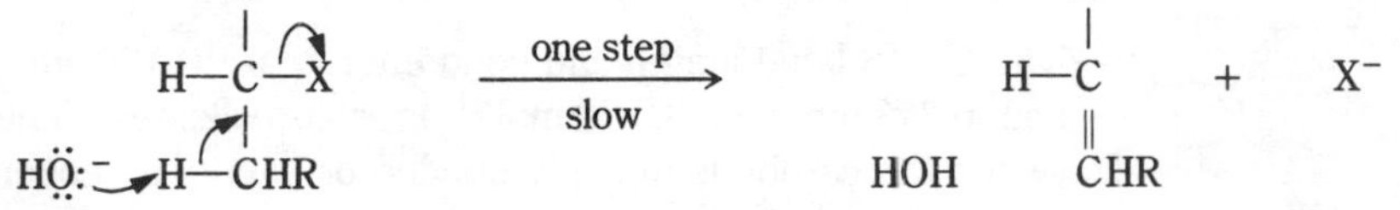

Fig. 16.5 E_N2 reaction mechanism

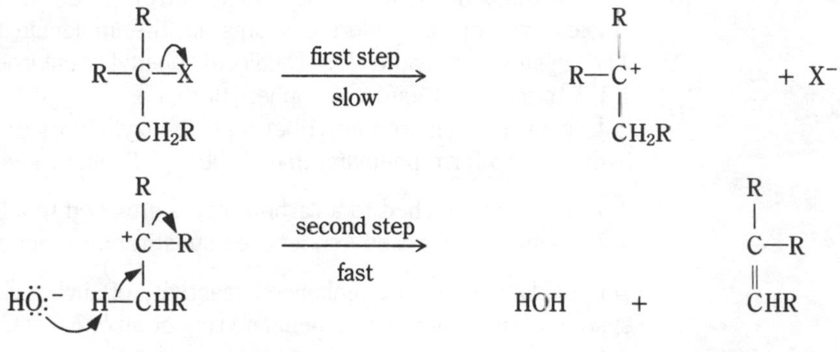

Fig. 16.6 E_N1 reaction mechanism

" Learn the different conditions for substitution and elimination. "

■ Halogenoalkanes can undergo both substitution and elimination reactions if they contain — C — C — in their structure with H and X on adjacent carbons

NUCLEOPHILE	REAGENTS AND CONDITIONS	SUBSTITUTION PRODUCTS	
HO: $^-$	sodium hydroxide aqueous; heat under reflux	RCH_2OH alcohol	
CH_3O: $^-$	sodium methoxide (CH_3ONa)* in methanol; under reflux	RCH_2OCH_3 ether	
: CN: $^-$	potassium cyanide (KCN) in ethanol; under reflux	RCH_2CN nitrile	
AgCN:	silver cyanide (AgCN) in ethanol; under reflux	RCH_2NC isonitrile	
: NH_3	ammonia (NH_3) aqueous or gaseous; heat in sealed tube under pressure	RCH_2NH_2 amine	– (primary)
: NH_2CH_2R		$(RCH_2)_2NH$	– (secondary)
: $NH(CH_2R)_2$	with excess RCH_2Br	$(RCH_2)_3N$	– (tertiary)
: $N(CH_2R)_3$		$(RCH_2)_4N^+$ quaternary ammonium bromide	

* made by adding sodium metal to methanol: $Na + CH_3OH \rightarrow CH_3ONa + \frac{1}{2}H_2$

Fig. 16.7 Substitution reactions of a typical halogenoalkane RCH_2Br

Nucleophilic substitution reactions with primary and secondary (but not tertiary) halogeno-alkanes provide useful ways of putting certain functional groups into a molecule. The more important substitutions are listed in Figure 16.7.

LABORATORY PREPARATION

You may be asked to make a halogenoalkane as part of your practical work. For A-level chemistry the starting material is usually an alcohol. The laboratory preparation of 1-bromobutane from butan-1-ol is often used as a typical example of an organic preparation and it gives quite a good yield. Details of this experiment and of the standard practical techniques involved are given in the unit on alcohols.

HALOGENO-ALKENES AND HALOGENOARENES

In general, chlorobenzene or bromobenzene only react with nucleophilic reagents e.g. aqueous NH_3, OH^- or CN^-, at high temperatures and/or pressures.

■ Halogen attached to a benzene ring or a multiply-bonded carbon atom does not readily undergo substitution reactions

The C—Cl bond length and bond energy are 0.170 nm and 400 kJ mol^{-1} in chlorobenzene and 0.177 nm and 346 kJ mol^{-1} in chloroalkanes. The shorter, stronger C—Cl bond (when the carbon is multiply-bonded or part of a benzene ring) may be explained by lone pairs of electrons on the halogen atom participating in the π-electron system of the double bond or the benzene ring. The shorter, stronger C—Cl bond is one reason why DDT is so stable: two of the chlorine atoms in the molecule are attached to benzene rings. The systematic name for DichloroDiphenylTrichloroethane (Cl—C_6H_4)$_2$CH—CCl$_3$ is 1,1,1-trichloro-2,2-di(4-chlorophenyl)ethane.

In contrast, (chloromethyl)benzene (benzyl chloride) reacts very readily with aqueous hydroxide to form phenylmethanol (benzyl alcohol): see Fig. 16.8.

■ Halogen attached to a carbon that is attached to a benzene ring or a multiply-bonded carbon atom readily undergoes substitution reactions

One explanation for the enhanced reactivity of such a halogen may be that the π-electron system of the neighbouring benzene ring or alkene C=C bond participates in the formation and stabilisation of the carbocation.

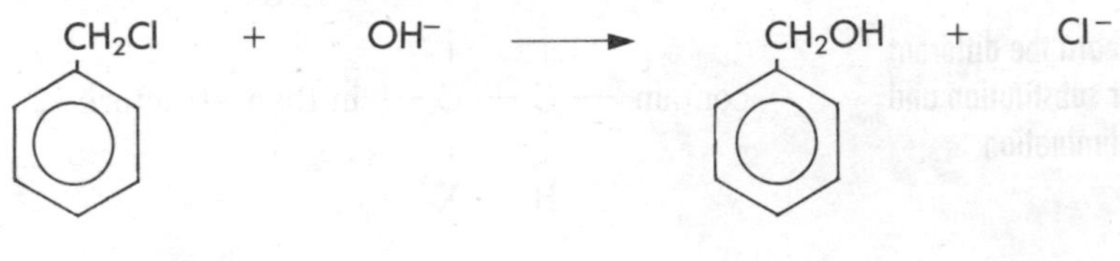

Fig. 16.8 Hydrolysis of (chloromethyl)benzene

USES OF HALOGENOHYDROCARBONS

Many but not all of the uses of these organic halogen compounds depends upon their chemical and thermal stability. Fig. 16.9 illustrates the wide variety of uses of the halogenohydrocarbons.

HALOGENOHYDROCARBON	USES
chloromethane CH_3Cl	making silicone polymers
chloroethane C_2H_5Cl	local anaesthetic; making tetraethyllead(IV) – petrol additive
dichloromethane CH_2Cl_2	solvent; paint stripper; making viscose fibres
tri- & tetrachloromethane $CHCl_3$ & CCl_4	to make chlorofluoromethanes ('FREONS'): $CHClF_2$, CCl_2F_2, CCl_3F
Freons	aerosol propellants; refrigerants; solvents
1,2-dibromomethane CH_2BrCH_2Br	petrol additive
Halothane $CF_3CHBrCl$	anaesthetic
bromochlorodifluoromethane $CBrClF_2$	BCF in fire extinguishers
tetrafluoroethene $F_2C{=}CF_2$	making PTFE addition polymer
chloroethene $CH_2{=}CHCl$	making PVC addition polymer
trichloroethene $CHCl{=}CCl_2$	drycleaning; degreasing metal surfaces
1,2,3,4,5,6-hexachloro-cyclohexane: $C_6H_6Cl_6$	gamma-BHC insecticide
1,1,1-trichloro-2,2-di(4-chlorophenyl)ethane	DDT insecticide

Fig. 16.9 Uses of some halogenohydrocarbons

ACYL CHLORIDES

These are extremely reactive compounds. The two you are most likely to encounter are ethanoyl chloride (acetyl chloride) and benzenecarbonyl chloride (benzoyl chloride). They are colourless, corrosive liquids that fume in moist air to form the corresponding organic acid and hydrogen chloride:

$$R - C \overset{O}{\underset{Cl}{<}} \ + H_2O \ \rightarrow \ R - C \overset{O}{\underset{OH}{<}} \ + HCl$$

The electronegative oxygen and chlorine atoms makes the carbon atom to which they are attached strongly electrophilic. So acyl chlorides readily react with nucleophilic reagents e.g. H_2O, ROH, NH_3, RNH_2.

REACTION MECHANISM

The activation energy of the first step in Fig. 16.10 may be quite low because the energy required for breaking only one of the bonds in the $C{=}O$ group is more than adequately provided by the simultaneous formation of the $\overset{+}{O}{-}C$ bond. An ionising solvent and alkaline conditions would facilitate the expulsion of the stable chloride ion and the proton in the second step which would also have a low activation energy. Hence the substitution would be fast.

■ The substitution of an acyl group into a molecule using an acyl chloride is known as an acylation reaction

As part of your practical work you may have to make a phenyl ester. As you will see in the next unit, phenols do not react directly with carboxylic acids. However, you can, for example, prepare phenyl benzoate in good yield just by shaking benzenecarbonyl (benzoyl) chloride with benzoic acid in aqueous alkali. The more important acylation reactions are listed in Figure 16.11.

The first step could be considered as a nucleophilic addition:

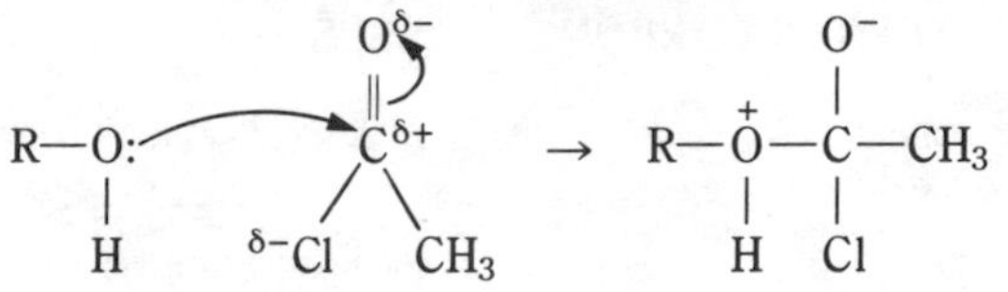

The second step could then be regarded as an elimination:

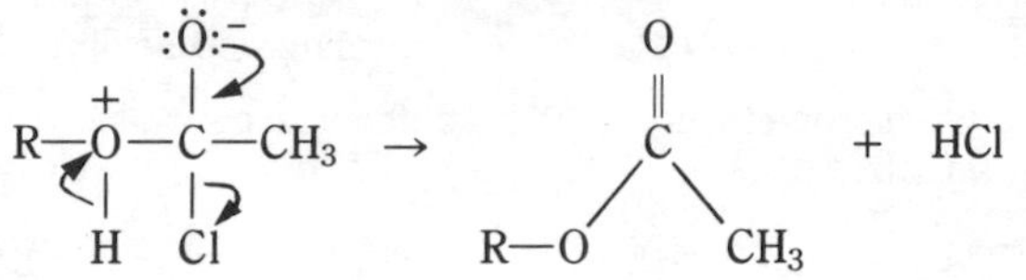

The overall effect is a nucleophilic substitution of the H in ROH by the acyl group: $(CH_3C{=}O)$.

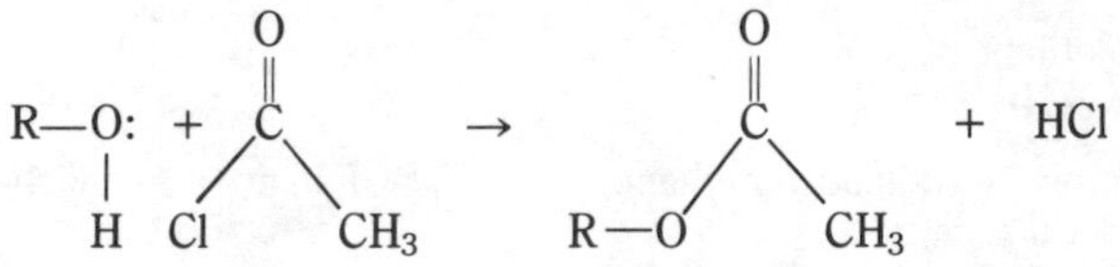

Fig. 16.10 Nucleophilic substitution with acyl chlorides

REACTANT MOLECULE	PRODUCT(S)	CLASS OF COMPOUND
water H_2O	$R-\overset{O}{\overset{\|\|}{C}}-OH$	carboxylic acid
methanol CH_3OH	$R-\overset{O}{\overset{\|\|}{C}}-OCH_3$	methyl ester
phenol C_6H_5OH	$R-\overset{O}{\overset{\|\|}{C}}-OC_6H_5$	phenyl ester
ammonia NH_3	$R-\overset{O}{\overset{\|\|}{C}}-NH_2$	amide
methylamine CH_3NH_2	$R-\overset{O}{\overset{\|\|}{C}}-NHCH_3$	substituted amide
diethylamine $(C_2H_5)_2NH$	$R-\overset{O}{\overset{\|\|}{C}}-N(C_2H_5)_2$	substituted amide
$R-\overset{O}{\overset{\|\|}{C}}-O^-Na^+$	$R-\overset{O}{\overset{\|\|}{C}}-O-\overset{O}{\overset{\|\|}{C}}-R$	acid anhydride
benzene C_6H_6 (+ Friedel-Craft catalyst – $AlCl_3$)	$R-\overset{O}{\overset{\|\|}{C}}-C_6H_5$	phenyl ketone

Fig. 16.11 Nucleophilic acyl substitutions (acylations) with RCOCl

EXAMINATION QUESTIONS AND OUTLINE ANSWERS

Q1 a) Give the *name* and *structure* of:
   i) a primary alkyl halide;
   ii) an acyl halide.
   Give the equations and conditions for the hydrolysis of each of these two halides and comment briefly on any differences. *(10)*

b) Write down the equation for the **ethanoylation of benzene**; give the formula of the catalyst used and explain briefly how the catalyst works. *(4)*
   With the aid of equations, show how the organic product of this reaction can be used to prepare:

   i) *(3)*

$$C_6H_5- \overset{\overset{\displaystyle H}{|}}{\underset{\underset{\displaystyle CH_3}{|}}{C}} - OH$$

   ii) $C_6H_5CO_2^- Na^+$ *(3)*

(ODLE 1988)

Q2 a) Draw the structure of 2,4-dimethylpentane and identify the different types of C—H bonds present. Write down the structures of all the possible monochloro derivatives of 2,4-dimethylpentane and state with reasons which one of the isomers will be produced most readily by direct chlorination. *(9)*

b) Explain in detail why the reaction conditions required to prepare (chloromethyl)-benzene and chloro-4-methylbenzene from methylbenzene are very different. *(15)*

c) Account for the fact that (chloromethyl)benzene produces a white precipitate when shaken with warm aqueous silver nitrate whereas chloro-4-methylbenzene does not. *(6)*

(JMB 1988)

Q3 Write an equation for the preparation of propanoyl chloride from propanoic acid. Give the structures of the products of the reactions between propanoyl chloride and
   i) ammonia,
   ii) phenol.
   Give a mechanism for the reaction between propanoyl chloride and ammonia. *(10)*

(JMB 1988)

Outline answers

Q1

(a) See text (PP247-248,251)

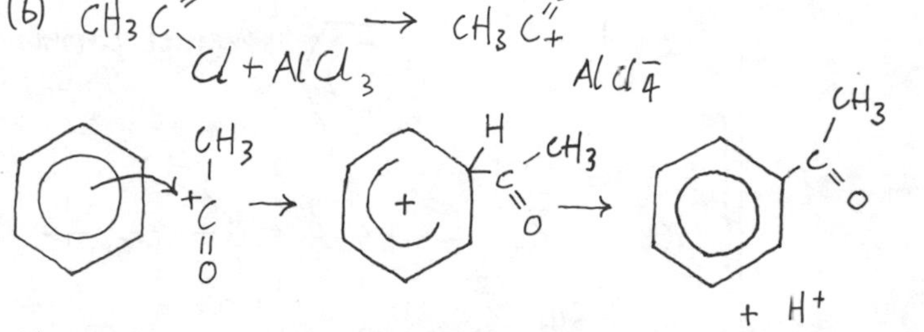

(b) $CH_3C\overset{\displaystyle =O}{\underset{\displaystyle Cl}{}} + AlCl_3 \rightarrow CH_3 C_+^{\displaystyle O} \quad AlCl_4^-$

[reaction scheme with benzene rings]

+ H⁺

Aluminium chloride catalyst promotes formation of the electrophile $CH_3 \overset{+}{C}O$

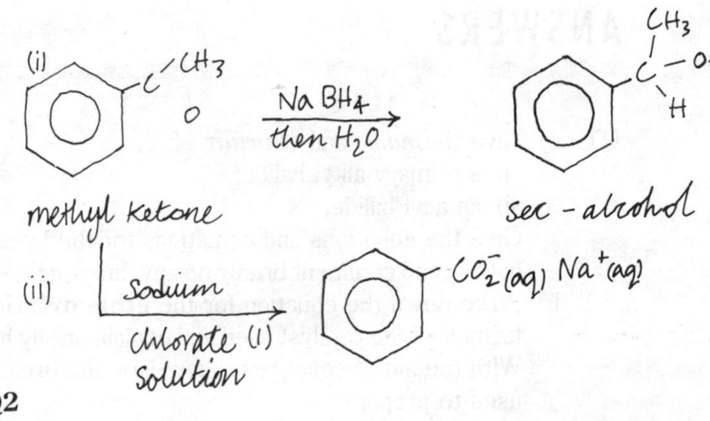

(i)

methyl ketone $\xrightarrow[\text{then } H_2O]{Na BH_4}$ sec - alcohol

(ii)

$\xrightarrow{}$ Sodium Chlorate (I) Solution $\rightarrow$ $CO_2^-(aq) \ Na^+(aq)$

Q2

(a)

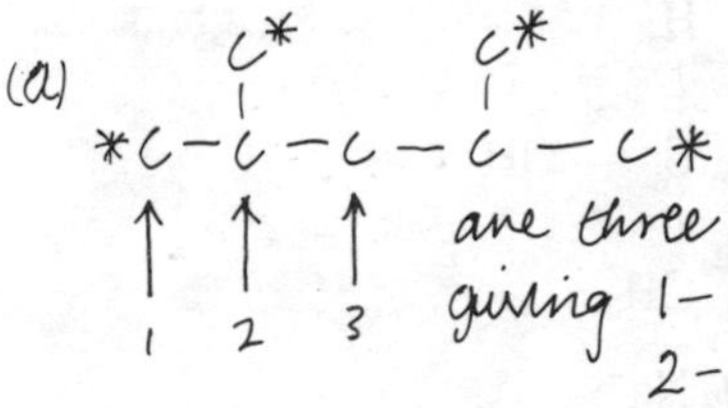

are three places for chlorine to substitute giving 1- chloro - 2,4 - dimethyl pentane
2- chloro - · · ·
3- chloro - · · ·

The 1 chloro will be most readily produced because there are four carbon atoms to attack (✻). Two carbons can be attacked to form the 2-chloro product but only one carbon can be attacked to form the 3-chloro product.

(b) (Chloro methyl) benzene means substitution of the alkyl (CH_3) side chain by free - radical mechanism. Chloro-4- methyl- benzene means electrophilic substitution onto the benzene ring.

(c) hydrolysis of halogen in an alkyl group is easier because the C–Cl bond is weaker than that for chlorine attached to a benzene ring.

Q3

$C_2H_5 - C{\overset{O}{\underset{OH(l)}{\big\langle}}} + SOCl_2(l) \rightarrow C_2H_5 - C{\overset{O}{\underset{Cl(l)}{\big\langle}}} + HCl(g) + SO_2(g)$

(i) 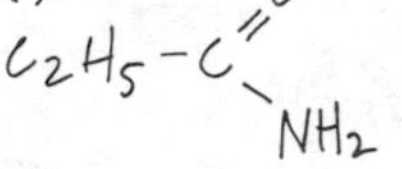
$C_2H_5 - C{\overset{O}{\underset{NH_2}{\big\langle}}}$ propanamide

(ii)
$C_2H_5 - C{\overset{O}{\underset{O-}{\big\langle}}}$ ⬡ phenyl propanoate 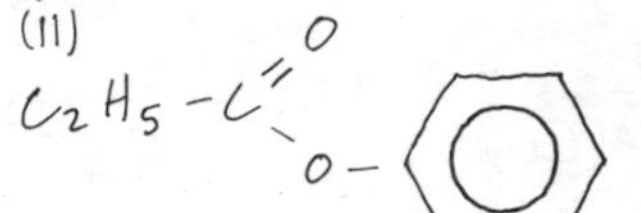

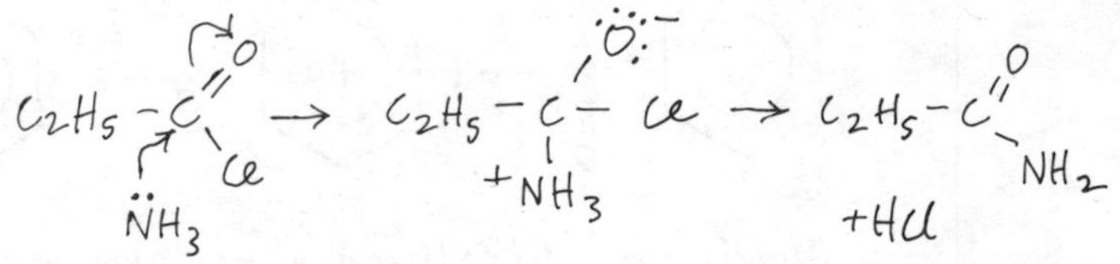

$C_2H_5 - C{\overset{O}{\underset{\ddot{N}H_3}{\overset{|}{\big\langle}}}}Cl \rightarrow C_2H_5 - C - Cl \rightarrow C_2H_5 - C{\overset{O}{\underset{NH_2}{\big\langle}}}$
$\qquad\qquad\qquad +NH_3 \qquad\qquad +HCl$

EXAMINATION QUESTION AND TUTOR'S ANSWER

Q4   a)   State the reagent(s) and write an equation for the formation of each of the following compounds from 1-bromopropane.

     i) $CH_3CH_2CH_2OCH_3$

        Reagent(s) <u>Sodium methoxide (sodium reacted with methanol)</u>

        Equation <u>$CH_3CH_2CH_2Br + CH_3O^-Na^+ \rightarrow CH_3CH_2CH_2OCH_3 + Na^+Br^-$</u>

     ii) $CH_3CH_2CH_2CN$

        Reagent(s) <u>ethanolic potassium cyanide</u>

        Equation <u>$CH_3CH_2CH_2Br + KC^-N^- \rightarrow CH_3CH_2CH_2CN + K^+Br^-$</u>

     iii) $CH_3CH=CH_2$

        Reagent(s) <u>ethanolic potassium hydroxide</u>

        Equation <u>$CH_3CH_2CH_2Br + KOH \rightarrow CH_3CH=CH_2 + K^+Br^- + H_2O$</u>   *(7)*

  b)   Outline a mechanism for the reaction taking place in a)ii).

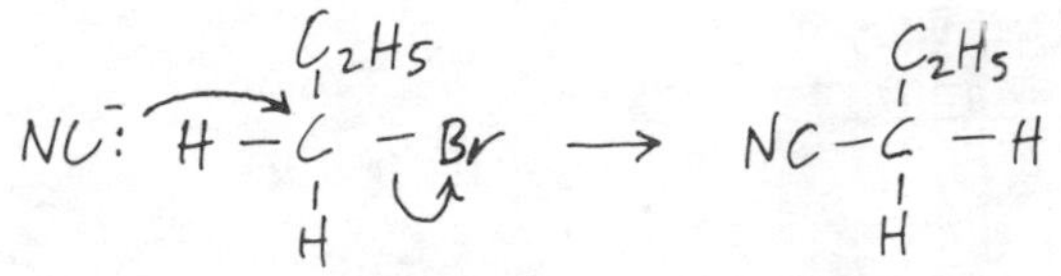

                                                  (3)

                                               (JMB 1988)

PRACTICE QUESTIONS

Q5   Study the section in the text on S_N2 and E_N2 reactions

Describe what products may be formed when a haloalkane (such as 1-bromobutane) reacts with a nucleophile (such as the hydroxide ion). Explain how the reaction conditions may influence the nature of the products.   *(6)*

                                                (WJEC 1988)

Q6   a)   Discuss the mechanism of the formation of chloromethane from methane and chlorine in the presence of ultra-violet light. Explain briefly why the chloromethane obtained in this way is impure.   *(12)*

    b)   Describe, giving a mechanism, the nucleophilic substitution reactions of haloalkanes and the uses of these reactions in synthesis.   *(18)*

                                                (JMB (AS-level specimen))

CHAPTER

17

ALCOHOLS
AND PHENOLS

GETTING STARTED

Alcohols and phenols contain the —OH functional group. The two classes of compounds show both similarities and differences in properties. In alcohols the —OH is attached directly to a carbon atom that is part of an aliphatic chain or an alicyclic ring. The emphasis is upon comparing the effect of the organic molecular structure upon the reactions of the —OH group. The influence of the —OH group upon the reactivity of the benzene ring is also important.

ESSENTIAL PRINCIPLES

Organic compounds containing the hydroxy group (—OH) are normally classed as alcohols
or phenols.

- In alcohols the —OH is attached directly to a carbon atom that is part of an
  aliphatic chain or an alicyclic ring
- In phenols the —OH is attached directly to a carbon atom that is part of a
  benzene ring

GENERAL PROPERTIES

Hexan-1-ol ($C_6H_{13}OH$) and phenol (C_6H_5OH) are two hydroxy-compounds whose properties are typical of alcohols and phenols respectively.

**PHYSICAL
PROPERTIES**

The physical properties are typical of substances with covalent molecular structures but
their melting and boiling points are higher than those of aliphatic and aromatic hydrocarbons of similar molar mass: see Fig. 17.1.

NAME	M_r	T_b/°C	FORMULA AND TYPE OF STRUCTURE		INTERMOLECULAR FORCES
triptane*	100	81	$(CH_3)_3C$—$CH(CH_3)_2$	branched chain	van der Waals
heptane	100	98	$CH_3(CH_2)_5CH_3$	straight chain	van der Waals
hexan-1-ol	102	158	$CH_3(CH_2)_5CH_2$—OH	straight chain	van der Waals dipole-dipole H-bonding
phenol	94	182	⬡—OH	planar ring	van der Waals dipole-dipole H-bonding

* systematic name is 2,2,3-trimethylbutane

Fig. 17.1 Physical properties of
some hydroxycompounds

Methanol (CH_3OH) and ethanol (C_2H_5OH) can be used as ionising solvents (for KOH,
for example) and they mix with water in all proportions. In general, however, the solubility
in water of the higher alcohols and other hydroxyhydrocarbons decreases as the proportion
of hydrocarbon in the molecular structure increases and they are not good solvents for
ionic or polar solutes. Hexanol and phenol are not very soluble but behave as surfactants
by lowering the surface tension of water. Hexadecan-1-ol (cetyl alcohol) has been spread
on Australian lakes and reservoirs to reduce evaporation and conserve water.

The —OH group makes the molecules polar, so permanent dipole-dipole attractions
contribute to the intermolecular forces. Hydrogen bonding between the OH-groups
increases the intermolecular forces of attraction even more. These dipole-dipole and
hydrogen-bonding forces account for the differences in the physical properties of the
alcohols and phenols compared to the hydrocarbons. The hydrophilic —OH group and
hydrophobic character of the alkyl or aryl group together account for the surfactant
behaviour.

**CHEMICAL
PROPERTIES**

COMBUSTION

Alcohols will burn in excess air or oxygen to form carbon dioxide and water: e.g.

$$C_2H_5OH(l) + 3O_2(g) \rightarrow 2CO_2(g) + 3H_2O(l)$$

Phenols ignite less readily and tend to burn with very smoky flames.

REACTION WITH SODIUM

Metallic sodium reduces the —OH group and liberates hydrogen: e.g.

$$C_2H_5\text{—}O\text{—}H(l) + Na(s) \rightarrow C_2H_5\text{—}O^-Na^+(alc) + \tfrac{1}{2}H_2(g)$$

The (alc) indicates that the product, sodium ethoxide, is dissolved in the excess liquid reactant, ethanol (alcohol). This reaction with alcohol provides a safe way of disposing of sodium residues. The ethoxide ion ($C_2H_5O^-$) is a powerful base and the ethanol molecule, its conjugate acid, is extremely weak, so aqueous wide-range indicator turns purple (pH 14 because $[OH^-] \gg [H^+]$) when you add it to the alcoholic sodium ethoxide:

$$C_2H_5O^-(alc) + H_2O(l) \rightarrow C_2H_5OH(aq) + OH^-(aq)$$

Phenol reacts with sodium in a similar way to ethanol but only if it is heated and the sodium is added to the molten phenol:

$$C_6H_5O{-}H(l) + Na(l) \rightarrow C_6H_5O^-Na^+(l) + \tfrac{1}{2}H_2(g)$$

REACTION WITH ACYL CHLORIDES

Acyl chlorides react with the —OH group to produce esters: e.g.

$$
\begin{array}{ccccc}
& & \overset{\displaystyle O}{\underset{\displaystyle \parallel}{}} & & \overset{\displaystyle O}{\underset{\displaystyle \parallel}{}} \\
C_3H_7O{-}H(l) & + & CH_3C{-}Cl(l) & \rightarrow & CH_3C{-}OC_3H_7(l) + HCl(g) \\
\text{propanol} & & \text{ethanoyl chloride} & & \text{propyl ethanoate}
\end{array}
$$

You may use this method in the laboratory to prepare a phenyl ester because phenol does not react directly with a carboxylic acid. It is usually conducted in aqueous alkali which neutralises the HCl(aq) formed:

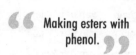
Making esters with phenol.

$$
\begin{array}{ccccc}
& & \overset{\displaystyle O}{\underset{\displaystyle \parallel}{}} & & \overset{\displaystyle O}{\underset{\displaystyle \parallel}{}} \\
C_6H_5O{-}H(aq) & + & C_6H_5C{-}Cl(l) & \rightarrow & C_6H_5C{-}OC_6H_5(s) + HCl(aq) \\
\text{phenol} & & \text{benzoyl chloride} & & \text{phenyl benzoate}
\end{array}
$$

REACTION WITH ACID ANHYDRIDES

These reactions are similar to those with acyl chlorides but usually less vigorous: e.g.

$$
\begin{array}{lll}
C_6H_5OH(aq) + (CH_3CO)_2O(l) & \rightarrow & CH_3CO_2C_6H_5(l) + CH_3CO_2H(aq) \\
\text{phenol} \qquad \text{ethanoic anhydride} & & \text{phenyl ethanoate} \quad \text{ethanoic acid}
\end{array}
$$

The reaction is carried out in aqueous alkali e.g. NaOH and the ethanoic acid converted into the ethanoate salt $CH_3CO_2^-Na^+$.

- ROH reacts with (i) Na (metal) and (ii) $R'COCl$ (acyl chlorides) or $(R'CO)_2O$ (acid anhydrides) to form (i) RO^-Na^+ and (ii) $R'CO_2R$

ALCOHOLS

Alcohols constitute a broad class of organic compounds containing one or more hydroxyl (—OH) groups not attached directly to a benzene ring. They may be grouped into a variety of homologous series based on the number of hydroxyl groups and the nature of the carbon skeleton.

NOMENCLATURE

- A monohydric, dihydric, trihydric or polyhydric alcohol contains one, two, three or more than three —OH groups respectively in its molecule

The systematic method of naming alcohols is based on the method of naming the alkanes.

- The root name is based on the longest continuous carbon chain and the name of the straight-chain alkane with the same number of carbons
- For monohydric alcohols the letter 'e' at the end of the root name is replaced by the ending 'ol' with a number, when necessary, to show the position of the —OH group on the carbon skeleton

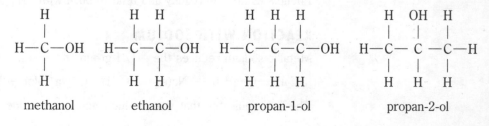

The traditional (unsystematic) names are methyl alcohol, ethyl alcohol, propyl alcohol and isopropyl alcohol. These names may be found in some textbooks and are still used by some chemists. Make sure that you use the correct systematic names.

■ For dihydric, trihydric, etc. alcohols the letter 'e' is NOT replaced but the letters 'diol', 'triol', etc. and numbers 1,2,3 etc. are added to the ending to show how many —OH groups and their positions

$$
\begin{array}{ccc}
& \overset{\displaystyle OH}{\underset{\displaystyle |}{}} \ \ \overset{\displaystyle OH}{\underset{\displaystyle |}{}} & \qquad \overset{\displaystyle OH}{\underset{\displaystyle |}{}} \ \ \overset{\displaystyle OH}{\underset{\displaystyle |}{}} \ \ \overset{\displaystyle OH}{\underset{\displaystyle |}{}} \\
H - C - C - H & H - C - C - C - H \\
| \quad | & | \quad | \quad | \\
H \quad H & H \quad H \quad H
\end{array}
$$

ethane-1,2-diol   propane-1,2,3-triol
(glycol)   (glycerol)

The terms primary, secondary and tertiary are used to describe the type of alcohol by the alkyl group to which the —OH group is attached

$$
\begin{array}{ccc}
\overset{\displaystyle H}{\underset{\displaystyle |}{}} & \overset{\displaystyle H}{\underset{\displaystyle |}{}} & \overset{\displaystyle CH_3}{\underset{\displaystyle |}{}} \\
C_3H_7 - C - OH & C_2H_5 - C - OH & CH_3 - C - OH \\
| & | & | \\
H & CH_3 & CH_3
\end{array}
$$

primary   secondary   tertiary

butan-1-ol   butan-2-ol   2-methylpropan-2-ol
(butyl alcohol)   (sec-butyl alcohol)   (tert-butyl alcohol)

COMBUSTION OF ALCOHOLS

You may have to measure the enthalpy change upon combustion of an alcohol as part of your practical work. This important experiment is discussed in Chapter 6.

OXIDATION OF ALCOHOLS

Primary alcohols are readily oxidised to aldehydes by heating with sodium dichromate(VI) in aqueous sulphuric acid:

Primary alcohols give aldehydes.

Secondary alcohols give ketones.

Tertiary alcohols don't oxidise.

$$
\begin{array}{c}
\overset{\displaystyle H}{\underset{\displaystyle |}{}} \\
R - C - OH \\
| \\
H
\end{array}
\quad \xrightarrow[\text{distillation}]{\overset{Cr_2O_7^{2-}(aq)/H^+(aq)}{\text{continuous}}} \quad
R - C \overset{\displaystyle O}{\underset{\displaystyle \diagdown H}{\diagup\!\!\!\parallel}}
$$

primary alcohol   aldehyde

You may have to prepare an aldehyde as part of your laboratory work. To improve the yield, you should drip the primary alcohol onto the hot oxidising agent at about the same rate as you are collecting in the cooled receiver the aqueous aldehyde distillate from the reaction mixture: see Fig. 17.2.

Even using continuous distillation, some aldehyde will be oxidised further to the less volatile and more water-soluble carboxylic acid:

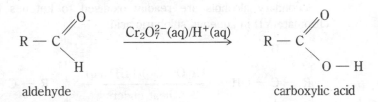

aldehyde   carboxylic acid

You may also have to prepare a carboxylic acid by oxidising a primary alcohol. In this case you must heat the mixture under reflux to prevent the escape of the volatile aldehyde and return it to the oxidising agent to ensure complete oxidation: see Fig. 17.3.

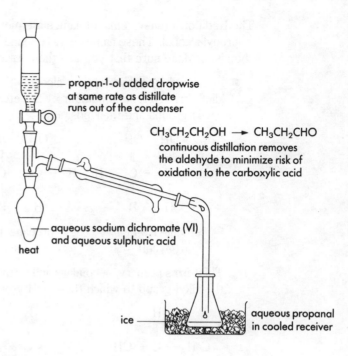

propan-1-ol added dropwise at same rate as distillate runs out of the condenser

$$CH_3CH_2CH_2OH \longrightarrow CH_3CH_2CHO$$

continuous distillation removes the aldehyde to minimize risk of oxidation to the carboxylic acid

aqueous sodium dichromate (VI) and aqueous sulphuric acid

heat

ice

aqueous propanal in cooled receiver

Fig. 17.2 Laboratory preparation of an aldehyde

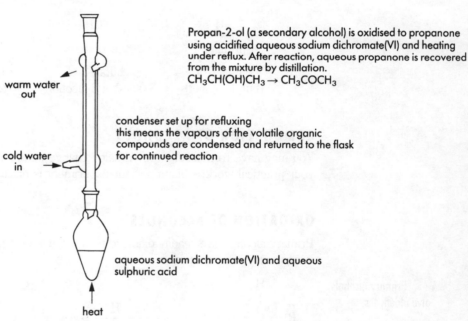

warm water out

Propan-2-ol (a secondary alcohol) is oxidised to propanone using acidified aqueous sodium dichromate(VI) and heating under reflux. After reaction, aqueous propanone is recovered from the mixture by distillation.
$$CH_3CH(OH)CH_3 \rightarrow CH_3COCH_3$$

condenser set up for refluxing this means the vapours of the volatile organic compounds are condensed and returned to the flask for continued reaction

cold water in

aqueous sodium dichromate(VI) and aqueous sulphuric acid

Fig. 17.3 Laboratory preparation of a ketone

heat

$$R - \underset{\underset{H}{|}}{\overset{\overset{H}{|}}{C}} - OH \xrightarrow[\substack{\text{heat under} \\ \text{reflux}}]{Cr_2O_7^{2-}(aq)/14H^+(aq)} R - C\overset{\displaystyle O}{\underset{\displaystyle O-H}{\big\langle}}$$

primary alcohol   carboxylic acid

If you warm a few drops of a primary alcohol with acidified sodium dichromate(VI) in a test-tube you can usually detect the pungent smell of the aldehyde.

Secondary alcohols are readily oxidised to ketones by refluxing with sodium dichromate(VI) in aqueous sulphuric acid:

$$R - \underset{\underset{R}{|}}{\overset{\overset{H}{|}}{C}} - OH \xrightarrow[\substack{\text{heat under} \\ \text{reflux}}]{Cr_2O_7^{2-}(aq)/14H^+(aq)} R - C\overset{\displaystyle O}{\underset{\displaystyle R}{\big\langle}}$$

secondary alcohol   ketone

Further oxidation of the ketone usually does not occur. However, you can oxidise cyclohexanol (secondary alcohol) beyond cyclohexanone (cyclic ketone) to hexanedioic acid (adipic acid – a straight-chain dicarboxylic acid) by using more drastic oxidising conditions e.g. refluxing with potassium manganate(VII) in aqueous sulphuric acid:

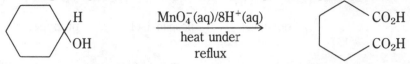

- Primary alcohols oxidise to aldehydes and carboxylic acids
- Secondary alcohols oxidise to ketones
- Tertiary alcohols oxidise only under drastic conditions

DEHYDRATION

The elements of water can be removed from an alcohol by passing the vapour over a heated catalyst such as pumice or aluminium oxide granules or by heating the liquid with concentrated phosphoric or sulphuric acid. At least two products could be obtained depending upon the conditions:

$$RCH_2—CH_2OH \xrightarrow[\textit{intra}\text{molecular dehydration}]{\text{excess sulphuric acid } 170\,°C} RCH=CH_2$$

alkene

$$2RCH_2—CH_2OH \xrightarrow[\textit{inter}\text{molecular dehydration}]{\text{excess alcohol } 140\,°C} RCH_2—CH_2—O—CH_2—CH_2R$$

ether

For example, ethanol reacts with conc. sulphuric acid to form ethyl hydrogensulphate (perhaps by electrophilic attack of the sulphur atom in the Lewis acid, SO_3, upon a lone electron-pair on the oxygen atom in the nucleophilic —OH group acting as a very weak Lewis base):

$$H—CH_2—CH_2OH \xrightarrow[\delta+SO_3{}^{\delta-}]{\text{excess sulphuric acid}} H—CH_2—CH_2O—SO_2—OH$$

ethanol   ethyl hydrogensulphate

The ethyl hydrogensulphate, which is not isolated from the reaction, undergoes an elimination reaction to form ethene and regenerate the sulphuric acid catalyst:

$$H—CH_2—CH_2O—SO_2—OH \xrightarrow{170\,°C} CH_2=CH_2 + H_2SO_4$$

The reverse of these reversible reactions provides a way of hydrating an alkene to make an alcohol.

- Propene can be reacted with conc. sulphuric acid followed by water to form propan-2-ol (but NOT propan-1-ol)

If excess ethanol (a nucleophile) is added to the ethyl hydrogensulphate a nucleophilic substitution occurs to form ethoxyethane:

$$CH_3CH_2OH + CH_3CH_2O—SO_2—OH \xrightarrow{140\,°C} CH_3CH_2—O—CH_2CH_3$$

↑   ↑
nucleophilic   electrophilic
oxygen atom   carbon atom

- Dehydration of alcohols gives mixtures of alkenes and ethers in varying proportions depending upon the conditions of the reaction

ESTERIFICATION

Alcohols react reversibly with carboxylic acids to form esters and water. The reaction is catalysed by hydrogen ions, usually from a few drops of conc. sulphuric acid, and an equilibrium mixture is obtained: e.g.

| ethanoic acid | propan-1-ol | propyl ethanoate |

This reaction is discussed in more detail in the unit on carboxylic acids.

Esters are formed when alcohols react with acyl chlorides and with acid anhydrides but these reactions are generally called acylations. For example, alcohols react with ethanoyl chloride (or ethanoic anhydride) to form esters of ethanoic acid:

$$CH_3COCl(l) + ROH \rightarrow CH_3CO_2R + HCl$$
$$(CH_3CO)_2O(l) + 2ROH \rightarrow 2CH_3CO_2R + H_2O$$

FORMATION OF HALOGENOALKANES

The —OH group can be replaced by a halogen atom. For example, alcohols react with phosphorus pentachloride, phosphorus trichloride or sulphur dichloride oxide to form chloroalkanes: e.g.

$$C_4H_9OH(l) + PCl_5(s) \rightarrow C_4H_9Cl(l) + POCl_3(l) + HCl(g)$$
butan-1-ol $\qquad$ 1-chlorobutane
$$(CH_3)_3COH(l) + SCl_2O(l) \rightarrow (CH_3)_3CCl(l) + SO_2(g) + HCl(g)$$
2-dimethylpropan-2-ol $\qquad$ 2-chloro-2-methylpropane

You can easily detect the hydrogen chloride fumes by the dense white smoke of ammonium chloride formed with ammonia. Consequently, we can use the reaction with, for example, phosphorus pentachloride to test for an hydroxyl group in a molecule.

You can convert alcohols into bromoalkanes (or iodoalkanes) by refluxing with concentrated hydrobromic (or hydroiodic acid): e.g.

$$CH_3CH_2-OH + HBr \rightarrow CH_3CH_2-OH_2^+ + Br^-$$
$$\qquad\qquad\qquad\qquad\qquad\uparrow$$
ethanol $\qquad\qquad$ electrophilic $\qquad$ nucleophilic
$$\qquad\qquad\qquad\qquad centre \qquad\qquad anion$$
$$CH_3CH_2-OH_2^+ + Br^- \rightarrow CH_3CH_2Br + H_2O$$
$$\qquad\qquad\qquad\qquad\qquad bromoethane$$

Concentrated hydrochloric acid is less effective because HCl is not as strong a protonating agent as HBr or HI. This is so because Cl^- is a weaker nucleophile than Br^- or I^-.

You can also use a mixture of potassium bromide (or iodide) and concentrated phosphoric (or sulphuric) acid to generate the hydrogen halide in situ. Phosphoric acid is usually preferred because, unlike sulphuric acid, it does not oxidise the hydrogen halide to halogen. Similarly we can use red phosphorus and iodine (or bromine but not usually chlorine) to generate phosphorus triiodide or tribromide in situ.

PREPARATION OF 1-BROMOBUTANE

As part of your practical work you may have to prepare and determine the yield of 1-bromobutane from butan-1-ol, potassium bromide and conc. sulphuric acid:

$$C_4H_9OH(l) + KBr(s) + H_2SO_4(l) \rightarrow C_4H_9Br(l) + KHSO_4(s) + H_2O(l)$$

You can measure the mass (m_a) of alcohol accurately by differential weighing directly into your reaction flask or by finding its volume (v_a) with a measuring cylinder and using its density (ρ_a) (from a data book) to calculate the mass ($m_a = v_a \times \rho_a$). From this mass (m_a) and the molar mass (M_a) (from a data book) you can find the amount ($X_a = m_a/M_a$) of the alcohol. You can determine in the same way the amount (X_b) of 1-bromobutane you have purified and collected. According to the equation above, one mole of alcohol should produce one mole of bromoalkane to give a 100% yield. This is never achieved. Loss of yield is caused by such things as incomplete reaction, side-reactions and experimental

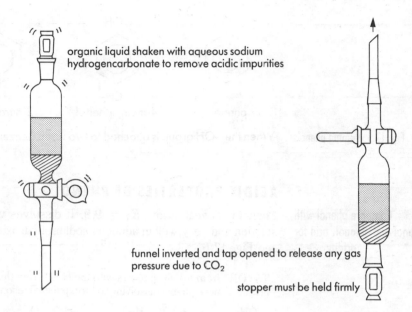

organic liquid shaken with aqueous sodium
hydrogencarbonate to remove acidic impurities

funnel inverted and tap opened to release any gas
pressure due to CO_2

stopper must be held firmly

Fig. 17.4 Purification of
1-bromobutane in a separating
funnel

techniques. When the theoretical ratio of reactant to product is 1:1 as in this example, you calculate your actual percentage yield by $100 \times X_b/X_a$.

You put the potassium bromide and butan-1-ol into the reaction flask. Make sure none of the crystals gets into the glass joint to spoil the seal to the condenser. You connect up the apparatus for heating under reflux as in Fig. 17.3, add and mix in the sulphuric acid, a little at a time, down through the condenser and then reflux the mixture for at least 45 minutes. You allow the apparatus to cool to minimise loss of product by evaporation before reassembling the apparatus for distillation: see Chapter 5, Fig. 5.12. The product steam distils and you stop distilling when no more oily drops leave the condenser. You carry out most of the purification in a separating funnel: see Fig. 17.4.

The steps are:

1. Separate the dense product from the less dense water
2. Shake with conc. hydrochloric acid to remove unreacted butan-1-ol, which becomes protonated and soluble in the aqueous acid, and separate the two layers
3. Shake with aqueous sodium hydrogencarbonate to remove any acid (sodium hydroxide is not used because it is too alkaline and likely to hydrolyse the product back to an alcohol) and separate the two layers
4. Shake with distilled water to remove the alkali and separate the two layers
5. Shake the cloudy product with anhydrous sodium sulphate to remove moisture – this may take some time
6. Separate the clear, colourless liquid from the solid drying agent and distil in a dry apparatus, collecting your purified product distilling at the appropriate temperature

THE IODOFORM REACTION

Alcohols containing the structure $CH_3CH(OH)$— will, when warmed with iodine and aqueous sodium hydroxide, produce a yellow precipitate of triiodomethane CHI_3 (iodoform):

$$CH_3CH(OH)—R \xrightarrow[NaOH]{I_2} CH_3CO—R \xrightarrow[NaOH]{I_2} CI_3CO—R \xrightarrow[NaOH]{H_2O} CHI_3 + R—CO_2^-$$

The alcohol is first oxidised to a methyl ketone, so the iodoform reaction also takes place with ethanal CH_3CHO and ketones containing CH_3CO— in their structure.

PHENOLS

Phenols are a class of aromatic compounds with one or more —OH groups attached directly to a benzene ring.

NOMENCLATURE

The simplest compound with only one —OH group attached to an unsubstituted benzene ring is called phenol. The name of a substituted phenol contains the name and position on the ring of the substituent group: see Fig. 17.5.

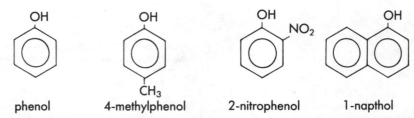

phenol   4-methylphenol   2-nitrophenol   1-napthol

Fig. 17.5 Naming phenols   When the —OH group is attached to two fused benzene rings the compound is called a naphthol.

ACIDIC PROPERTIES OF PHENOL

Compare phenol with ethanol and ethanoic acid for acidity.

Phenol is a weak acid: $pK_a = 9.9$. It dissolves very sparingly in water to give an acidic solution and very well in aqueous sodium hydroxide to give a solution of sodium phenate: see Fig. 17.6.

The OH^- from the alkali reacts with the H_3O^+ from the phenol and disturbs the equilibrium in favour of more phenol dissolving to form phenoxide ions:

| OH | + | H_2O | ⇌ | O^- | + | H_3O^+ |

| Na^+OH^- | + | H_3O^+ | → | Na^+ | + | $2H_2O$ |

| C_6H_5OH | + | Na^+OH^- | → | $C_6H_5O^-Na^+$ | + | H_2O |

Fig. 17.6 Phenols dissolve in aqueous alkali

$C_6H_5O^-Na^+$ is also known as sodium phenolate, sodium phenoxide and even sodium carbolate (carbolic acid being the old name for phenol).

We explain phenol's ability to donate a proton and behave as an acid by the stabilisation of its conjugate base, the $C_6H_5O^-$ anion. The negative charge does not reside only on the oxygen but is delocalised by interaction with the π-electrons of the benzene ring: see Fig. 17.7.

Ethanol (a solvent for some acid-base indicators) is not an acid because the $C_2H_5O^-$ cannot be stabilised by delocalisation.

- Phenol is a weaker acid than carbonic acid (H_2CO_3) and so it does not react with carbonates and hydrogencarbonates to give off carbon dioxide
- Ethanoic acid is stronger than carbonic acid and therefore does react with carbonates and hydrogencarbonates to give off carbon dioxide

Carboxylic acids are stronger than unsubstituted phenols because more effective delocalisation of the charge makes the conjugate base carboxylate anion more stable than the phenolate anion: see the unit on carboxylic acids. This difference between phenol and carboxylic acids (and the explanation for it) is often tested in A-level examinations.

negative charge on the oxygen atom becomes part of the delocalised π-electrons of the benzene ring and helps to stabilise the phenoxide ion

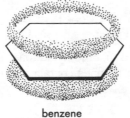

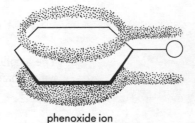

Fig. 17.7 Delocalised structure of the phenoxide anion

benzene   phenoxide ion

COMPLEXING PROPERTIES OF PHENOL

- Phenol turns 'neutral' aqueous iron(III) chloride from yellow-orange to purple in colour

This is a simple diagnostic test for many phenols. You may use it in your practical work. It often turns up in exams. The phenolate ion $C_6H_5O^-$ probably acts as a ligand with a lone electron-pair on the oxygen atom forming a dative bond with the iron(III) cation: e.g.

$$[Fe(H_2O)_6]^{3+}(aq) + 2C_6H_5O^-(aq) \rightarrow [Fe(H_2O)_4(C_6H_5O)_2]^+(aq) + 2H_2O(l)$$
$$\text{purple colour}$$

It is important that the test be carried out in neutral or near-neutral conditions. If the conditions are alkaline, there will be plenty of $C_6H_5O^-$ ions but the iron(III) ions will precipitate as iron(III) hydroxide:

$$[Fe(H_2O)_6]^{3+}(aq) + 3OH^-(aq) \rightarrow Fe(OH)_3(s) + 6H_2O(l)$$

If the conditions are acidic the iron(III) ions will not precipitate but the strongly basic electron-pair donating $C_6H_5O^-$ ions will accept protons to become the weakly acidic C_6H_5OH molecules, which are weaker ligands.

NUCLEOPHILIC PROPERTIES OF PHENOL

■ The —OH group activates the benzene ring so that phenol undergoes very rapid electrophilic substitutions in the ortho/para (2,4,6) positions

When you shake phenol with aqueous bromine, the orange-brown solution rapidly decolourises and a white precipitate forms: see Fig. 17.8

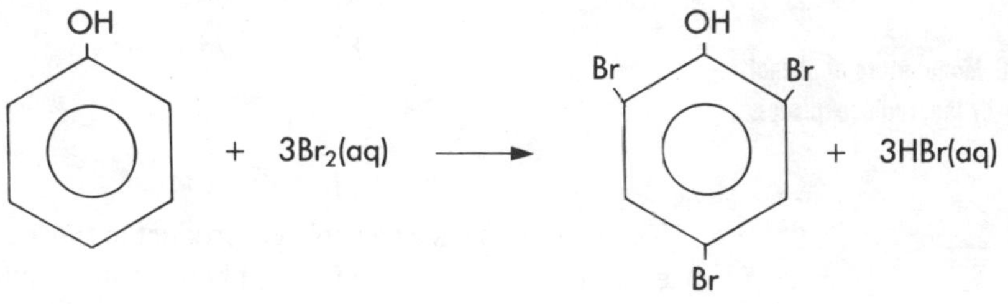

Fig. 17.8 Electrophilic substitution of phenol

2,4,6-tribromophenol
(white solid with an
'antiseptic' smell)

■ Phenols and naphthols act as good nucleophilic coupling agents with electrophilic diazonium ions to form strongly coloured 'azo' dyes

When you add a solution of phenol in aqueous sodium hydroxide to ice-cold aqueous benzene diazonium chloride, an orange precipitate forms: see Fig. 17.9

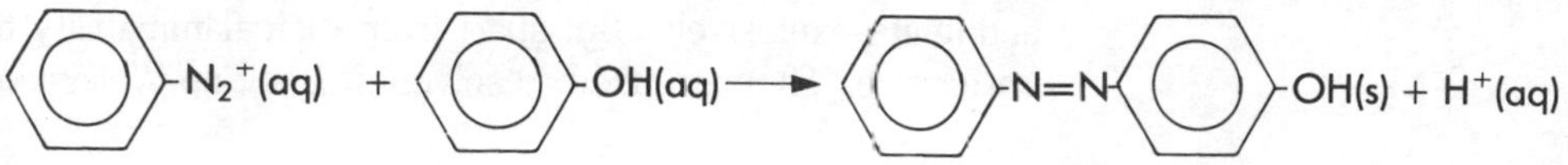

Fig. 17.9 Nucleophilic phenols couple with electrophilic diazonium ion

4-hydroxyazobenzene
an orange solid

MANUFACTURE AND USES OF ETHANOL AND PHENOL

Pure ethanol is manufactured by hydration of ethene produced by the petrochemical industry from oil and natural gas: e.g.

$$CH_2{=}CH_2 + H_2O \xrightarrow[\substack{\text{phosphoric acid catalyst} \\ \text{on a silica support}}]{60\text{--}70\,\text{atm at }300\,°C} CH_3{-}CH_2OH$$

Fermentation is still a very important biochemical process for ethanol production, especially in the 'drinks' industry. The essential reaction is:

$$C_6H_{12}O_6(aq) \xrightarrow[\text{enzyme catalyst}]{35\,°C} 2C_2H_5OH(aq) + 2CO_2(g)$$

In some countries, e.g. Brazil, it is economic to ferment carbohydrates as a method of manufacturing ethanol for use as a motor fuel. Some lead-free petrols may contain 10–20% ethanol. Ethanol is an important industrial solvent.

Phenol is manufactured by the 'cumene' process: see Fig. 17.10.

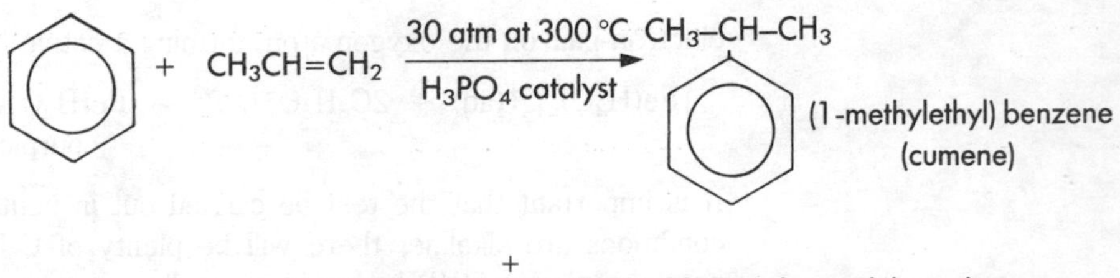

$$\text{benzene} + CH_3CH=CH_2 \xrightarrow[\text{H}_3\text{PO}_4 \text{ catalyst}]{30 \text{ atm at } 300\,°C} \begin{array}{c} CH_3\text{–}CH\text{–}CH_3 \\ \text{(1-methylethyl) benzene} \\ \text{(cumene)} \end{array}$$

The catalyst may protonate the alkene (CH₃–$\overset{+}{C}$H–CH₃) to give an electrophilic carbocation to attack the benzene ring.

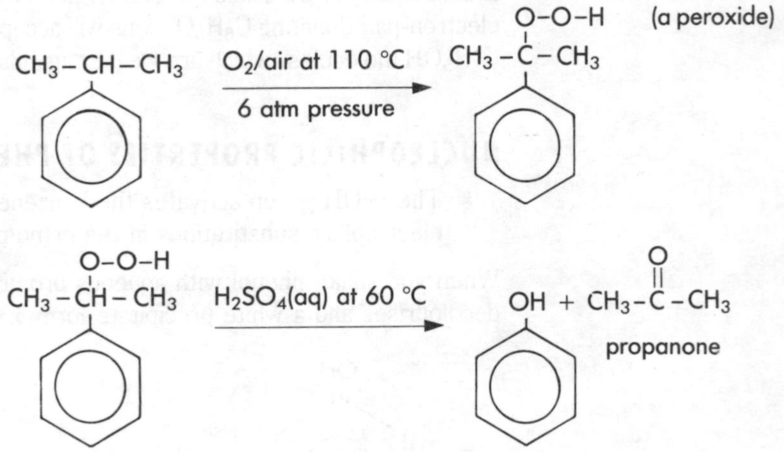

$$CH_3\text{–}CH\text{–}CH_3 \xrightarrow[6 \text{ atm pressure}]{O_2/\text{air at } 110\,°C} CH_3\text{–}C\text{–}CH_3 \quad (\text{a peroxide})$$

$$CH_3\text{–}CH\text{–}CH_3 \text{ (O–O–H)} \xrightarrow{H_2SO_4(aq) \text{ at } 60\,°C} OH + CH_3\text{–}C\text{–}CH_3 \quad \text{propanone}$$

Fig. 17.10 Manufacture of phenol by the 'cumene' process

The propanone is a valuable co-product in the 'cumene' manufacturing process. Phenol is used in the production of epoxy-, phenolic and poly(carbonate) resins. Another major use is to make caprolactam – an intermediate in the manufacture of nylon-6.

ETHERS

The ethers are a class of compounds that are isomeric with alcohols and phenols but the oxygen atom is connected to two carbon atoms: e.g.

CH_3—O—CH_3 methoxymethane (dimethyl ether) C_6H_5—O—CH_3 methoxybenzene (anisole or methylphenyl ether)

CH_3CH_2—OH ethanol CH_3—C_6H_4—OH methylphenol(s)

Some ethers, like ethoxyethane (diethyl ether or just 'ether'), have very low flash-points and ignite explosively. But apart from their flammability the ethers are chemically rather unreactive. They are used as solvents. Look up Williamson's synthesis in your textbook.

EXAMINATION QUESTIONS AND OUTLINE ANSWERS

Q1 a) Outline the laboratory preparation of a pure sample of phenyl benzoate. (6)

b) How, and under what conditions, would you expect a compound with the formula

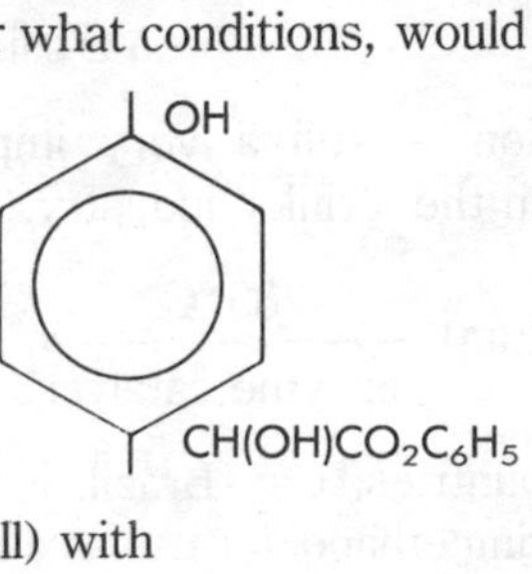

to react (if at all) with
  i) sodium
  ii) sodium hydroxide

iii) phosphorus pentachloride
iv) potassium dichromate(VI)
v) lithium tetrahydridoaluminate, $LiAlH_4$? *(3+4+2+2+3)*

(UCLES 1988)

Q2 a) Give the structural formula of each of the isomeric alcohols represented by C_4H_9OH.
b) Discuss the extent to which it is possible to distinguish between the isomers of C_4H_9OH by oxidation with acidified potassium dichromate(VI) solution.
c) What relevance has the formation of carboxylic acids from alcohols to dental decay?

(ULSEB (AS-level specimen))

Q3 a) Describe with brief experimental details, how phenol may be prepared, starting from benzene.
b) i) Arrange the following in order of increasing acidity:

ethanoic acid;   methanol;   phenol.

Increasing acidity sequence: _____
ii) Comment upon the position of phenol in the sequence in b)i).

(WJEC 1988)

Outline answers

Q1

(a)

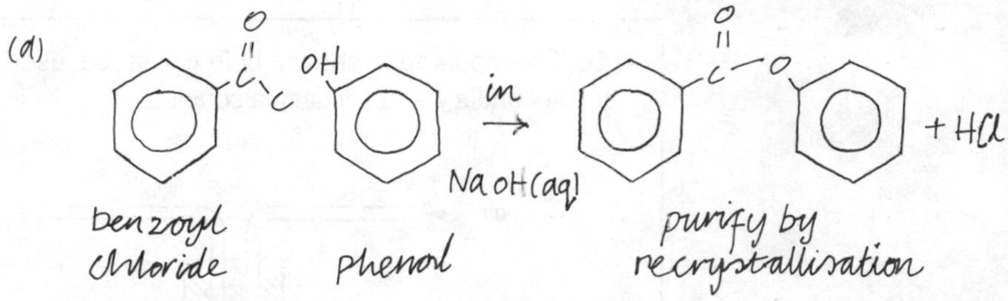

benzoyl chloride   phenol   purify by recrystallisation

(b) (i) −OH on ring attacked by molten sodium
−OH in side chain attacked without heating
to give $-O^-Na^+ + \frac{1}{2}H_2(g)$

(ii) −OH + NaOH (aq) → $-O^-Na^+$ (aq) + H_2O
on ring

(iii) −OH in side chain → − Cl + HCl

(iv) $-CH(OH)CO_2C_6H_5 \xrightarrow[aqueous]{acidified}$ $-C-CO_2C_6H_5$ with $\underset{O}{\overset{\|}{}}$

sec − alcohol   ketone

in anhydrous
(v) $-CO_2C_6H_5 \xrightarrow{ethoxyethane}$ $-CH_2OH$ and HOC_6H_5

Q2

(a) C−C−C−C−OH   butan−1−ol

C−C−C−C
        |
       OH   butan−2−ol

C−C−C−OH   2−methyl propan−1−ol
    |
    C

       OH
       |
C−C−C   2−methyl propan−2−ol
    |
    C

(b) (i) 2-methyl propan-2-ol is unaffected by acidified potassium dichromate

(ii) 2-methyl propan-1-ol and butan-1-ol will oxidise to aldehyde (that will give an orange-red ppᵛ with Fehlings) and also to carboxylic acids that give CO_2 with carbonates

(iii) butan-2-ol will oxidise to a ketone (that has no effect on Fehlings).

(c) Acids formed by bacterial attack on carbohydrates in mouth. The acids attack the enamel on teeth

Q3

(a) 'Cumene' process - see Fig. 17.10
(b) (i) Methanol → phenol → ethanoic acid (ii) see text and fig 17.7.

EXAMINATION QUESTION AND TUTOR'S ANSWERS

Q4 The apparatus shown below can be used to distinguish between primary, secondary and tertiary alcohols.

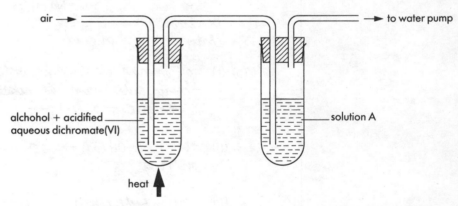

a) State the classes of compound, if any, which are produced by the oxidation of:
   i) primary alcohols   *aldehydes and carboxylic acids*

   ii) secondary alcohols   *ketones*
   iii) tertiary alcohols   *not oxidised*   (3)

b) i) What would be a suitable reagent as solution A to distinguish between primary and secondary alcohols in the first tube?
                                              Fehling's solution

   Describe what would be seen if the alcohol were
   ii) primary   *orange-red precipitate would form and the solution would turn from blue to green in colour*
                                                                          (3)
   iii) secondary   *no change*
                                              (WESSEX (specimen))

PRACTICE QUESTION

Q5 a) i) Describe the main industrial method for the manufacture of ethanol. *(4)*

ii) Describe the preparation of a secondary alcohol using methylmagnesium iodide (Grignard reagent), and state the mechanism of the reaction. *(5)*

b) Compare the behaviour of primary and secondary alcohols on oxidation, stating which oxidising agents are used. *(3)*

c) State how **both** (i) bromine water, **and** (ii) the iodoform (triiodomethane) reaction may be used to distinguish between ethanol and phenol. *(4)*

d) Given that the acid dissociation constant of phenol is $1 \times 10^{-10}\,mol\,dm^{-3}$, calculate the pH of aqueous phenol of concentration $1 \times 10^{-2}\,mol\,dm^{-3}$. *(4)*

(WJEC 1988)

GETTING STARTED

Aldehydes and ketones contain the carbonyl functional group $>C=O$ and have distinctive infra-red absorption spectra. The two classes of compounds show both similarities and differences in properties. In aldehydes the carbonyl group is always attached to one hydrogen atom to give the functional group —CHO. In ketones the carbonyl group is always attached to two carbon atoms. The emphasis is upon comparing the similarities and differences in the reactions of the —CHO and $>CO$ functional groups. The reactions of the $>C=O$ group are also compared to the reactions of the $>C=C<$ group.

ESSENTIAL PRINCIPLES

- Aldehydes contain the functional group — $C\begin{smallmatrix}H\\ \| \\ O\end{smallmatrix}$ (attached to a C-atom) and the name ends with the suffix -al for aliphatic compounds and with the word aldehyde for aromatic compounds

- Ketones contain the functional group $>C=O$ (attached to C-atoms but not to H-atoms) and the name ends with the suffix -one for aliphatic compounds and with the word ketone for aromatic compounds

GENERAL PROPERTIES

Propanal CH_3CH_2CHO and propanone $(CH_3)_2CO$ are two isomeric compounds whose properties are typical of the two classes of carbonyl compounds – aldehydes and ketones.

PHYSICAL PROPERTIES

The physical properties are typical of substances with polar covalent molecular structures. Their melting and boiling points are higher than those of aliphatic hydrocarbons and ethers but lower than those of aliphatic alcohols of similar molar mass: see Fig. 18.1.

NAME	M_r	$T_b/°C$	DIPOLE MOMENT/D	STRUCTURE	INTERMOLECULAR FORCES
butane*	58	0	0	$CH_3(CH_2)_2CH_3$	van der Waals
methoxyethane	60	11	1.23	$C_2H_5OCH_3$	dipole-dipole
propanal	58	49	2.52	CH_3CH_2CHO	dipole-dipole
propanone	58	56	2.88	$(CH_3)_2CO$	dipole-dipole
propan-1-ol	60	98	1.68	$CH_3CH_2CH_2OH$	dipole-dipole and H-bonding
propan-2-ol	60	83	1.66	$(CH_3)_2CHOH$	dipole-dipole and H-bonding

* non-polar molecule included for comparison with polar molecules

Fig. 18.1 Physical properties of some polar compounds

The $>C=O$ group makes the molecules polar, so permanent dipole-dipole attractions contribute to the intermolecular forces. Consequently only methanal and ethanal are gases at 25 °C and 1 atm. All the ketones and all the other aldehydes you will meet are liquids. Propanone and butanone are good solvents in which to carry out many organic reactions because they will disolve both ionic and covalent substances. Aldehydes are not such useful solvents because they are more reactive than ketones.

Aldehyde and ketone molecules do NOT hydrogen-bond to each other because they do not have a hydrogen atom attached to a sufficiently electronegative atom (such as N, O or F). Consequently nucleophilic substitution reactions occur readily in, for example, propanone as solvent (and faster than in water or alcohols as solvents) because the nucleophiles do not become solvated by hydrogen-bonding to the propanone molecules.

Carbonyl compound molecules may not hydrogen-bond to each other but they will form hydrogen bonds with water molecules and with the molecules of other hydrogen-bonded substances such as alcohols and carboxylic acids: see Fig. 18.2.

Consequently the lower aldehydes and ketones are soluble in water. So you may handle methanal and ethanal as aqueous solutions. Aqueous methanal (formalin) is a powerful

$$>C=O\cdots H—O \begin{smallmatrix} H\cdots O=C< \\ \ \end{smallmatrix}$$

hydrogen bonding between
the water molecules and
the carbonyl groups

Fig. 18.2 Lower aldehydes and ketones are water-soluble

disinfectant that has been used for preserving biological specimens. Propanone (acetone) is often used to rinse apparatus prior to drying. In general, as you should expect, the solubility of the higher members decreases as the proportion of hydrocarbon in the molecular structure increases.

CHEMICAL PROPERTIES

COMBUSTION

Carbonyl compounds will burn in excess air or oxygen to form carbon dioxide and water:
e.g. $(CH_3)_2CO(l) + 4O_2(g) \rightarrow 3CO_2(g) + 3H_2O(l)$

The lower members are extremely flammable and their vapours ignite explosively. Take great care when you are handling propanone (acetone). Keep it away from naked flames.

OXIDATION

■ Aldehydes are readily oxidised to carboxylic acids

When you warm in a test tube a few colourless drops of an aldehyde with acidified aqueous sodium dichromate(VI), you may expect to see the pale orange solution turn deep green. The unpleasant smell of the aldehyde may disappear and you may detect the smell of the carboxylic acid formed:

$$3CH_3{-}C{\overset{O}{\diagup}}_{\diagdown H} + Cr_2O_7^{2-}(aq) + 8H^+(aq) \rightarrow CH_3{-}C{\overset{O}{\diagup}}_{\diagdown OH} + 2Cr^{3+}(aq) + 4H_2O(l)$$
orange solution   green solution

■ Most aldehydes give a positive test with Fehling's solution

> **Fehling's test is positive for aldehydes but not ketones.**

Fehling's solution is alkaline but copper(II) hydroxide does not precipitate because the Cu^{2+} ion is held in solution as a complex cuprate(II) anion with 2,3-dihydroxybutanedioate (tartrate) anions acting as bidentate ligands: see Chapter 13.

When you warm in a test-tube a few colourless drops of aldehyde with a few cubic centimetres of Fehling's solution, you may expect to see the blue solution turn pale green or even colourless and an orange-red ppt. form:

$$3CH_3{-}C{\overset{O}{\diagup}}_{\diagdown H} + 2Cu^{2+}(aq) + 4OH^-(aq) \rightarrow CH_3{-}C{\overset{O}{\diagup}}_{\diagdown OH} + Cu_2O(s) + 2H_2O(l)$$
blue solution   orange-red ppt.
copper(II) complex   copper(I) oxide

C_6H_5CHO (usually called benzaldehyde rather than benzenecarbaldehyde) does not readily reduce Fehling's solution whereas HCHO (usually called methanal rather than formaldehyde) reduces the copper(II) complex to a metallic copper mirror on the inside of the test tube!

■ Most ketones are not readily oxidised to carboxylic acids

Some ketones (often cyclic) will degrade to a mixture of carboxylic acids with fewer carbon atoms than the original ketone when refluxed with vigorous oxidising agents such as acidified potassium manganate(VII).

REACTION WITH HALOGEN

Oxygen is more electronegative than carbon so the $C{=}O$ bond is polar and the C-atom is an electrophilic centre open to attack by nucleophiles: see Fig. 18.3.

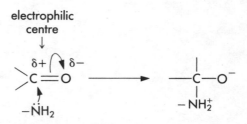

Fig. 18.3 Nucleophiles attack the carbon atom in the $>C{=}O$ group

Moreover, the $C=O$ bond is polarisable and the O-atom can accept a displaced π-electron pair to form a fairly stable intermediate oxoanion and make the α-carbon atom a nucleophilic centre open to attack by electrophiles: see Fig. 18.4.

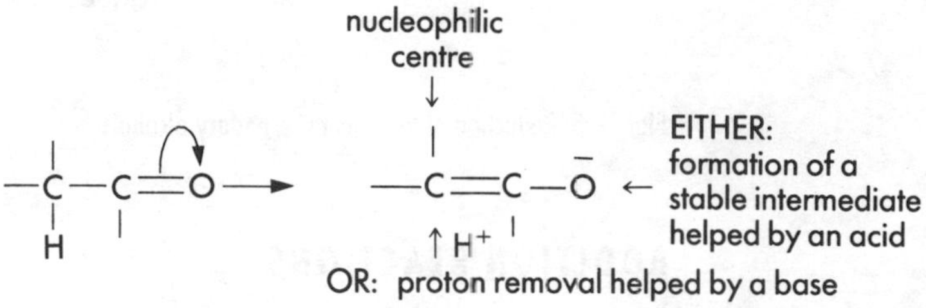

EITHER:
formation of a
stable intermediate
helped by an acid

OR: proton removal helped by a base

Fig. 18.4 Electrophiles attack the α-carbon atom next to the $\diagdown CO$ group

You should contrast this with the $C=C$ bond in which a C-atom is attacked initially by electrophiles. If nucleophiles were to attack the $C=C$ bond initially, the intermediate carbanion would be relatively unstable.

- A hydrogen atom on the α-carbon atom (i.e., the carbon atom next to the $C=O$ group) of an aldehyde or ketone is readily replaced by a halogen atom

Propanone reacts with iodine in aqueous acid (or alkali) according to the following equation:

$$CH_3-\overset{\overset{\displaystyle \|}{O}}{C}-CH_3 + I_2(aq) \xrightarrow[OH^-(aq)]{H^+(aq) \text{ or}} CH_3-\overset{\overset{\displaystyle \|}{O}}{C}-CH_2I + H^+(aq) + I^-(aq)$$

You may well investigate the kinetics of the acid-catalysed reaction in your practical work: see Chapter 9.

Iodoform test

- Ethanal and methyl ketones contain the $CH_3-\overset{\overset{\displaystyle H}{|}}{C}=O$ group and therefore give a positive result (yellow ppt. of CHI_3) with the iodoform test

You warm a small sample of the compound with aqueous alkali and a large amount of iodine (e.g. 2 drops of propanone need about 1 g of iodine) in a test-tube. The pale yellow crystalline ppt. of triiodomethane (iodoform) appears on cooling:

$$\overset{R}{\underset{CH_3}{\diagdown}}C=O + 3I_2(aq) + 4OH^-(aq) \rightarrow CHI_3(s) + R-CO_2^-(aq) + 3I^-(aq)$$

The hydrocarbon parts of the aldehyde and ketone molecules may undergo free-radical substitution reactions with chlorine and bromine under conditions similar to those for substitution reactions of alkanes: e.g.

$$CH_3-CHO \xrightarrow{Cl_2 \text{ \& UV light}} CCl_3-CHO \quad \text{trichloroethanal (chloral)}$$

And the oxygen in the carbonyl group may be replaced by chlorine by reaction with phosphorus pentachloride under anhydrous conditions: e.g.

$$(CH_3)_2CO \xrightarrow[\text{conditions}]{PCl_5 \text{ in anhydrous}} (CH_3)_2CCl_2 \quad 2,2\text{-dichloropropane}$$

REDUCTION

- Carbonyl compounds are reduced to alcohols
- Aldehydes are reduced to primary alcohols
- Ketones are reduced to secondary alcohols

$\diagdown C=O$ is polar and is attacked by nucleophiles.

$\diagdown C=C \diagup$ is non-polar and attacked by electrophiles.

In the Nuffield course you will use sodium tetrahydridoborate to reduce 1,2-diphenylethanedione to 1,2-diphenylethane-1,2-diol as part of your practical work: see Fig. 18.5.

Lithium tetrahydridoaluminate ($LiAlH_4$) is used in dry ethoxyethane. It is a powerful reducing agent that needs careful handling. It will reduce carboxylic acids to primary alcohols. We do not use it for practical work in schools.

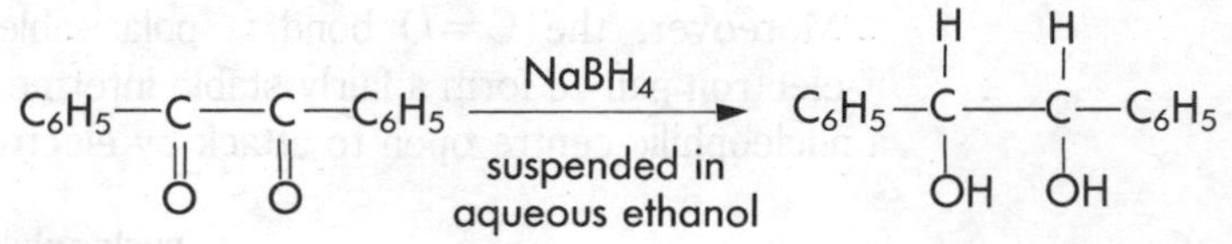

$$C_6H_5-\underset{\underset{O}{\|}}{C}-\underset{\underset{O}{\|}}{C}-C_6H_5 \xrightarrow[\substack{\text{suspended in} \\ \text{aqueous ethanol}}]{NaBH_4} C_6H_5-\underset{\underset{OH}{|}}{\overset{\overset{H}{|}}{C}}-\underset{\underset{OH}{|}}{\overset{\overset{H}{|}}{C}}-C_6H_5$$

Fig. 18.5 Reduction of ketones to secondary alcohols

ADDITION REACTIONS

The reduction of carbonyl compounds by sodium tetrahydridoborate or lithium tetra-hydridoaluminate may be seen as the addition of hydrogen across the $C=O$ double bond. As a possible mechanism, the BH_4^- and AlH_4^- ions could supply hydride anions, $:H^-$, for nucleophilic attack upon the electrophilic carbon atom in the carbonyl group and the water could supply hydrogen (ions) for electrophilic attack upon the nucleophilic oxygen atom in the carbonyl group: see Fig. 18.6.

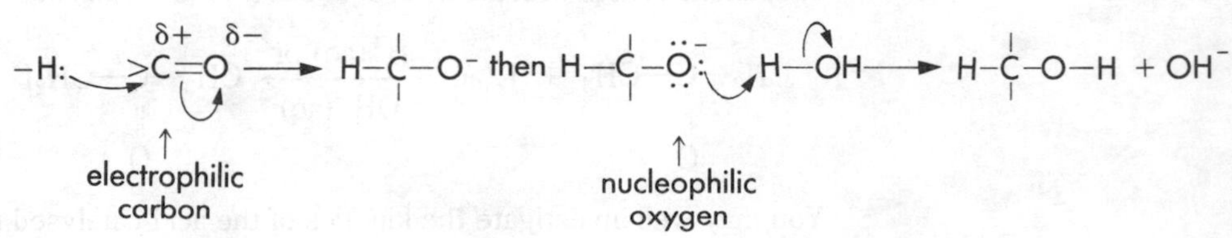

Fig. 18.6 Reduction of carbonyl compounds by metal hydrides

When you shake a few drops of an aliphatic aldehyde or ketone with a few cubic centimetres of freshly prepared aqueous sodium hydrogensulphite, you may expect to see colourless or white crystals form. The carbonyl compound undergoes a nucleophilic *addition* reaction to form the sodium hydroxysulphonate: see Fig. 18.7.

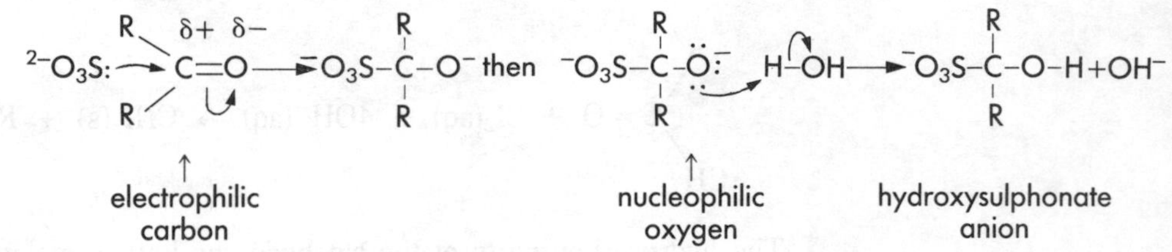

Fig. 18.7 Addition reaction with aqueous hydrogensulphite

You can purify the addition compound by recrystallisation then regenerate the original aldehyde or ketone by adding acid or alkali to the crystals. This gives you a way of purifying liquid or gaseous carbonyl compounds that would be difficult to purify by direct recrystallisation.

The addition of hydrogen cyanide to propanone is the first stage of a multistage process for the manufacture of a poly(methyl 2-methylpropenoate) and other acrylic polymers. You will find a summary of the nucleophilic addition reactions of aldehydes and ketones in Figure 18.8.

CONDENSATION REACTIONS

Compounds containing $-NH_2$ may initially undergo an addition reaction between the nucleophilic N-atom in the $-NH_2$ and the electrophilic C-atom in the $C=O$, but the product usually undergoes an elimination reaction to lose a water molecule and form a $N=C$ double bond: see Fig. 18.9.

■ A condensation reaction is an addition reaction followed by an elimination reaction

NUCLEOPHILE	REAGENTS AND CONDITIONS	FORMULA AND NAMES OF ADDITION PRODUCTS	
: H⁻	NaBH₄(aq) or LiAlH₄ in dry ethoxyethane then add water	OH | R'—C—H | R''	If R' = CH₃ & R'' = H then ethanol If R' = R'' = CH₃ then propan-2-ol
: CN⁻	aqueous sodium cyanide (KCN) then excess mineral acid	OH | R'—C—CN | R''	If R' = CH₃ & R'' = H then 2-hydroxy-propanenitrile If R' = R'' = CH₃ then 2-hydroxy-2-methylpropane-nitrile
: SO₃⁻	freshly made saturated aqueous sodium hydrogensulphite	OH | R'—C—SO₃⁻Na⁺ | R''	If R' = CH₃ & R'' = H then sodium 2-hydroxy-propane sulphonate
: C₂H₅⁻	Grignard reagent CH₃CH₂—Mg—Br in dry ethoxyethane then add water	OH | R'—C—CH₂CH₃ | R''	If R' = CH₃ & R'' = H then butan-2-ol If R' = R'' = CH₃ then 2-methylbutan-2-ol

The glucose reaction:

: O—H
H—C—CH₂OH

← nucleophilic oxygen atom

glucose sugar molecule forms a six-membered heterocyclic ring structure

H—C—OH
H—C—OH
H—C—OH
H—C=O ← carbonyl group

Fig. 18.8 Addition reactions of carbonyl compounds

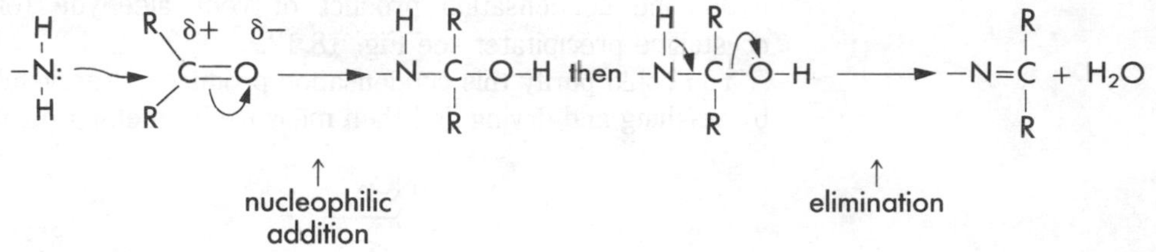

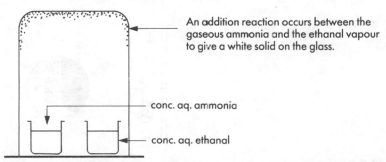

Fig. 18.9 Condensation as addition followed by elimination

nucleophilic addition → elimination

REACTION WITH AMMONIA

■ Aldehydes and ketones usually react in a complicated way with ammonia often to give resinous polymeric condensation products

When you place side by side two miniature beakers – one containing conc. aqueous ammonia, the other ethanal – and cover them with a large lipless beaker, you may expect to see a whity crystalline solid form on the inside wall of the large beaker within about one hour: see Fig. 18.10.

An addition reaction occurs between the gaseous ammonia and the ethanal vapour to give a white solid on the glass.

— conc. aq. ammonia

— conc. aq. ethanal

Fig. 18.10 Reaction of ammonia with ethanal

Ammonia molecules ($M_r = 17$) diffuse faster than the ethanal molecules ($M_r = 44$) (Graham's law) so reaction eventually occurs between the ammonia and ethanal at or in the aqueous ethanal surface.

Over a period of several days the colourless aqueous ethanal diminishes in volume, turns yellow-orange and eventually becomes a dark red-brown resin. The dark red-brown resin is probably a polymer formed by extensive addition and elimination reactions between the ethanal and the 'simple' addition products such as 1-amino-ethanal because they still have a nucleophilic N-atom to attack the electrophilic C-atom. The white crystalline solid is a simple nucleophilic addition product: see Fig. 18.11.

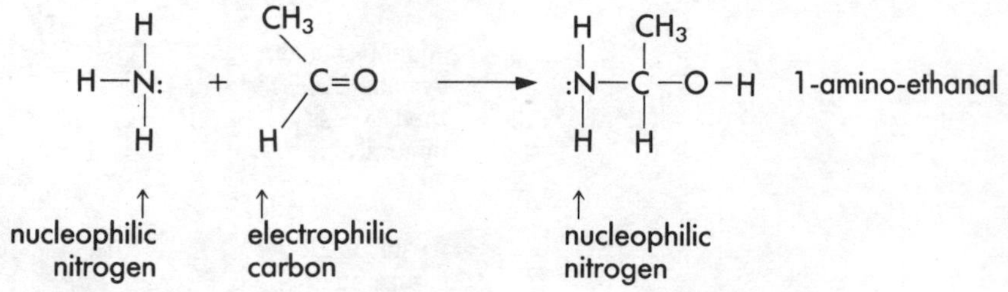

Fig. 18.11 Addition of ammonia to ethanal

Six moles of methanal react with four moles of ammonia probably by a sequence of nucleophilic additions and eliminations to produce hexamethylenetetramine:

$$6HCHO + 4NH_3 \rightarrow (CH_2)_6N_4 + 6H_2O$$
$$\text{hexamethylenetetramine}$$

This is an interesting heterocyclic fused ring compound (see Fig. 18.16) with a variety of industrial uses e.g. corrosion inhibitor, fungicide, rubber-to-textile adhesives, shrink-proofing textiles, and is used in high explosive cyclonite.

REACTION WITH OTHER —NH₂ COMPOUNDS

- Aldehydes and ketones react with compounds derived from ammonia usually to give simple, well-defined condensation products

In your practical work you may use a solution of 2,4-dinitrophenylhydrazine ('Brady's reagent') to help you identify an aldehyde or ketone. You add the reagent to a methanolic solution of your unknown carbonyl compound and you expect to obtain a 2,4-dinitrophenyl-hydrazone condensation product of your aldehyde (or ketone) as a yellow-orange crystalline precipitate: see Fig. 18.12.

You could purify this condensation product by recrystallisation from methanol followed by washing and drying and then measure its melting point: see Fig. 18.13.

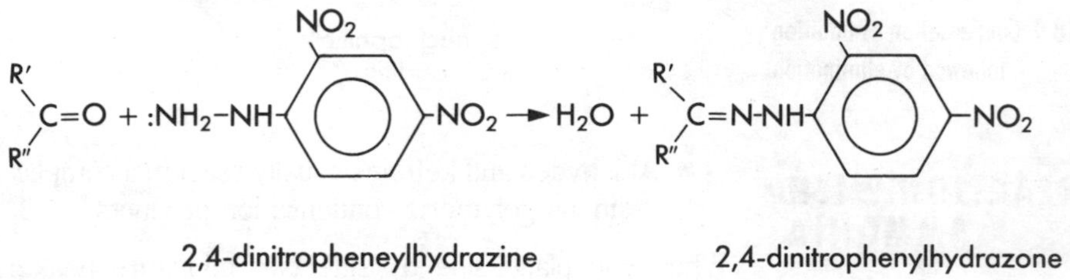

Fig. 18.12 Formation of 2,4-dinitrophenylhydrazone derivations

2,4-dinitropheneylhydrazine   2,4-dinitrophenylhydrazone

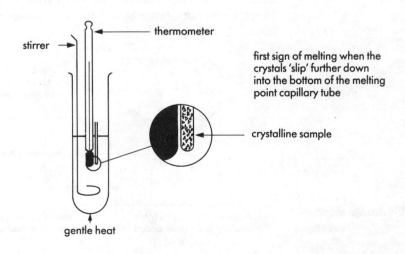

Fig. 18.13 Melting-point determination

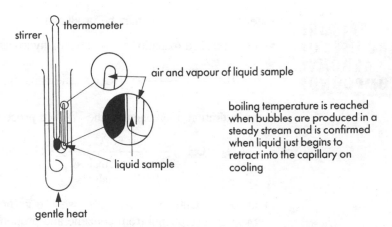

Fig. 18.14 Boiling-point determination

You could also measure the boiling point of your original unknown liquid carbonyl compound but not of its condensation product: see Fig. 18.14.

In spite of your experimental errors in making these two measurements, you should be able to use your values to distinguish between propanal and propanone for example: see Fig. 18.15.

There is a summary of the condensation reactions of carbonyl compounds in Fig. 18.16.

	PROPANAL	PROPANONE
formula	CH$_3$CH$_2$—C(=O)H	(CH$_3$)$_2$C=O
boiling point/°C	49	56
reaction with Fehling's solution	positive: forms orange-red precipitate	negative: does not form an orange-red ppt.
reaction with Brady's reagent	positive: forms yellow-orange crystalline ppt.	positive: forms yellow-orange crystalline ppt.
formula of the yellow-orange crystalline ppt.	C$_2$H$_5$(H)C=N—NH—C$_6$H$_3$(NO$_2$)—NO$_2$	(CH$_3$)$_2$C=N—NH—C$_6$H$_3$(NO$_2$)—NO$_2$
melting point/°C	156	128

Fig. 18.15 Distinguishing between an aldehyde and a ketone

NUCLEOPHILIC REAGENT	CARBONYL COMPOUND AND ITS CONDENSATION PRODUCT	
:NH$_2$—NHC$_6$H$_4$(NO$_2$)$_2$ 2,4-dinitro-phenylhydrazine	C$_2$H$_5$—C(CH$_3$)=O	C$_2$H$_5$—C(CH$_3$)=N—NH—C$_6$H$_3$(NO$_2$)—NO$_2$ butanone 2,4-dinitrophenylhydrazone
:NH$_2$—NHC$_6$H$_5$ phenylhydrazine	C$_3$H$_7$—C(H)=O	C$_3$H$_7$—C(H)=N—NHC$_6$H$_5$ butanal phenylhydrazone
:NH$_2$—NHCONH$_2$ semicarbazide	CH$_3$—C(CH$_3$)=O	CH$_3$—C(CH$_3$)=N—NHCONH$_2$ propanone semicarbazone
:NH$_2$OH hydroxylamine	CH$_3$—C(H)=O	CH$_3$—C(H)=N—OH ethanal oxime
:NH$_3$ ammonia	H—C(H)=O	(CH$_2$)$_6$N$_4$ heterocyclic fused rings hexamethylenetetramine 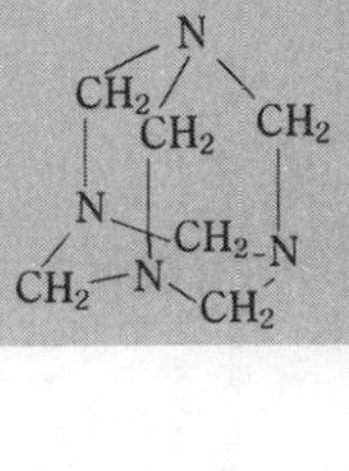

Fig. 18.16 Condensation reactions of some carbonyl compounds

MANUFACTURE AND USES OF CARBONYL COMPOUNDS

Aldehydes and ketones may be produced

a) by controlled oxidation of alcohols or hydrocarbons: e.g.

$$CH_4 \xrightarrow[\text{Ag catalyst}]{\text{air } 500\,°C} HCHO \quad \text{methanal (formaldehyde)}$$

b) by hydration of alkenes by the Wacker process: e.g.

$$CH_2{=}CH_2 \xrightarrow[\substack{\text{aqueous } CuCl_2 \,\& \\ PdCl_2 \text{ catalysts}}]{\text{air } 50\,°C} CH_3CHO \quad \text{ethanal (acetaldehyde)}$$

c) as co-products of other processes: e.g. propanone from the cumene process to manufacture phenol from benzene and propene (see unit on alcohols and phenols).

Methanal (formaldehyde) is used with phenol and with urea to make a variety of plastics (sold under trade names such as bakelite, formica and melamine) which have a wide range of applications e.g. electrical plugs and sockets, chipboard, kitchen worktops, light-weight picnic utensils, heat shields and rocket nose cones. These phenol-formaldehyde and urea-formaldehyde plastics are all thermosetting condensation copolymers: see Fig. 18.16.

Their formation probably involves initial attack of the electrophilic carbon in the $C{=}O$ group of the methanal upon the nucleophilic nitrogen atoms in the NH_2 groups of urea or of melamine, a trimeric condensation product of urea.

Propanone (acetone) is reacted with phenol to produce 2,2-bis(4-hydroxyphenyl)propane (Bisphenol 'A') which is used in the manufacture of epoxy and poly(carbonate) resins. Another important use of propanone is in the manufacture of methyl 2-methylpropenoate (methyl methacrylate) for the production of 'perspex': see Fig. 18.18; and other acrylic polymers: see Figs. 18.17 and 18.18.

Propanone is also an important solvent.

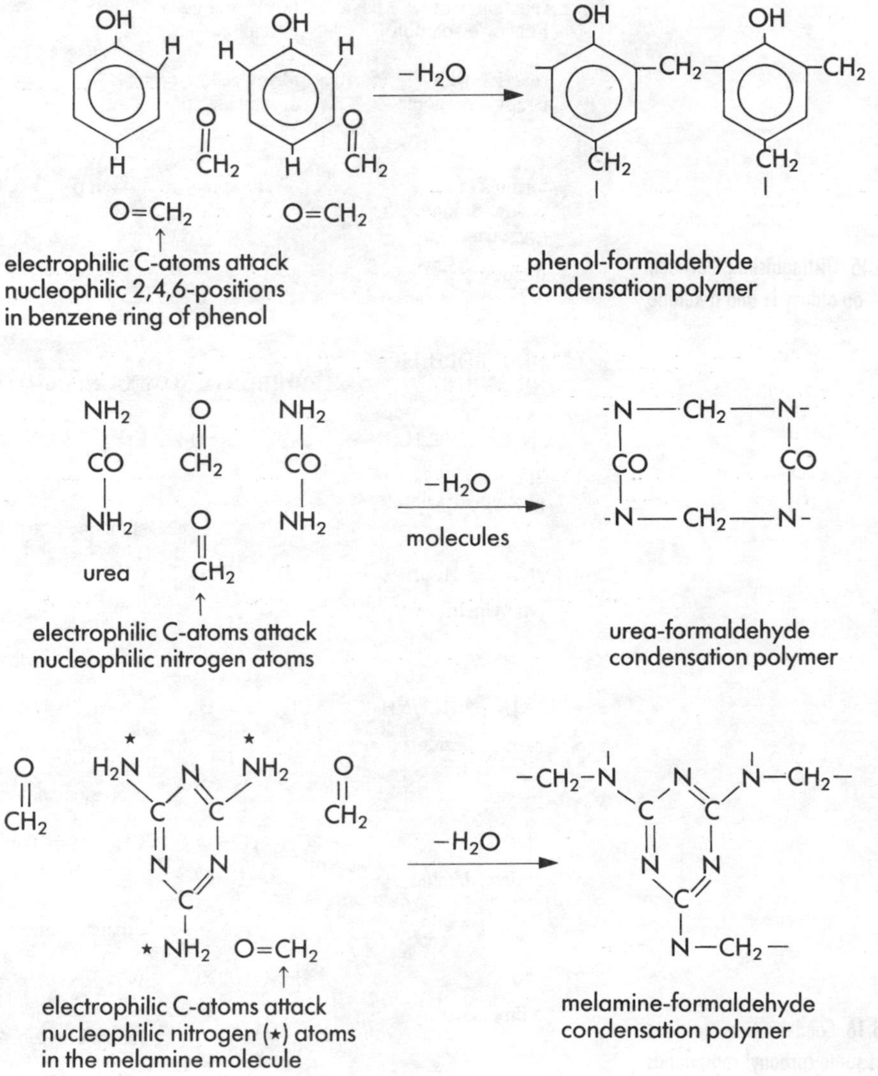

electrophilic C-atoms attack nucleophilic 2,4,6-positions in benzene ring of phenol

phenol-formaldehyde condensation polymer

electrophilic C-atoms attack nucleophilic nitrogen atoms

urea-formaldehyde condensation polymer

Fig. 18.17 Methanal plastics

electrophilic C-atoms attack nucleophilic nitrogen (⋆) atoms in the melamine molecule

melamine-formaldehyde condensation polymer

Fig. 18.18 Propanone is used in
making perspex

nucleophilic
addition of
HCN to C=O group
in propanone

dehydration of
hydroxy compound

hydrolysis of nitrile (CN)
to carboxylic acid (–CO_2H);
formation of methyl ester

methyl 2-methylpropenoate
forms an addition polymer

poly(methyl 2-methylpropenoate)
[poly(methyl methacrylate) or 'perspex']

EXAMINATION QUESTIONS AND OUTLINE ANSWERS

Q1 a) Indicate, giving essential reagents and conditions, how propanal may be
prepared, in the laboratory, from propan-1-ol. *(7)*

b) Explain briefly why aldehydes and ketones are susceptible to attack by
nucleophiles. *(4)*

c) Give **two** different types of reaction in which propanal and propanone behave
similarly and **one** reaction in which only propanal takes part. Write balanced
equations wherever possible. *(9)*

d) Write an equation for the preparation of propanoyl chloride from propanoic acid.
Give the structures of the products of the reactions between propanoyl chloride
and
i) ammonia,
ii) phenol.
Give a mechanism for the reaction between propanoyl chloride and ammonia. *(10)*
(JMB 1988)

Q2 Synthesis gas, a mixture of hydrogen and carbon monoxide, is prepared as shown
below. Nickel is known to catalyse the reaction.

$$CH_4(g) + H_2O(g) \rightleftharpoons 3H_2(g) + CO(g)$$

a) i) An increase in temperature increases the yield of synthesis gas. What
information does this give about the enthalpy change in the forward reaction?
(1)

ii) Using Le Chatelier's Principle, explain how a change in pressure will
affect the composition of the equilibrium mixture. *(2)*

iii) State how the rate of formation of synthesis gas will be affected by
the use of the catalyst. *(1)*

iv) State how the composition of the equilibrium mixture will be affected
by the use of the catalyst. *(1)*

b) A reaction sequence involving an addition reaction between synthesis
gas and propene is shown below.

$$2H_2(g) + 2CO(g) + 2C_3H_6(g) \rightarrow CH_3CH_2CH_2CHO(l) + \text{compound B}$$

compound A

↓

butan-1-ol

↓

alkanol C

Compound A and compound B are isomers and belong to the same class of organic compounds.

 i) Name the class of organic compounds to which A and B belong.   *(1)*

 ii) Draw the full structural formula of compound B and name it.   *(2)*

 iii) If hexan-1-ol was required as a product instead of butan-1-ol, which reagent would be used in place of propene?   *(1)*

 iv) Will alkanol C be primary, secondary or tertiary?   *(1)*

c) i) Which type of chemical reaction occurs when butan-1-ol reacts with ethanoic acid in the presence of concentrated sulphuric acid?   *(1)*

 ii) Draw the full structural formula for the organic product c) i).   *(1)*

                              (12)

                            (SEB 1988)

Outline answers

Q1

(a) see fig 17.2 and text page 260

$$C_2H_5CH_2OH \xrightarrow[\substack{H_2SO_4 \ (aq) \\ distil}]{K_2Cr_2O_7 \ (aq)} C_2H_5CHO$$

(b) Carbonyl group is polar $\overset{\delta+}{C}=\overset{\delta-}{O}$ and carbon atom is an electrophilic centre :

$$Nu: \longrightarrow \overset{}{C}=\overset{\curvearrowright}{O} \longrightarrow Nu-\overset{+}{\underset{|}{C}}-O^-$$

(c)
$$\begin{matrix} R \\ \searrow \\ R \end{matrix} C=O \xrightarrow{NaBH_4} \begin{matrix} R & H \\ \searrow & / \\ R & \nearrow \\ & OH \end{matrix} C$$

$$\begin{matrix} R \\ \searrow \\ R \end{matrix} C=O + H_2N-NH-\bigcirc-NO_2$$ (with NO_2 below)

$$\longrightarrow \begin{matrix} R \\ \searrow \\ R \end{matrix} C=N-NH-\bigcirc-NO_2 + H_2O$$ (with NO_2 below)

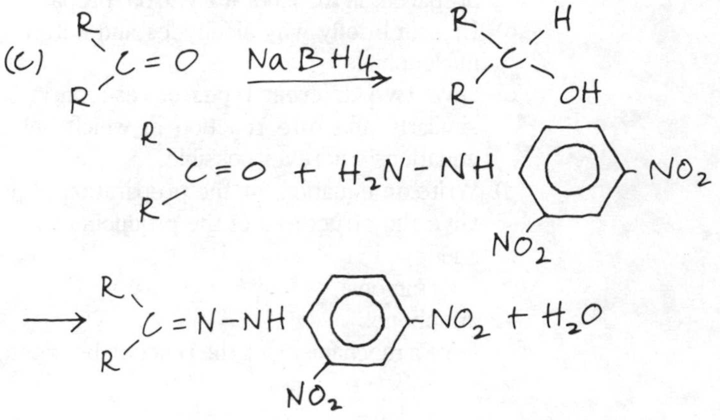

(d) See chapter 16 Examination Questions

Q2

(a) (i) Endothermic

 (ii) 2 volumes reactants give 4 volumes products so increasing pressure will increase percentage of reactants at equilibrium

 (iii) Equilibrium reached faster but

 (iv) Composition is *not* changed.

(b) (i) A & B = aldehydes

 (ii)

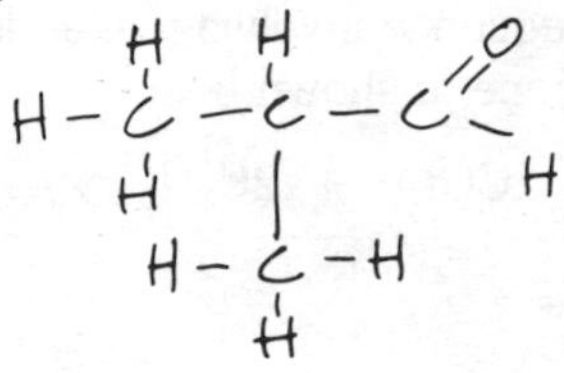

$$H-\overset{\overset{H}{|}}{C}-\overset{\overset{H}{|}}{\underset{\underset{\displaystyle H-\overset{\overset{H}{|}}{\underset{\underset{H}{|}}{C}}-H}{|}}{C}}-C\overset{\displaystyle O}{\underset{\displaystyle H}{\diagup\!\!\diagdown}} \qquad 2-methylpropanal$$

(iii) pent-1-ene

(iv) primary

(c) (i) esterification

$$H-\overset{\overset{\displaystyle H}{|}}{\underset{\underset{\displaystyle H}{|}}{C}}-\overset{\displaystyle O}{\overset{\|}{C}}\underset{O}{\diagdown}-\overset{\overset{\displaystyle H}{|}}{\underset{\underset{\displaystyle H}{|}}{C}}-\overset{\overset{\displaystyle H}{|}}{\underset{\underset{\displaystyle H}{|}}{C}}-\overset{\overset{\displaystyle H}{|}}{\underset{\underset{\displaystyle H}{|}}{C}}-\overset{\overset{\displaystyle H}{|}}{\underset{\underset{\displaystyle H}{|}}{C}}-H$$

STUDENT'S ANSWER WITH EXAMINER'S COMMENTS

Q3 a) i) Write a mechanism for the addition reaction between ethanal
and hydrogen cyanide.   (3)
   ii) Explain how and why the addition reactions of aldehydes differ from
the addition reactions of alkenes.   (5)
   b) Explain how you would use 2,4-dinitrophenylhydrazine to identify an
unknown aldehyde as ethanal. Give an equation for the reaction.   (6)

(AEB 1988)

Student answer

(A)

(i) The reaction is nucleophilic addition

> **The nucleophile is :CN⁻ ion _not_ the HCN molecule!**

$$CH_3-\overset{\overset{\displaystyle O^{\delta-}}{\|}}{\underset{\underset{\displaystyle \delta+}{}}{C}}-H + \overset{\delta+\quad\delta-}{H-CN} \rightarrow \left[CH_3-\overset{\overset{\displaystyle H}{|}}{\underset{\underset{\displaystyle CN}{|}}{C}}-O\right]^{\ominus} + H^+$$

ethanol   hydrogen cyanide   intermediate

> **The C-atoms are joined together thus C**
> **|**
> **CN**

$$CH_3-\overset{\overset{\displaystyle H}{|}}{\underset{\underset{\displaystyle CN}{|}}{C}}-O^- + H^+ \longrightarrow CH_3-\overset{\overset{\displaystyle H}{|}}{\underset{\underset{\displaystyle CN}{|}}{C}}-OH$$

(ii) The main difference is that the $\diagup C=C\diagdown$ bond in alkenes is (un) polar and hence only electrophiles are drawn towards the double bond to form addition products. In aldehydes however the $\diagup C^{\delta\pm}=O^{\delta-}\diagdown$ bond is quite polar as the more electronegative oxygen draws electrons towards itself and gives the carbon atom a positive charge. Hence nucleophilles like $NH_3^{\circ\circ}$, OH^-, etc are attracted towards the carbon atom and it is susceptible to nucleophillic attack.

> **non-polar**

> **nucleophiles**

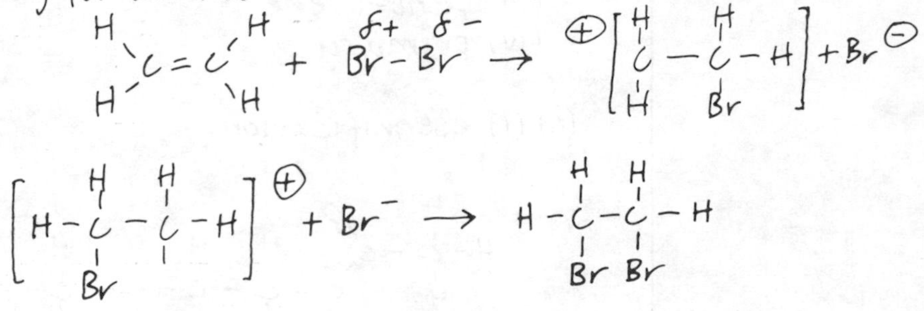

This is an important mechanism to learn: study the section on reaction mechanisms in Chapter 14 and on addition reactions of alkenes in Chapter 15.

eg for alkenes

Aldehydes do not undergo this type of addition.

(b) consider the reaction between the two, which is a condensation reaction.

This could be left out. It is not a mechanism.

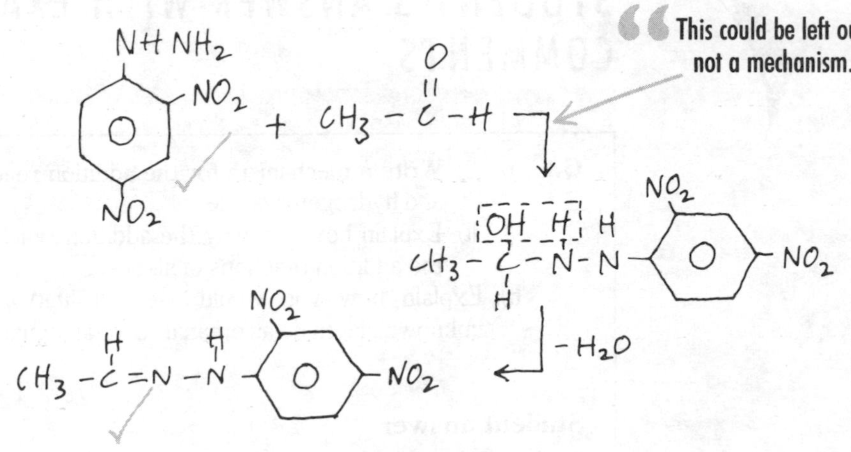

(orange brown ppr)

Aldehydes do not form 'salts' – they are not acids or bases.

The derivative is ethanal 2,4-dinitrophenylhydraz*one*. It must be purified by recrystallisation and the crystals dried before their melting point is measured.

Whereas in any other aldehydes and there salts do not have sharp or fixed melting points, those which are derivatives of 2,4-dinitrophenylhydrazine have sharp melting points, and reference to tabulated values could enable one to identify the parent aldehyde.

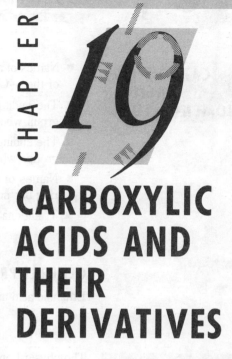

CARBOXYLIC ACIDS AND THEIR DERIVATIVES

GETTING STARTED

Carboxylic acids are compounds containing the structural unit

This functional group is called a carboxyl group and is usually represented by —CO_2H. Some of their properties and reactions may be compared to those of alcohols, phenols, aldehydes and ketones in order to note the few similarities and the many differences. Acid anhydrides, acyl chlorides, acid amides, esters and nitriles are regarded as derivatives of carboxylic acids. They are usually studied with the carboxylic acids.

CARBOXYLIC ACIDS: NOMENCLATURE

PHYSICAL PROPERTIES

CHEMICAL PROPERTIES

ESTERS

ACID ANHYDRIDES

ACID AMIDES

NITRILES

ESSENTIAL PRINCIPLES

CARBOXYLIC ACIDS: NOMENCLATURE

- Names of aliphatic carboxylic acids are based on the alkanes usually with the C-atom of the —CO_2H taken as part of the hydrocarbon chain
- The ending -ane is replaced by -anoic acid for carboxylic acids having one carboxyl group whose C-atom is part of the hydrocarbon chain
- The ending -ane is replaced by -anedioic acid for carboxylic acids having two —CO_2H groups whose C-atoms are part of the hydrocarbon chain
- Names of aromatic acids, with the C-atom of the —CO_2H attached to the benzene ring but not part of it, end with the words carboxylic acid
- Traditional names for some acids are still widely used

GENERAL PROPERTIES

Ethanoic acid and benzenecarboxylic acid (benzoic acid) are two typical carboxylic acids.

PHYSICAL PROPERTIES

The physical properties are typical of substances with covalent molecular structures. The molecules are polar and capable of hydrogen bonding. Consequently, their melting and boiling points are higher than those of aliphatic and aromatic hydrocarbons of similar molar mass, but not as high as you might expect: see Fig. 19.1.

	ETHANOIC ACID	BENZENECARBOXYLIC ACID
traditional name	acetic acid	benzoic acid
formula	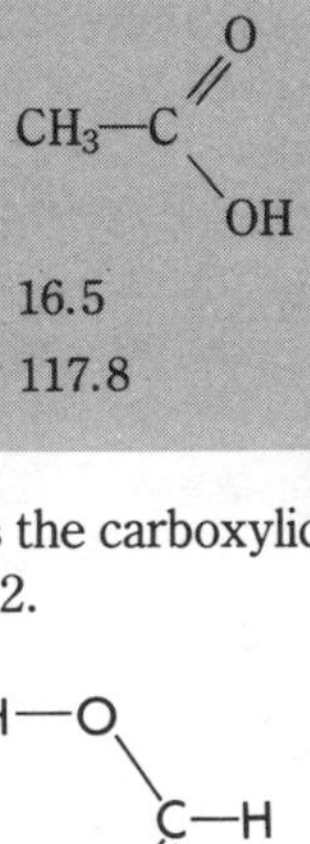 $CH_3{-}C\overset{O}{\underset{OH}{}}$	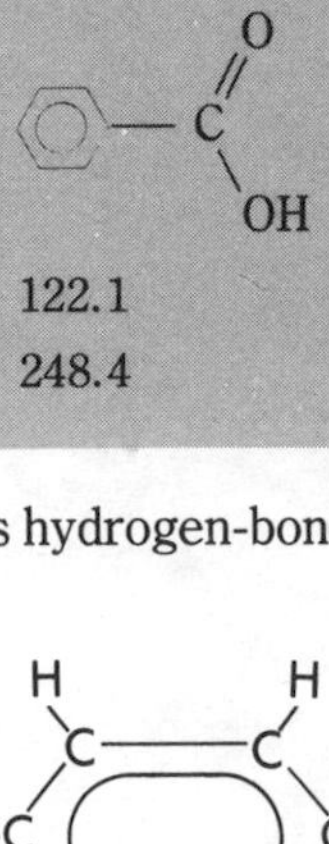 benzene ring—$C\overset{O}{\underset{OH}{}}$
melting point/°C	16.5	122.1
boiling point/°C	117.8	248.4

Fig. 19.1 Physical properties of two carboxylic acids

In pure crystals or liquids the carboxylic acid molecules hydrogen-bond together in pairs, called dimers: see Fig. 19.2.

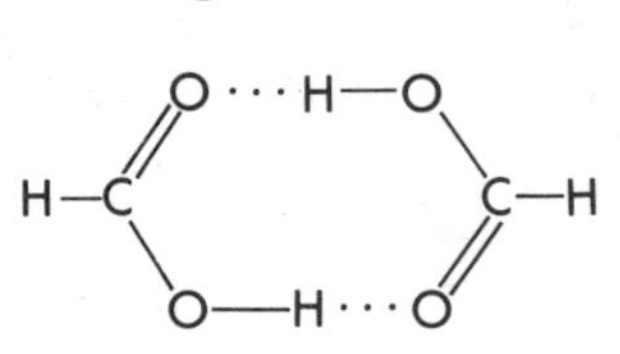

Fig. 19.2 Comparison of methanoic acid and benzene

dimer of methanoic acid
m.pt./°C = 8.3
b.pt./°C = 100.5
M_r (dimer) = 92

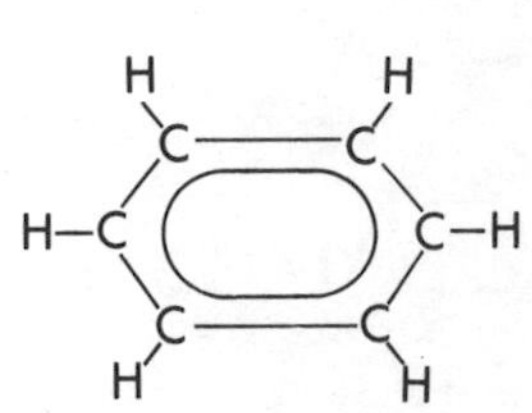

benzene
m.pt./°C = 5.4
b.pt./°C = 80.0
M_r = 78

These dimers are non-polar and not hydrogen-bonded to each other but because of their symmetrical shape they can pack tightly together to give plate-like crystals with fairly high melting points. This explains the unexpected similarity in the physical properties of methanoic acid and benzene: see Fig. 19.2.

The lower fatty acids have characteristic but often unpleasant smells: see Fig. 19.3.

SYSTEMATIC NAME	TRADITIONAL NAME	LATIN ORIGIN	SMELL
ethanoic acid	acetic acid	acetum: vinegar	vinegar
butanoic acid	butyric acid	butyrum: butter	rancid butter
hexanoic acid	caproic acid	caper: goat	old goat

Fig. 19.3 Smells of some carboxylic acids

You should contrast this with the equally characteristic but usually pleasant fruity smells of the esters (see Fig. 19.6).

Ethanoic acid mixes with water in all proportions. As you should expect, the solubility of the higher fatty acids decreases as the proportion of hydrocarbon in the molecular structure increases. Benzoic acid is soluble in hot water but crystallises readily when the solution is cooled. This point about benzoic acid is worth remembering. It crops up frequently in both practical and theory papers.

CHEMICAL PROPERTIES

COMBUSTION

Carboxylic acids will burn in excess air or oxygen to form carbon dioxide and water. In the Granada Schools Television programme on the Bomb Calorimeter, benzoic acid is used to calibrate the apparatus:

$$C_6H_5CO_2H(s) + 6\tfrac{1}{2}O_2(g) \rightarrow 6CO_2(g) + 3H_2O(l); \quad \Delta H_c^\ominus = -3227.0\,\text{kJ}\,\text{mol}^{-1}$$

OXIDATION

- Carboxylic acids are the oxidation products of primary alcohols and aldehydes and are not readily oxidised

- Methanoic (formic) acid and ethanedioic (oxalic) acid are exceptions and react with oxidising agents to form carbon dioxide and water

You should find that methanoic acid, whose structure seems to contain an aldehyde functional group —CHO, will give a positive result (orange-red ppt.) on warming with Fehling's solution:

$$H-C\overset{\displaystyle O}{\underset{\displaystyle OH(aq)}{{\Big\backslash}}} + 2Cu^{2+}(aq) + 4OH^-(aq) \rightarrow CO_2(g) + 3H_2O(l) + Cu_2O(s)$$

As part of your practical work you could well be asked to titrate aqueous ethanedioic acid with acidified aqueous potassium manganate(VII) solution:

$$5\begin{array}{c} CO_2H \\ | \\ CO_2H(aq) \end{array} + 2MnO_4^-(aq) + 6H^+(aq) \xrightarrow{\text{warm}} 10CO_2(g) + 8H_2O(l) + 2Mn^{2+}(aq)$$

You warm the solution in the conical flask to speed up the oxidation. You might expect the reaction to have a high activation energy and be slow because a C—C bond must break and ethanedioate anions (negatively charged) would repel the manganate(VII) anions. The manganese(II) cations catalyse the reaction, so it gets faster as you approach the end-point. This is a popular example of an auto-catalytic reaction and worth remembering: see the unit on kinetics.

CHLORINATION

- The alkyl chain of aliphatic carboxylic acids can undergo a free radical substitution reaction with chlorine in UV light

Chlorine can be bubbled into boiling ethanoic acid in sunlight to form mono-, di- or trichloroethanoic acid. The reaction can be monitored by measuring the increase in mass of the liquid:

$$CH_3CO_2H \rightarrow CH_2ClCO_2H \rightarrow CHCl_2CO_2H \rightarrow CCl_3CO_2H$$

- The —OH in the carboxyl group can be replaced by —Cl by reaction of carboxylic acids with phosphorus chlorides or sulphur dichloride oxide

When you add a small amount of phosphorus pentachloride powder to anhydrous ethanoic (glacial acetic) acid, you should expect to observe a vigorous reaction in which the mixture becomes warm and steamy fumes of hydrogen chloride are formed:

$$CH_3-C \underset{OH(l)}{\overset{O}{\Vert}} + PCl_5(s) \rightarrow CH_3-C \underset{Cl(l)}{\overset{O}{\Vert}} + PCl_3O(l) + HCl(g)$$

ethanoyl   steamy
chloride   fumes

REDUCTION

■ Carboxylic acids and their esters can be reduced to primary alcohols by a powerful reducing agent such as LiAlH$_4$ in dry ethoxyethane

This principle is quite important and may be tested in your theory papers. You could also note that the reduction may well involve the nucleophilic attack of H$^+$ upon the electrophilic C-atom in the carboxyl group of the acid and subsequently in the carbonyl group of the aldehyde:

$$\xrightarrow[\text{followed by hydrolysis with } H_2O]{\text{LiAlH}_4 \text{ in dry ethoxyethane}}$$

$$CH_3CH_2-C \overset{O}{\underset{OH}{\Vert}} \longrightarrow CH_3CH-C \overset{O}{\underset{H}{\Vert}} \longrightarrow CH_3CH_2-\overset{H}{\underset{H}{C}}-OH$$

propanoic acid   propanal   propan-1-ol

Note that this method CANNOT be used to prepare aldehydes because they are reduced to the primary alcohol even more readily than the carboxylic acids.

■ Aqueous carboxylic acids react with calcium, sodium, magnesium and some other active metals to produce hydrogen gas

Most textbooks regard these reactions as typical of any acid. This does not mean they are simply acid-base or proton-transfer reactions!

$$CH_3CO_2H(aq) + Na(s) \rightarrow CH_3CO_2^-(aq) + Na^+(aq) + \tfrac{1}{2}H_2(g)$$

The metal atom is oxidised to the cation and the hydrogen cation is reduced to the hydrogen molecule. In this reaction with sodium, the proton is extracted from the weak ethanoic acid which would be only partially ionised. This reaction is a redox but the carboxylic acid is merely involved in proton-transfer.

ACID-BASE REACTIONS

■ Most carboxylic acids are weak acids that in aqueous solution partially ionise into hydrogen cations and carboxylate anions

Although ethanoic acid is a weak acid (pK$_a$ = 4.8) it is stronger than carbonic acid (pK$_a$ = 6.4) and phenol (pK$_a$ = 9.9) because delocalisation of the negative charge makes the ethanoate anion more stable than the hydrogencarbonate anion and the phenoxide anion:

> ❝Carboxylic acids are good examples of weak acids.❞

$$CH_3-C \overset{O}{\underset{OH}{\Vert}} \rightleftharpoons CH_3-C \overset{O}{\underset{O}{\diagup}} - + H^+$$

■ Most carboxylic acids displace CO$_2$ from carbonates and hydrogencarbonates but most phenols do not

$$CH_3CO_2H(aq) + CO_3^{2-}(aq) \rightarrow CH_3CO_2^-(aq) + H_2O(l) + CO_2(g)$$

This is an important test to distinguish between carboxylic acids and phenols. You will meet it in your practical work and theory papers.

■ Carboxylic acids dissolve in aqueous sodium hydroxide to form solutions of their sodium salts

- Soapy detergents (soaps) are sodium salts of long, straight-chain aliphatic (fatty) acids

When you warm benzoic crystals with aqueous sodium hydroxide you should expect a colourless solution to form that does NOT crystallise on cooling. The ion-dipole interactions of the sodium and benzoate ions with the water molecules are stronger than the hydrogen-bonding interactions of benzoic acid molecules with the water molecules. So the sodium benzoate stays dissolved when you cool the solution

$$C_6H_5CO_2H(s) + NaOH(aq) \rightarrow C_6H_5CO_2^-(aq) + Na^+(aq) + H_2O(l)$$

However, if you add a strong acid (e.g. HCl or H_2SO_4) to the solution, you can expect to see white crystals of benzoic acid form as the strong acid displaces the weaker carboxylic acid:

$$C_6H_5CO_2^-(aq) + Na^+(aq) + H^+(aq) + Cl^-(aq) \rightarrow$$
$$C_6H_5CO_2H(s) + Na^+(aq) + Cl^-(aq)$$

or more simply: $C_6H_5CO_2^-(aq) + H^+(aq) \rightarrow C_6H_5CO_2H(s)$

The effect of aqueous alkali on benzoic acid and acid upon the resulting aqueous solution is a simple illustration of Le Chatelier's principle and the law of chemical equilibrium applied to the following reversible reaction:

$$C_6H_5CO_2H(s) \rightleftharpoons C_6H_5CO_2^-(aq) + H^+(aq)$$

- Aqueous carboxylic acids react with basic metal oxides and with ammonia to form salts and water or just salts

Aqueous ethanoic acid ($pK_a = 4.8$) and ammonia ($pK_b = 4.8$) react to form a neutral solution (pH = 7) because they are equally weak:

$$CH_3CO_2H(aq) + NH_3(aq) \rightarrow CH_3CO_2^-(aq) + NH_4^+(aq)$$

Don't forget that you CANNOT titrate aqueous ethanoic acid with aqueous ammonia because the pH does not change sharply enough at the end-point: see the unit on acid-base equilibria.

- The strength of a carboxylic acid is affected by the structure of the molecule attached to the carboxyl group

Electron-attracting groups attached to the carboxyl group help further to delocalise the negative charge and stabilise the carboxylate anion. This makes the carboxylic acid stronger. Trichloroethanoic acid could be regarded as a strong acid! Electron-donating groups have the opposite effect: see Fig. 19.4.

NAME	FORMULA	pK_a
trichloroethanoic acid	CCl_3CO_2H	0.7
dichloroethanoic acid	$CHCl_2CO_2H$	1.3
chloroethanoic acid	CH_2ClCO_2H	2.9
benzoic acid	$C_6H_5CO_2H$	4.2
ethanoic acid	CH_3CO_2H	4.8
propanoic acid	$C_2H_5CO_2H$	4.9

Fig. 19.4 Strengths of some carboxylic acids

ESTERIFICATION

- Carboxylic acids react with alcohols (but not directly with phenols) and an acid catalyst (HCl or H_2SO_4) to form esters and water

These reactions are reversible and come to equilibrium. No matter how long you reflux a mixture of propanoic acid, ethanol and sulphuric acid catalyst, you will NOT obtain a 100% yield of ethyl propanoate:

Link this with chemical equilibrium in Chapter 7.

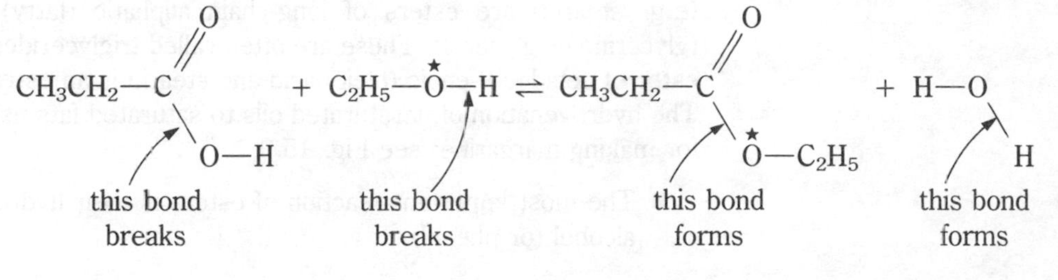

$$CH_3CH_2-C + C_2H_5-O-H \rightleftharpoons CH_3CH_2-C + H-O$$

this bond breaks   this bond breaks   this bond forms   this bond forms

Experiments with isotopes (^{18}O marked by ★) tell us which bonds break.

Notice that the bonds that break are very similar but not identical to the bonds that form. So you should expect the enthalpy changes for these esterifications to be quite small: the reactions are endothermic with values between 40 and 100 kJ mol^{-1}.

Notice that the reactant and product molecules are quite similar and equal in number. So you should expect the changes in entropy and free energy also to be quite small, making the values of the equilibrium constants close to 1 (the value of K for the esterification of ethanoic acid by ethanol is about 4).

You should expect the reaction to have a high activation energy and be quite slow because strong bonds must break. The reaction is extremely slow but can be catalysed by mineral acids: see Fig. 19.5.

An acid catalyst could speed up the reaction by protonating the O-atom of the C=O in the carboxyl group:

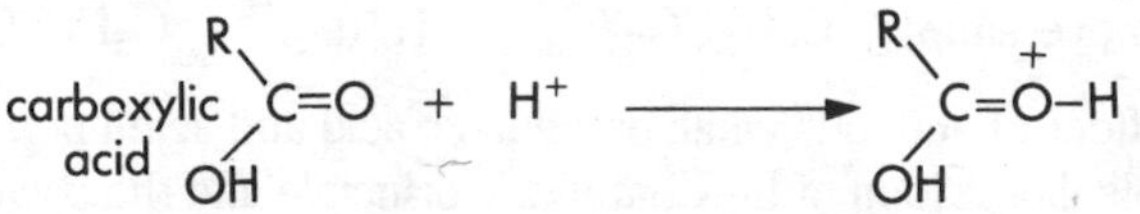

so that the nucleophilic O-atom of the OH in the alcohol can begin to form a bond before the C—O bond breaks:

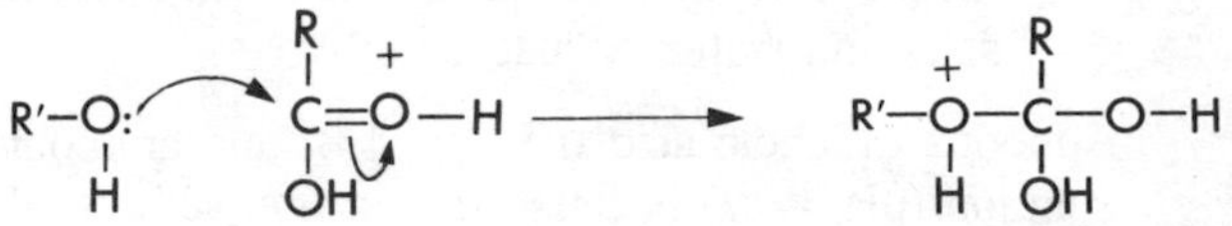

and then the OH can be eliminated as a water molecule and the H$^+$ catalyst regenerated:

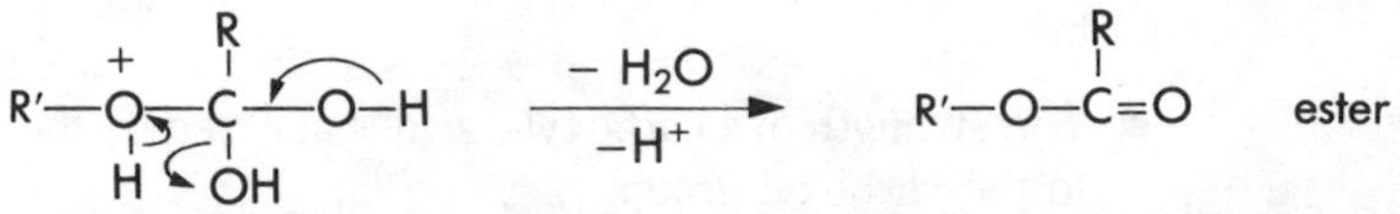

Fig. 19.5 A mechanism for acid-catalysed esterification

[Compare this mechanism with that for acyl chlorides on page 258.]

CARBOXYLIC ACID DERIVATIVES

ESTERS

If you put some ethanoic acid, 3-methylpentan-1-ol and 1 drop of conc. sulphuric acid in a test-tube, warm the mixture in a hot-water bath for a few minutes and then pour the contents of the tube into a beaker of cold water, you should expect to observe the strong 'pear drops' smell of the 3-methylpentyl ethanoate ester floating as oily drops on the top of the water:

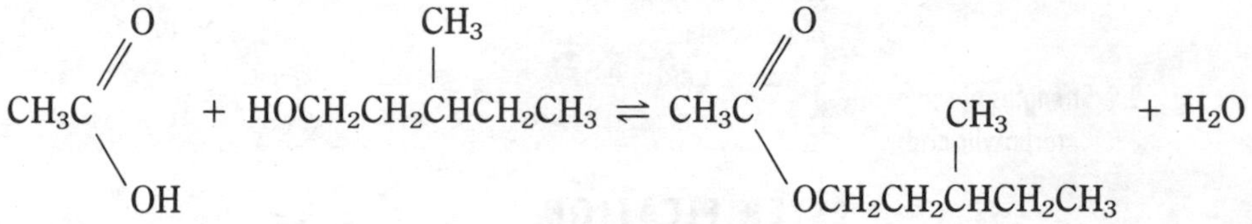

3-methylpentyl ethanoate
ester with 'pear drops' smell

■ Esters are usually volatile liquids that are insoluble in water and can often be identified by their strong fruity smells

Many esters occur naturally in plants. Naturally occurring oils (e.g. olein) and fats (e.g. stearin) are esters of long-chain aliphatic (fatty) acids and propane-1,2,3-triol (glycerine or glycerol). These are often called triglycerides or triacylglycerols. Olein is an ester of octadec-9-enoic (oleic) acid and stearin is an ester of octadecanoic (stearic) acid. The hydrogenation of unsaturated oils to saturated fats using a nickel catalyst is the basis for making margarine: see Fig. 15.9.

■ The most important reaction of esters is their hydrolysis to the carboxylic acid and alcohol (or phenol)

■ Refluxing with aqueous sodium hydroxide hydrolyses an ester completely but produces the sodium salt of the carboxylic acid

Hydrolysis of oils and fats by boiling with conc. aqueous sodium hydroxide is called saponification (Latin: sapo – soap) because the sodium salts of the long-chain fatty acids are soaps: e.g.

$$CH_3(CH_2)_{16}CO_2CH_2$$
$$|$$
$$CH_3(CH_2)_{16}CO_2CH \quad + \quad 3NaOH \quad \rightarrow \quad 3CH_3(CH_2)_{16}CO_2^-Na^+ \quad + \quad CHOH$$
$$|$$
$$CH_3(CH_2)_{16}CO_2CH_2$$

| stearin (fat) | caustic soda sodium hydroxide | sodium stearate (a soap) | glycerine or glycerol |

$$CH_2OH$$
$$|$$
$$CHOH$$
$$|$$
$$CH_2OH$$

Nowadays many esters are synthesised by industrial chemists for food flavourings: see Fig. 19.6.

NAME	FORMULA	SMELL
ethyl methanoate	$HCO_2C_2H_5$	rum
3-methylbutyl ethanoate	$CH_3CO_2CH_2CH(CH_3)C_2H_5$	pear
methyl butanoate	$C_3H_7CO_2CH_3$	apple
ethyl butanoate	$C_3H_7CO_2C_2H_5$	pineapple
pentyl butanoate	$C_3H_7CO_2C_5H_{11}$	apricot
3-methylbutyl butanoate	$C_3H_7CO_2CH_2CH(CH_3)C_2H_5$	plum
octyl ethanoate	$CH_3CO_2C_8H_{17}$	orange
methyl 2-hydroxybenzoate	CO_2CH_3 ⬡—OH	wintergreen

Fig. 19.6 Smells of some esters

Benzene-1,4-dicarboxylic acid (terephthalic acid) is used with ethane-1,2-diol (glycol) to make polyester fibres, 'Terylene' or 'Dacron', and with propane-1,2,3-triol (glycerine) to make polyester (alkyd) resins in paints: see Fig. 19.7.

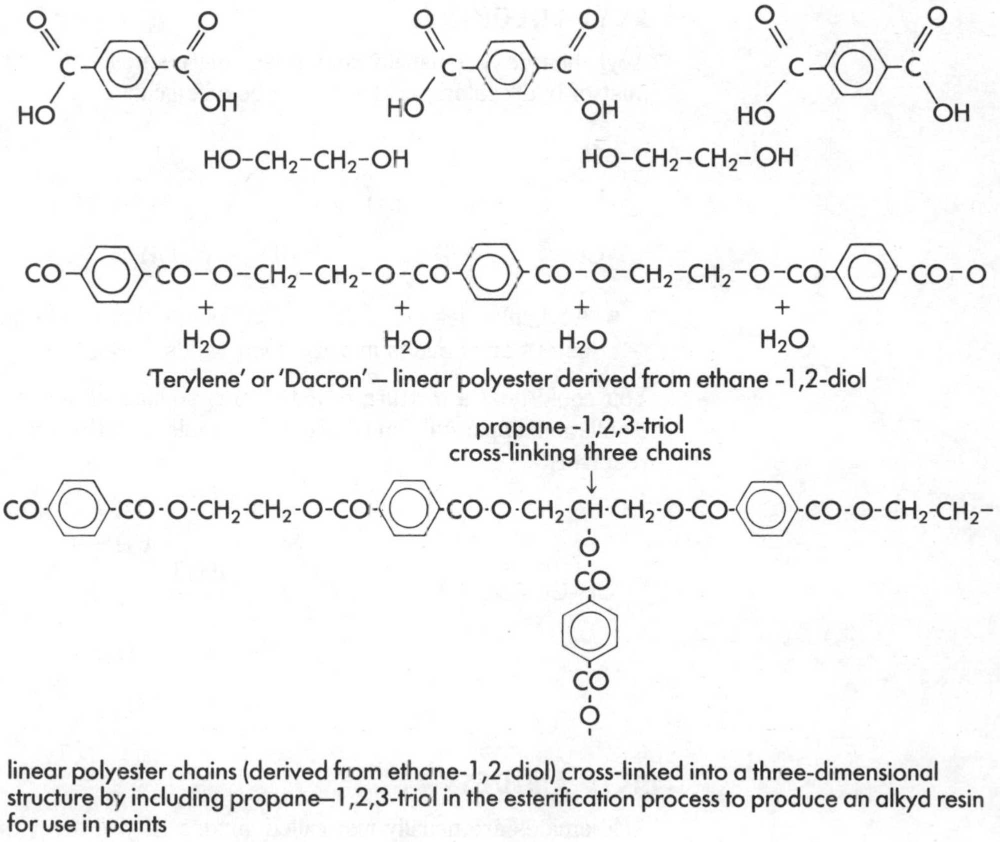

'Terylene' or 'Dacron' — linear polyester derived from ethane -1,2-diol

propane -1,2,3-triol
cross-linking three chains

Fig. 19.7 Polyesters of benzene-1, 4-dicarboxylic (terephthalic) acid

linear polyester chains (derived from ethane-1,2-diol) cross-linked into a three-dimensional structure by including propane—1,2,3-triol in the esterification process to produce an alkyd resin for use in paints

ACID ANHYDRIDES

Acid anhydrides are usually colourless liquids with sharp smells. They are like acyl chlorides in their reactions but they are less reactive. You can think of an acid anhydride as two molecules of carboxylic acid joined by the removal of a water molecule: e.g.

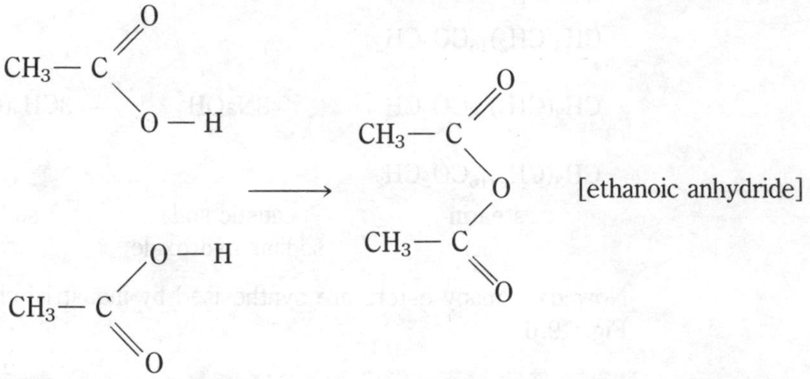

[ethanoic anhydride]

- Acid anhydrides can be used to prepare esters and amides

As part of your practical work you could use ethanoic anhydride to prepare aspirin from 2-hydroxybenzoic (salicylic) acid: see Fig. 19.8.

Notice that the phenolic OH attached to the benzene ring (unlike the OH in alcohols) CANNOT be esterified by reaction with a carboxylic acid. You have to use a carboxylic acid anhydride or an acyl chloride.

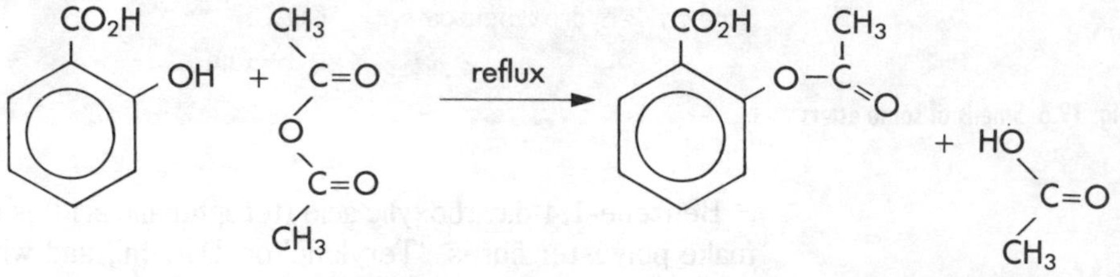

Fig. 19.8 Preparation of aspirin

| 2-hydroxybenzoic (salicylic) acid | ethanoic (acetic) anhydride | (aspirin) 2-ethanoyloxy-benzoic acid | ethanoic acid |

ACYL CHLORIDES

Acyl chlorides are usually colourless, sharp smelling liquids that fume in moist air to form a mist of hydrochloric acid and a carboxylic acid:

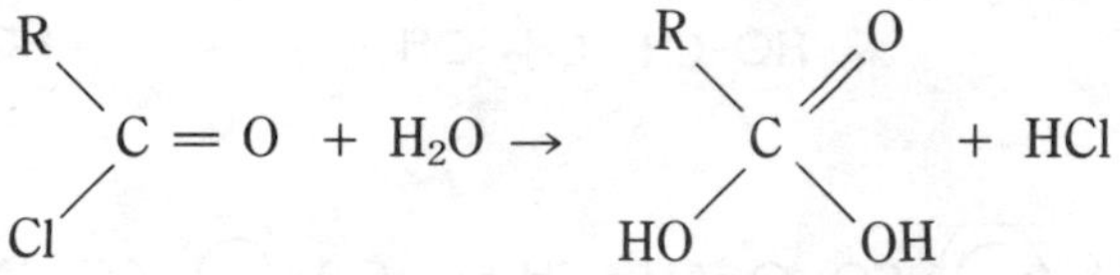

$$R{-}CCl{=}O + H_2O \rightarrow R{-}C({=}O)OH + HCl$$

- Acyl chlorides are reactive compounds that can be used to prepare acid anhydrides, esters and amides in very good yields

You could heat a mixture of anhydrous sodium ethanoate and ethanoyl chloride in a dry distillation apparatus and collect a sample of ethanoic anhydride as distillate in a dry receiver:

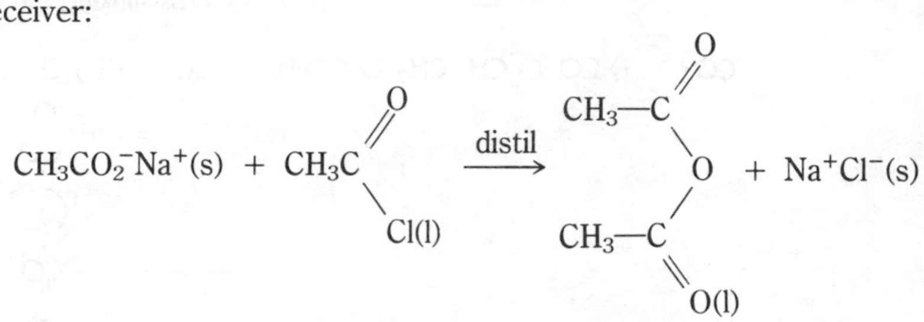

$$CH_3CO_2^- Na^+(s) + CH_3C({=}O)Cl(l) \xrightarrow{\text{distil}} (CH_3C({=}O))_2O(l) + Na^+Cl^-(s)$$

ACID AMIDES

Acid amides are usually just called amides. If you meet them it will normally be as white

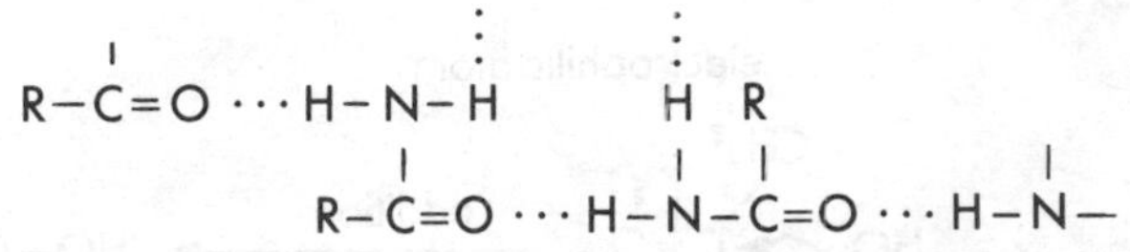

$$R-C=O \cdots H-N-H \qquad H \quad R$$
$$R-C=O \cdots H-N-C=O \cdots H-N-$$

Fig. 19.9 Hydrogen bonding in amides

Hydrogen bonding makes the lower aliphatic amides very soluble in water.

crystalline solids except for methanamide: m.pt. 2.5 °C. Compare ethanamide (molar mass 59 g mol^{-1}; m.pt. 82.2 °C) and ethanoic acid (molar mass 60 g mol^{-1}; m.pt. 16.5 °C). Amide molecules hydrogen-bond together but not predominantly as dimers; consequently, their melting points are usually higher than their parent carboxylic acids and the lower aliphatic amides are very water-soluble: see Fig. 19.9.

Ethanamide (acetamide) can be made by cautiously adding drops of ethanoyl (acetyl) chloride to some conc. aqueous ammonia in a beaker. The reaction is rapid but dangerously violent:

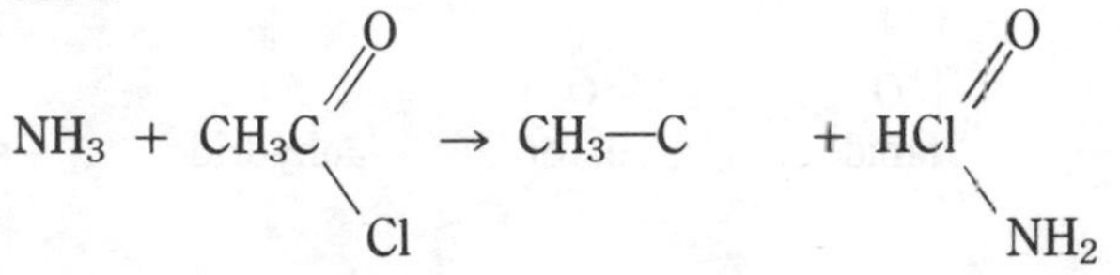

$$NH_3 + CH_3C\overset{O}{\underset{Cl}{\big\langle}} \rightarrow CH_3-C + HCl\overset{O}{\underset{NH_2}{\big\langle}}$$

Ethanamide can also be made by heating strongly under reflux an anhydrous mixture of ammonium ethanoate and ethanoic acid for several hours:

$$CH_3CO_2^-NH_4^+ \xrightarrow{-H_2O} CH_3CONH_2$$

Ethanamide has a strong smell of 'dead mice'! You are unlikely to prepare ethanamide in the laboratory because the first method is too dangerous and the second method too long and tedious. Moreover, ethanamide is now listed as a carcinogenic (i.e., cancer-inducing) compound.

Important in synthesis.

- When an amide is heated with bromine in alkaline conditions it loses a carbon atom as CO_2 and forms an amine

This Hofmann degradation is one way of precisely shortening a carbon chain:

$$RCH_2CONH_2 + Br_2 + 2NaOH \rightarrow RCH_2NH_2 + CO_2 + 2NaBr + H_2O$$

You will find more about amides in the chapter on organic nitrogen compounds.

NITRILES

These compounds are also known as cyanides e.g. CH_3CN methyl cyanide or ethanenitrile. Nitriles are toxic, dangerously flammable liquids. Strictly speaking, they are not carboxylic acid derivatives because their formulae do not conform to the pattern R—CO—R. However they can be made by dehydration of amides and are readily hydrolysed to carboxylic acids.

- When a nitrile is formed by heating alcoholic potassium cyanide with a halogeno-compound, a carbon atom is added to the chain

This nucleophilic substitution reaction is one way of precisely lengthening a carbon chain:

$$CH_3CH_2Br + K^+CN^- \xrightarrow[\text{alcoholic conditions}]{\text{reflux under}} CH_3CH_2CN + K^+Br^-$$

bromoethane   propanenitrile

There is more about nitriles in the chapter on organic nitrogen compounds.

HYDROLYSIS OF CARBOXYLIC ACID DERIVATIVES

The alkaline hydrolysis of the derivatives of ethanoic acid may be regarded in general as addition followed by elimination: see Fig. 19.10.

- Alkaline hydrolysis of carboxylic acid derivatives R— $\overset{O}{\overset{\|}{C}}$ —X is a nucleophilic addition of OH$^-$ to C=O followed by elimination of X$^-$
- The weaker X$^-$ is (as a nucleophile and base), the faster it is eliminated and the faster the hydrolysis takes place

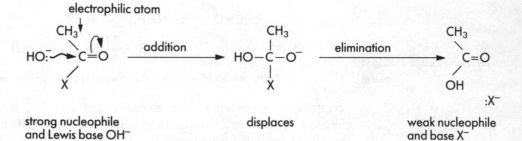

Fig. 19.10 Alkaline hydrolysis of carboxylic acid derivatives

strong nucleophile and Lewis base OH⁻   displaces   weak nucleophile and base X⁻

- In general the rate of hydrolysis of carboxylic acid derivatives increases from left to right as follows:

$$\text{slowest} \quad \underset{\substack{\| \\ O \\ \text{amide}}}{RC-NH_2} \rightarrow \underset{\substack{\| \\ O \\ \text{ester}}}{RC-OR'} \rightarrow \underset{\substack{\| \\ O \\ \text{anhydride}}}{RC-O-CR'} \rightarrow \underset{\substack{\| \\ O \\ \text{acyl chloride}}}{RC-Cl} \quad \text{fastest}$$

INTERCONVERSION OF CARBOXYLIC ACID DERIVATIVES

- Acyl chlorides can be simply converted into anhydrides, esters and amides but usually cannot be easily made from them
- Acid anhydrides can be simply converted into esters and amides but usually cannot be easily made from them
- In general the pattern for the simple conversion of one carboxylic acid derivative into another is as follows:

$$\underset{\substack{\| \\ O \\ \text{acyl chloride}}}{RC-Cl} \rightarrow \underset{\substack{\| \\ O \\ \text{anhydride}}}{RC-O-CR'} \rightarrow \underset{\substack{\| \\ O \\ \text{ester}}}{RC-OR'} \rightarrow \underset{\substack{\| \\ O \\ \text{amide}}}{RC-NH_2}$$

———————————— conversion possible in this direction ————————→

Compare this with the pattern for the hydrolysis of these derivatives.

EXAMINATION QUESTION AND OUTLINE ANSWER

Q1 a)   i) An organic compound, **A**, containing the elements carbon, hydrogen, oxygen and chlorine gave a mass spectrum in which the most prominent peaks corresponded to relative molecular masses of 92 and 94 in a 3:1 ratio. Hydrolysis of **A** produced a compound **B**. After isolation and purification of **B** it was found that:

        1) it did not contain chlorine;

        2) it reacted with sodium hydrogencarbonate to produce carbon dioxide;

        3) on heating with dry soda lime a gas, **C**, was produced which burned with a non-luminous flame.

        Name the functional group in **B** which can be deduced from this information, giving a brief explanation.

     ii) When 0.265 g of compound **A** was hydrolysed, the chloride ion released required 52.09 cm³ of 0.055 mol dm⁻³ aqueous silver nitrate solution for complete reaction. Calculate the number of chlorine atoms present in a molecule of **A** and identify **A**.

$$[A_r(H) = 1.0; \quad A_r(C) = 12.0; \quad A_r(O) = 16.0; \quad A_r(Cl) = 35.5; \quad A_r(Ag) = 107.9.]$$

   b) Write the formulae for structural isomers of **A** which have the following properties:

     i) **One** isomer which gives a yellow/orange precipitate with 2,4-dinitrophenyl-hydrazine, but which does not react with Fehling's solution.

ii) **Two** isomers which give a yellow/orange precipitate with 2,4-dinitro-phenylhydrazine and a red/brown precipitate with Fehling's solution.
iii) **Two** isomers which contain the methoxy (CH$_3$O·) functional group.
c) From the structural isomers of **A** in b) above choose
i) **one** which would exhibit optical isomerism,
ii) **one** which would exhibit geometrical isomerism.

(WJEC 1988)

Outline answer

Q1 (a)
(i) carboxylic acid) group $-C\overset{O}{\underset{OH}{\parallel}}$

(ii) 0.265g A contains $\dfrac{52.09 \times 0.055}{1000}$ mol Cl

$\rightarrow$ 92.5 g A $= \dfrac{92.5}{0.265} \times \dfrac{52.09 \times 0.055}{1000}$

$= 1$ mol Cl

A is $CH_3 \, CH_2 \, C\overset{O}{\underset{Cl}{\parallel}}$

propanoyl chloride

(b)
(i) $CH_3 \, \overset{\parallel}{\underset{O}{C}} \, CH_2 \, Cl$

(ii) $CH_3 \, CHCl \, C\overset{O}{\underset{H}{\parallel}}$ and $CH_2 Cl \, CH_2 \, C\overset{O}{\underset{H}{\parallel}}$

(iii) CH_3O, $C = C$ H; Cl, H / CH_3O, $C = C$ Cl; H, H

(c)
(i) $CH_3 \; C \; Cl \; H \; CHO$ (ii) CH_3O, $C = C$ H; H, Cl

STUDENT'S ANSWERS WITH EXAMINER'S COMMENTS

Q2 A pleasant smelling liquid, **A**, was hydrolysed in acidic conditions and gave two organic products.

One product, **B**, was readily soluble in water and was a monocarboxylic acid. Its mass spectrum is shown below:

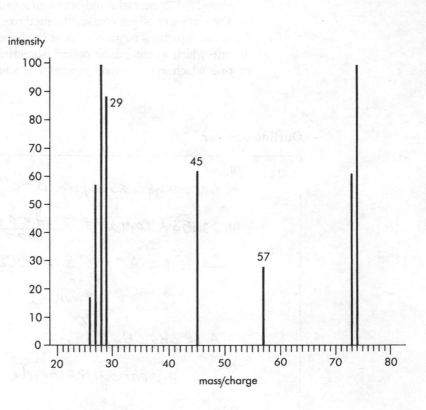

a) i) What is the relative molecular mass of **B**?

   74 ✓

ii) Suggest fragment ions which could be responsible for the peaks listed below. (Relative atomic masses: H = 1, C = 12, O = 16)

   29 _____ C_2H_5 _____

   45 _____ CO_2H _____

   57 _____ C_2H_5CO _____

iii) Hence deduce the displayed formula and name of **B**. $CH_3\,CH_2\,CO_2H$ ✗

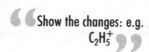

(6)

The second product, **C**, was also readily soluble in water and on analysis by mass was found to contain 60.0% carbon, 13.3% hydrogen and 26.7% oxygen.

The relative molecular mass of **C** was found to be 60.
(Relative atomic masses: H = 1, C = 12, O = 16)

b) i) Calculate the molecular formula of **C**.

	%	÷ RAM =	'mol' Ratio	Empirical
C	60·0	12	5	3
H	13·3	1	13·3	8
O	26·7	16	1·67	1

Empirical Formula is C_3H_8O

RMM of empirical formula is $3 \times 12 + 8 + 16 = 60$

→ molecular formula C is C_3H_8O ✓

Show the changes: e.g.
$C_2H_5^+$

Show *all* the bonds.

Good answer

ii) Draw three *possible* displayed formulae for isomers having this molecular formula.

"Remember ethers are isomeric with alcohols"

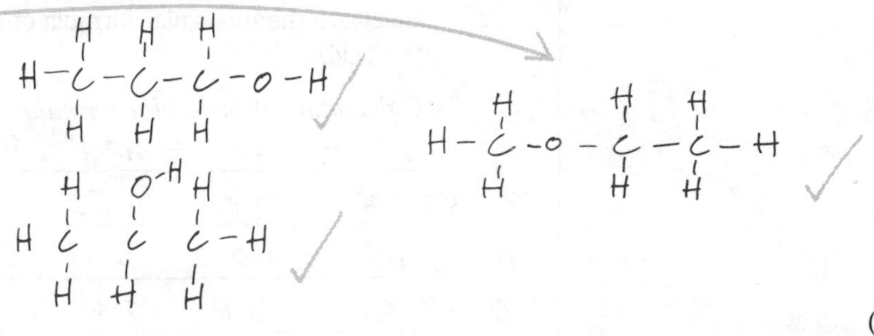

(5)

c) **C** was oxidized to **D** by refluxing with acidified sodium dichromate(VI). **D** was neither acidic, nor did it react on warming with Fehling's solution but it contained the same number of carbon atoms per molecule as **C**.

i) To what class of compound does **D** belong?

ketones

ii) Hence deduce the correct formula for **C** and name it.

"Answer *all* the question! Propan-2-ol"

secondary alcohols oxidise to ketones so C must be $CH_3 - CH(OH) - CH_3$

(2)

d) From your answers to parts a), b) and c) draw the displayed formula for **A**.

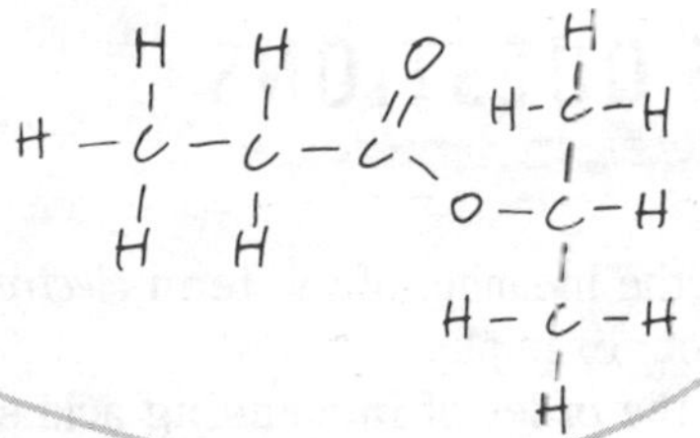

"The name was *not* asked for. Don't waste time in exams."

2-propyl propanoate

(1)

e) Outline two experiments you could do in your school laboratory to confirm the identity of **D**.

1. Measure its boiling point and compare it with the data book value.

2. React it with 2,4 - dinitrophenylhydrazine to make the 2,4-dinitrophenylhydrazone derivative. purify the crystals and find their melting point. Compare the value with the one in the data book.

"Good answer."

(3)

f) i) The enthalpy change of vaporization of **D** is 30.3 kJ mol^{-1} and it boils at 329 K. Calculate the molar entropy change of vaporization.

$\dfrac{30.3}{329} = 0.092$

"Give the units and appropriate significant figures. e.g. $92.1 \text{ Jk}^{-1} \text{ mol}^{-1}$"

ii) Is this molar entropy change of vaporization typical for a molecular substance? Explain your answer.

Rather high (typical value $88 \text{ Jk}^{-1} \text{ mol}^{-1}$) because D (propanone) is polar.

(4)

(*Total 21 marks*)
(NUFF 1988)

Q3 a) An aromatic carboxylic acid **A** ($M_r = 166$) contains 57.83% of carbon, 3.64% of hydrogen and 83.53% of oxygen by mass.

Calculate the molecular formula of **A** and draw **three** possible structures of the acid.

Calculation of molecular formula

	%	÷	Av	=	'mol'	Ratio	
C	57·83		12·0		4·82	2	→ Empirical formula $C_4H_3O_2$
H	3·64		1·0		3·64	1·5	$Mr(C_4H_3O_2) = 83$
O	38·53		16·0		2·41	1	→ molecular formula $C_8H_6O_4$

Structure 1 *Structure 2* *Structure 3*

" Very good answer. "

(JMB 1988)

PRACTICE QUESTIONS

Q4 a) Explain the meaning of the term *electron delocalisation* using the ethanoate ion as your example. (4)

b) Predict the order of **increasing** acid strength for the following compounds, and explain your reasoning:

CH_3CO_2H C_6H_5OH C_2H_5OH H_2SO_4 (6)

Q5 This question is about benzoic acid.

a) Give the reagent(s) and reaction conditions for each of the following methods of preparation of benzoic acid.

  i) $C_6H_5CH_3 \rightarrow C_6H_5CO_2H$
  ii) $C_6H_5CONH_2 \rightarrow C_6H_5CO_2H$
  iii) $C_6H_5CO_2C_2H_5 \rightarrow C_6H_5CO_2H$

  Name the functional group of the reactant in ii) and iii).

  Reactant in ii) _____

  Reactant in iii) _____ (8)

b) Benzoic acid dissolves readily in sodium hydroxide solution, but this solution turns milky when excess hydrochloric acid is added. Explain. (3)

c) The organic reactant in a)iii) was isotopically labelled at one of its oxygen atoms with oxygen—18. The resulting benzoic acid after reaction contained only normal oxygen—16. Write out the structure of the labelled reactant, showing the position of the label clearly. Write a mechanism for the reaction consistent with the benzoic acid being unlabelled. (4)

(*Total 15 marks*)
(ULSEB 1988)

CHAPTER 20

ORGANIC NITROGEN COMPOUNDS

GETTING STARTED

Amines are classes of organic bases related to ammonia. Primary amines contain the —NH_2 functional group and are the most important of the amines. In phenylamine (aniline) the —NH_2 group is attached to a benzene ring. Amides and amino-acids are two other classes of compound with —NH_2 in their structure. The emphasis is upon comparing the effect of the organic molecular structure upon the reactions of the —NH_2 group. The influence of the —NH_2 group upon the reactivity of the benzene ring is also important.

AMINES: NOMENCLATURE

PHYSICAL PROPERTIES

CHEMICAL PROPERTIES

AMIDES

POLYAMIDES

AMINO ACIDS

NITRILES (CYANIDES)

ESSENTIAL PRINCIPLES

$$\begin{array}{ccc} & \text{R}' & \quad \text{R}' \\ & | & \quad | \end{array}$$

- Compounds containing the groups R'—NH_2, R'—NH and R'—N—R' (where the R's may be different but R' $\neq$ H) are called primary, secondary and tertiary amines, respectively: see Fig. 20.1
- Compounds containing the ion $(R')_4N^+$ are classed as quaternary ammonium compounds and named as substituted ammonium compounds: e.g., $(CH_3)_4N^+Cl^-$, tetramethylammonium chloride
- The names of 'simple' amines containing one N in the formula are based on the alkyl and phenyl groups attached to the N-atom: see Fig. 20.1
- The names of 'more complicated' primary amines are based on the name of the hydrocarbon and —NH_2 as the amino group: see Fig. 20.1
- Carboxylic acids, aldehydes, ketones and other substances containing the —NH_2 group as a substituent are called amino-compounds

GENERAL PROPERTIES

> The differences between aliphatic and aromatic amines are important.

Butylamine (1-aminobutane) $C_4H_9NH_2$ and phenylamine (aniline) $C_6H_5NH_2$ are typical of aliphatic and aromatic primary amines.

PHYSICAL PROPERTIES

The physical properties of primary and secondary amines are typical of substances with covalent molecular structures. Nitrogen is more electronegative than carbon and in amines the N-atom has one lone-pair of electrons repelling the three bonded-pairs. This gives the molecules their trigonal pyramidal shape and makes them polar. Consequently, permanent dipole-dipole attractions contribute to the intermolecular forces to make the boiling points of tertiary amines higher than those of aliphatic and aromatic hydrocarbons of similar molar mass: see Fig. 20.1.

Hydrogen bonding between the —NH_2 groups and $>$NH groups in primary amines and secondary amines increases the intermolecular forces of attraction even more. However, nitrogen is less electronegative than oxygen so hydrogen-bonding in amines is weaker than in alcohols. So primary and secondary amines have lower boiling points than the corresponding alcohols of similar relative molecular mass. Tertiary amine molecules do NOT hydrogen-bond to each other because there are no H-atoms attached directly to the electronegative N-atom.

NAME	M_r	T_b/°C	FORMULA AND TYPE OF AMINE		INTERMOLECULAR FORCES
methylamine	31	–6	CH_3NH_2	primary	dipole-dipole and H-bonding
dimethylamine	45	7	$(CH_3)_2NH$	secondary	dipole-dipole and H-bonding
trimethylamine	59	3	$(CH_3)_3N$	tertiary	dipole-dipole
ethylamine	45	17	$C_2H_5NH_2$	primary	dipole-dipole and H-bonding
*ethanol	46	78	C_2H_5OH	primary	dipole-dipole and H-bonding
1-aminopropane	59	48	$CH_3CH_2CH_2NH_2$	primary	dipole-dipole and H-bonding
2-aminopropane	59	32	$CH_3CHNH_2CH_3$	primary	dipole-dipole and H-bonding
phenylamine	93	184	$C_6H_5NH_2$	primary	dipole-dipole and H-bonding
*methylbenzene	92	111	$C_6H_5CH_3$	–	van der Waals

Fig. 20.1 Physical properties of some amines

* included for comparison

The first two members of the homologous series of aliphatic primary amines, methylamine CH_3NH_2 and ethylamine $C_2H_5NH_2$, are gases at $18°C$. However, they are very soluble in water and you may well experiment with their aqueous solutions. In general the solubility in water of the higher amines decreases as the proportion of hydrocarbon in the molecular structure increases. Phenylamine (aniline) is not very soluble in water. Quaternary alkylammonium ions (e.g. as in trimethylpentadecylammonium bromide), behave as surfactants (by lowering the surface tension of water) and are used as germicidal cationic liquid detergents.

■ Dipole-dipole and hydrogen-bonding forces account for the differences in physical properties of primary amines compared to hydrocarbons and explain the solubility of amines in water, alcohol and ethoxyethane

CHEMICAL PROPERTIES

COMBUSTION

■ Many amines are a dangerous explosion and fire risk, being highly flammable and having very low flash points

Amines will burn in excess air or oxygen (unlike ammonia which will burn in oxygen but NOT air) to form carbon dioxide, water and nitrogen; e.g.

$$2C_3H_7NH_2(l) + 10\tfrac{1}{2}O_2(g) \rightarrow 6CO_2(g) + 9H_2O(l) + N_2(g)$$

■ Many amines are hazardous in attacking the skin and in being toxic when absorbed through the skin

I have not seen these points tested in the examinations but for your own safety's sake you should be aware of them when carrying out your practical work. Take particular care if you are handling phenylamine (aniline)!

ACID-BASE REACTIONS

■ Most amines are weak bases that in aqueous solutions partially ionise to form hydroxide anions and substituted ammonium cations

■ Primary alkylamines are stronger bases than ammonia

Although butylamine is a weak base ($pK_b = 3.2$) it is stronger than ammonia ($pK_b = 4.8$) because the electron-donating alkyl group C_4H_9 makes the butylammonium cation more stable than the ammonium cation:

$$C_4H_9-\overset{\overset{\textstyle H}{|}}{\underset{\underset{\textstyle H}{|}}{N}}: + H_2O \rightleftharpoons C_4H_9-\overset{\overset{\textstyle H}{|}}{\underset{\underset{\textstyle H}{|}}{N}}{}^+\!-H + OH^-$$

> The benzene ring affects the properties of the group attached to it.

■ Phenylamine (aniline) is a weaker base than ammonia

Although ammonia is a weak base ($pK_b = 4.8$), phenylamine ($pK_b = 9.4$) is even weaker because the lone pair of electrons on the N-atom participate in the delocalised π-electron system of the benzene ring. This stabilises the free base molecule and makes the lone electron-pair less readily available for dative bonding. So the phenylammonium cation is less readily formed: see Fig. 20.2.

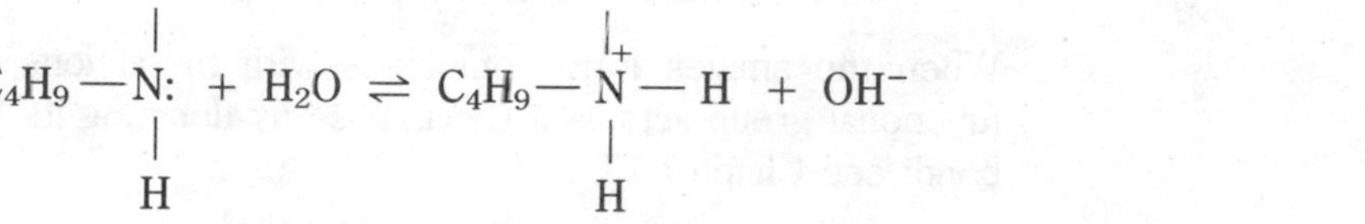

Fig. 20.2 Phenylamine is a weaker base than ammonia

■ Amines (like ammonia) turn moist red litmus paper blue and form a white smoke with hydrogen chloride

You can often detect aliphatic amines by their smells (methylamine and ethylamine smell like rotting fish) but chemical tests are usually safer:

$$R—NH_2(g) + HCl(g) \rightarrow R—NH_3^+Cl^-(s)$$ 　　white smoke of a substituted ammonium chloride

■ Amines dissolve in aqueous acids to form solutions of their substituted ammonium salts which can form white crystalline solids

When you add drops of hydrochloric acid to an aqueous emulsion of phenylamine you should expect the milky liquid to turn into a clear, colourless solution and the smell of the amine to disappear. The ion-dipole interactions of the phenylammonium and chloride ions with the water molecules are strong enough to overcome the hydrophobic character of the benzene ring. The phenylamine dissolves:

$$C_6H_5NH_2(l) + HCl(aq) \rightarrow C_6H_5NH_3^+(aq) + Cl^-(aq)$$

However, if you add a strong alkali e.g. NaOH to the solution, you can expect to see a white aqueous emulsion of phenylamine reform and the smell of the amine return as the strong base displaces the weaker amine:

$$C_6H_5NH_3^+(aq) + Cl^-(aq) + Na^+(aq) + OH^-(aq)$$
$$\rightarrow C_6H_5NH_2(l) + Na^+(aq) + Cl^-(aq) + H_2O(l)$$

or more simply 　$C_6H_5NH_3^+(aq) + OH^-(aq) \rightarrow C_6H_5NH_2(l) + H_2O(l)$

■ Amines (like ammonia) form complex ions (with cations of the d-block elements and other metals) whose solutions are often blue or green.

In general, if you add a few drops of an amine to blue aqueous copper(II) sulphate you may see a blue or blue-green precipitate of a basic copper(II) hydroxide form as the acidic solution becomes alkaline and then see it dissolve to give a deeper blue solution of a complex copper(II) cation as you add more drops of the amine. For simplicity these reactions may be represented as follows:

$$Cu^{2+}(aq) + 2OH^-(aq) \rightarrow Cu(OH)_2(s)$$

$$Cu(OH)_2(s) + 4:\!N\!\!-\!\!R(aq) \rightarrow [Cu(:N\!\!-\!\!R)_4]^{2+}(aq) + 2OH^-(aq)$$

In particular, if you use phenylamine (aniline) you can expect to see a green precipitate form but not subsequently dissolve. You could perform experiments with copper(II) sulphate solutions of known concentration and measured amounts of phenylamine to find the formula of the solid green copper-phenylamine complex.

■ EDTA (ethylenediaminetetraacetic acid) is an important complexing agent involving two N-atoms in tertiary amine groups

When the amines form complexes with metal ions, the nitrogen atom in the amine functional group acts as a Lewis base by donating its lone electron-pair to form a dative bond: see Chapter 13.

NUCLEOPHILIC SUBSTITUTIONS

■ Primary, secondary and tertiary amines (and ammonia) are nucleophiles

The lone electron-pair on the nitrogen atom in amines and ammonia makes the N-atom a nucleophilic centre. Consequently amines (and ammonia) react with halogenoalkanes, acyl chlorides and acid anhydrides.

REACTIONS WITH BROMOALKANES

■ Ammonia reacts with bromoethane to form a mixture of primary, secondary and tertiary ethylamines and/or the quaternary ethylammonium bromide salt depending upon the quantities and conditions used

The reactions may be carried out in aqueous or anhydrous conditions. They are not usually used for synthesis because they give mixtures of amines:

$$NH_3 + C_2H_5Br \rightarrow C_2H_5NH_2 + HBr \quad or \quad C_2H_5NH_3^+Br^-$$
$$C_2H_5NH_2 + C_2H_5Br \rightarrow (C_2H_5)_2NH + HBr \quad or \quad (C_2H_5)_2NH_2^+Br^-$$
$$(C_2H_5)_2NH + C_2H_5Br \rightarrow (C_2H_5)_3N + HBr \quad or \quad (C_2H_5)_3NH^+Br^-$$
$$(C_2H_5)_3N + C_2H_5Br \rightarrow (C_2H_5)_4N^+ + Br^- \quad or \quad (C_2H_5)_4N^+Br^-$$

Notice that the two products in each equation may react with each other and that the product in the first equation may become a reactant in the second equation. This illustrates the following important principle:

- A chemical change is often complicated because the products of one reaction may become reactants in another reaction

This is usually the cause of the unwanted side-reactions and by-products that are so frequently encountered, especially in organic chemistry.

ACYLATION REACTIONS

- Primary and secondary amines react with acyl chlorides (and acid anhydrides) to form substituted amides

These reactions occur readily in the cold but do not usually produce mixtures of amides because the N-atom is far less nucleophilic in the amide than in the original amine: see Fig. 20.3.

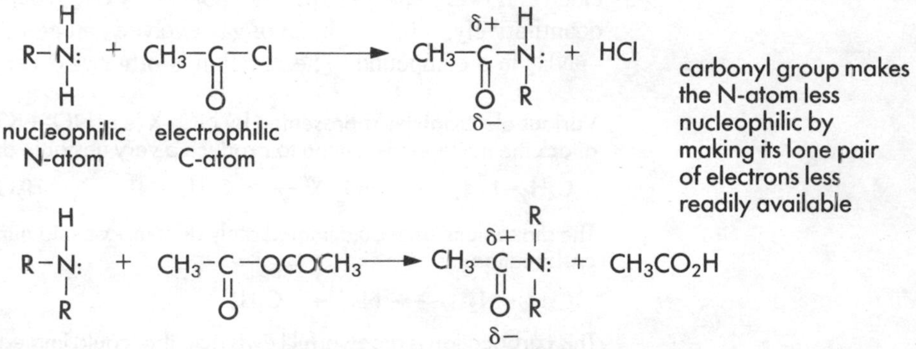

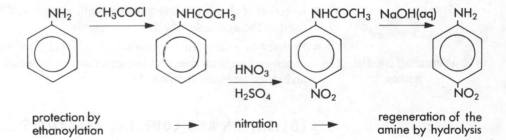

Fig. 20.3 Amides formed by acylation of amines

carbonyl group makes the N-atom less nucleophilic by making its lone pair of electrons less readily available

Ethanoylation of an amine produces an amide which is not attacked

We often use acylation to protect an amine functional group before we react the rest of the molecule: see Fig. 20.4.

Preparing 4-nitrophenylamine from phenylamine:

Fig. 20.4 Protecting an NH₂ group by acylation

protection by ethanoylation → nitration → regeneration of the amine by hydrolysis

Phenylamine is colourless when pure but on exposure to air and light it rapidly turns brown, presumably partly through oxidation and decomposition. If the amine is not protected by acylation, a nitrating mixture would have an even more destructive effect upon the molecule than air and light!

REACTION WITH NITROUS ACID

- A pale blue solution of nitrous acid is obtained by mixing cold aqueous sodium nitrite with hydrochloric acid

If you place a spatula measure of sodium nitrite (a pale yellow solid) into a test-tube and add about $3\,cm^3$ of hydrochloric acid (with a concentration of about $1\,mol\,dm^{-3}$), the liquid turns pale blue: see Fig. 12.22.

$$HCl(aq) + NaNO_2(s) \rightarrow HNO_2(aq) + NaCl(aq)$$

or solid dissolves: $\quad NaNO_2(s) \rightarrow Na^+(aq) + NO_2^-(aq)$

then weak acid forms: $\quad H^+(aq) + NO_2^-(aq) \rightarrow HNO_2(aq)$

The mixture gives off a colourless gas which turns brown just near the mouth of the tube. The bottom of the tube stays quite cold (endothermic dissolving of the solid and disproportionation of the aqueous nitrous acid):

$$3HNO_2(aq) \rightarrow H^+(aq) + NO_3^-(aq) + 2NO(g) + H_2O(l)$$

but the top becomes warm (exothermic spontaneous oxidation of the colourless nitrogen monoxide to the brown nitrogen dioxide):

$$2NO(g) + O_2(g) \rightarrow 2NO_2(g)$$

- Primary aliphatic amines react with ice-cold aqueous nitrous acid to give very unstable, aqueous diazonium salts. These salts decompose to give a variety of products depending upon the reaction conditions.

If some cold, aqueous sodium nitrite is added to a cold solution of butylamine in hydrochloric acid, the mixture gives off a colourless gas (nitrogen) and sometimes turns cloudy and very pale yellow. One mole of $-NH_2$ groups produce one mole of N_2 gas almost quantitatively, so the volume of gas evolved can be measured to determine the amount of $-NH_2$ in a compound. The reaction is otherwise very complicated: see Fig. 20.5.

Various electrophiles represented by NO–X (e.g. $NO–NO_2$, $NO–\overset{+}{O}H_2$ $NO–Cl$ NO^+) may attack the nucleophilic amine to produce a very unstable diazonium ion:

$$C_4H_9–NH_2 + NO–X \longrightarrow C_4H_9–N_2^+ + H_2O + X^-$$

The diazonium ion would immediately decompose into nitrogen and a very reactive carbocation:

$$C_4H_9–N_2^+ \longrightarrow N_2 + C_4H_9^+$$

This carbocation is a powerful Lewis acid that could immediately combine with any available Lewis base (e.g. OH^-, Cl^-, NO_2^-): e.g.

$$C_4H_9^+ + Cl^- \longrightarrow C_4H_9Cl$$

or simply lose a proton to form an alkene:

$$C_4H_9^+ \longrightarrow C_4H_8 + H^+$$

Fig. 20.5 Reaction of nitrous acid on primary aliphatic amines

Butan-1-ol, 1-chlorobutane, 1-nitrobutane, butyl nitrite and but-1-ene have all been detected and identified as products of the reaction of butylamine and nitrous acid

- Reaction of aliphatic primary amines with nitrous acid is complicated and not usually used in the synthesis of other compounds

 An important colourful reaction.

- Primary aromatic amines react with ice-cold aqueous nitrous acid to give stable aqueous diazonium salts that are important intermediates in the synthesis of azo-dyes and other aromatic compounds

DIAZOTISATION AND COUPLING REACTIONS

You may well prepare an azo dye as part of your practical work in the laboratory: see Fig. 20.6.

- Diazotisation followed by coupling with naphthalen-2-ol to produce an insoluble red ppt. is a simple test for aromatic primary amines

Notice that some coloured products may be formed if the coupling agent is just added to aqueous nitrous acid because the nucleophilic naphthalen-2-ol would be attacked by electrophiles in the acid. However, you are unlikely to confuse these coloured substances with the scarlet ppt. of the azo dye.

- Aromatic diazonium ions are stabilised by delocalisation, the π-electrons of the nitrogen-nitrogen bond participating in the delocalised π-electron system of the benzene ring: see Fig. 20.7.

1   Make an ice-cold solution of aniline in hydrochloric acid:

$$\langle\!\!\bigcirc\!\!\rangle\text{---}NH_2(l) + HCl(aq) \longrightarrow \langle\!\!\bigcirc\!\!\rangle\text{---}NH_3^+ (aq) + Cl^-(aq)$$

2   Make an ice-cold solution of sodium nitrite in water:

$$NaNO_2(s) \rightarrow Na^+(aq) + NO_2^-(aq)$$

3   Add the aqueous sodium nitrite to the phenylammonium chloride solution drop by drop with vigorous stirring and adding pieces of ice to the mixture as required to keep the solution cold:

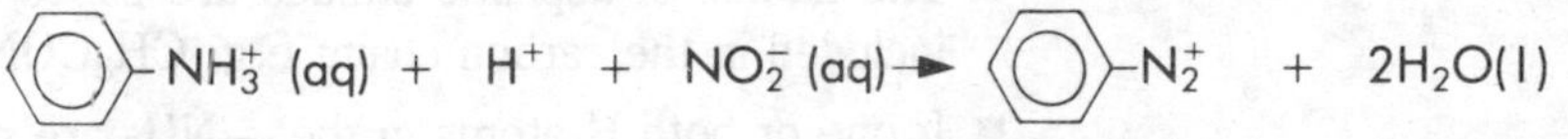

$$\langle\!\!\bigcirc\!\!\rangle\text{---}NH_3^+ (aq) + H^+ + NO_2^- (aq) \rightarrow \langle\!\!\bigcirc\!\!\rangle\text{---}N_2^+ + 2H_2O(l)$$

This step to form the aqueous benzene dazonium ion is called the diazotisation reaction.

4   Make an ice-cold solution of napthalen-2-ol (2-naphthol or β-naphthol) in aqueous sodium hydroxide:

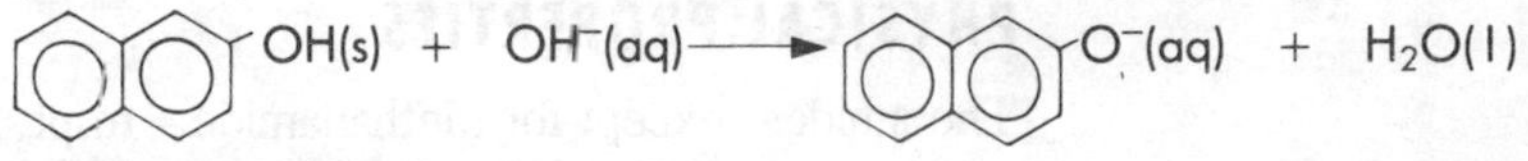

$$\langle\!\!\bigcirc\!\!\bigcirc\!\!\rangle\text{---}OH(s) + OH^-(aq) \longrightarrow \langle\!\!\bigcirc\!\!\bigcirc\!\!\rangle\text{---}O^-(aq) + H_2O(l)$$

This solution is called the coupling agent.

5   Mix together the ice-cold solutions of benzene diazonium ion and coupling agent to produce a scarlet precipitate of the azo dye by an electrophilic substitution called a coupling reaction:

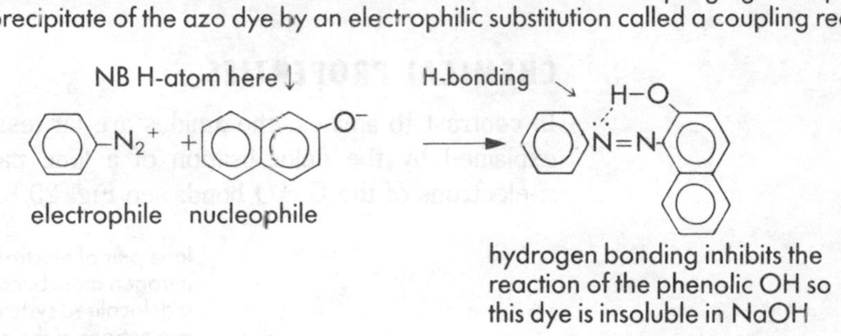

NB H-atom here ↓   H-bonding

$$\langle\!\!\bigcirc\!\!\rangle\text{---}N_2^+ + \langle\!\!\bigcirc\!\!\bigcirc\!\!\rangle\text{---}O^- \longrightarrow$$

electrophile   nucleophile

hydrogen bonding inhibits the reaction of the phenolic OH so this dye is insoluble in NaOH

Fig. 20.6 Preparing an azo-dye

π-electrons of the bond between the nitrogen atoms become part of the delocalised π-electron system of the benzene ring and help to stabilise the diazonium ion

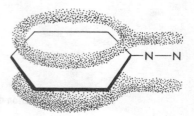

Fig. 20.7 Benzenediazonium ion is stabilised by delocalisation

REACTIONS OF AQUEOUS BENZENE DIAZONIUM ION

The benzene diazonium ion is stable only in ice-cold solution.

If you warm the solution nitrogen gas is evolved and the decomposition is first order with respect to the diazonium ion: see Chapter 9:

$$\langle\!\!\bigcirc\!\!\rangle\text{---}N_2^+ \rightarrow \langle\!\!\bigcirc\!\!\rangle + N_2(g)$$

Various other products are formed (including phenol, which you can detect by its smell) as the carbocation $C_6H_5^+$ combines with or attacks various Lewis bases: see Fig. 20.5.

If you warm the solution with aqueous potassium iodide, iodobenzene is formed:

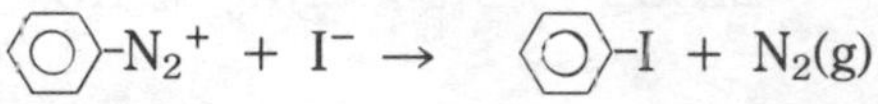

$$\langle\!\!\bigcirc\!\!\rangle\text{---}N_2^+ + I^- \rightarrow \langle\!\!\bigcirc\!\!\rangle\text{---}I + N_2(g)$$

This is important because it is one of the few ways of attaching iodine to a benzene ring.

If the solution is warmed with aqueous potassium cyanide and copper(I) cyanide as catalyst, benzonitrile is formed:

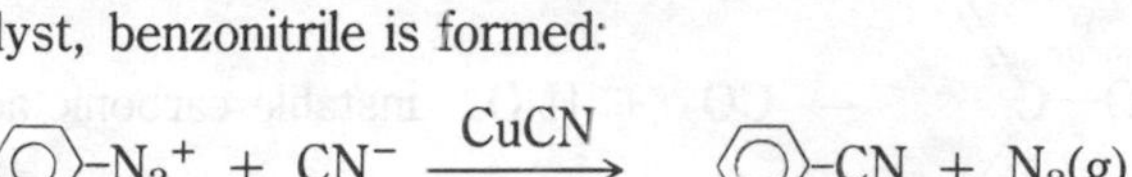

$$\langle\!\!\bigcirc\!\!\rangle\text{---}N_2^+ + CN^- \xrightarrow[\text{catalyst}]{CuCN} \langle\!\!\bigcirc\!\!\rangle\text{---}CN + N_2(g)$$

AMIDES

These compounds are usually considered as carboxylic acid derivatives because their

formulae contain the functional group $-C\!\!\begin{array}{c}O\\\|\\\diagdown\\NH_2\end{array}$: see chapter 19.

NOMENCLATURE

- The names of aliphatic amides are based on the alkanes with the C of the C=O included in the carbon chain: e.g. CH_3CONH_2; ethanamide
- If one or both H-atoms in the $-NH_2$ are replaced by alkyl groups the compound is named as N-alkyamide: e.g. $CH_3CONHCH_3$; N-methylethanamide
- $CO(NH_2)_2$ is a diamide known as urea or carbamide

PHYSICAL PROPERTIES

The amides (except for methanamide – m.pt. 2.5 °C) are white crystalline solids with high melting points. They dissolve well in water and organic liquids. The extensive hydrogen bonding between the molecules would account for the high melting points and solubilities in water: see Fig. 19.9.

CHEMICAL PROPERTIES

In contrast to amines, the amides are far less reactive and far less basic. This could be explained by the delocalisation of a lone pair of electrons (of the N-atom) with the π-electrons of the C=O bond: see Fig. 20.8.

lone pair of electrons on the nitrogen atom becomes part of a delocalised system with the π-electrons of the carbonyl group

Fig. 20.8 Electron delocalisation in amides

HYDROLYSIS

Amides are hydrolysed by refluxing with aqueous strong acids or alkalis:

$$R-C\!\!\begin{array}{c}O\\\|\\\diagdown\\NH_2\end{array} + H_2O \xrightarrow[\text{reflux}]{H^+ \text{ or } OH^-} NH_3 + R-C\!\!\begin{array}{c}O\\\|\\\diagdown\\OH\end{array} \quad \begin{array}{l}\text{carboxylic}\\\text{acid}\end{array}$$

Hydrolysis of urea is catalysed by specific enzymes found, for example, in soil bacteria and plants:

$$NH_2-C\!\!\begin{array}{c}O\\\|\\\diagdown\\NH_2\end{array} + 2H_2O \xrightarrow{\text{enzymes}} 2NH_3 + HO-C\!\!\begin{array}{c}O\\\|\\\diagdown\\OH\end{array} \quad \begin{array}{l}\text{carbonic}\\\text{acid}\end{array}$$

$$HO-C\!\!\begin{array}{c}O\\\|\\\diagdown\\OH\end{array} \rightarrow CO_2 + H_2O \quad \text{unstable carbonic acid decomposes}$$

The equation for the net hydrolysis reaction may be written as follows:

$$CO(NH_2)_2 + H_2O \rightarrow CO_2 + 2NH_3$$

The resulting ammonia molecules may be oxidised to nitrate ions which can be absorbed by plants. Note that urea is widely used as a fertiliser.

- Notice that two OH groups on the same C-atom (as in carbonic acid) are usually unstable and undergo elimination of a water molecule. This explains the formation of propanal and propanone by hydrolysis of 1,1-dichloropropane and 2,2-dichloropropane, respectively.

REACTION WITH NITROUS ACID

Nitrous acid replaces the —NH$_2$ groups in amides by —OH groups and liberates almost quantitatively one mole of N$_2$ gas for every one mole of —NH$_2$ group:

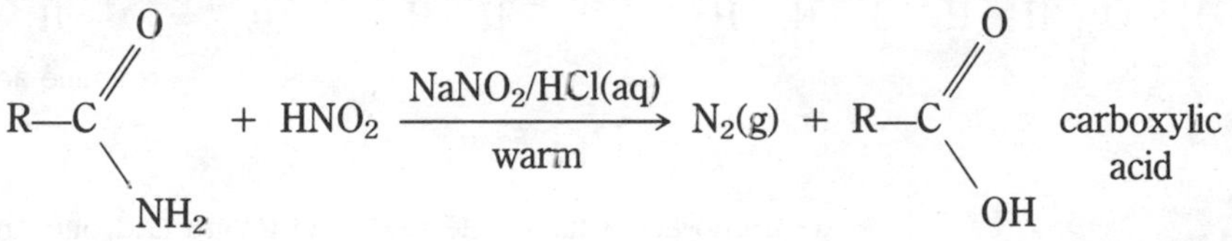

REDUCTION

Powerful reducing agents will reduce amides to amines:

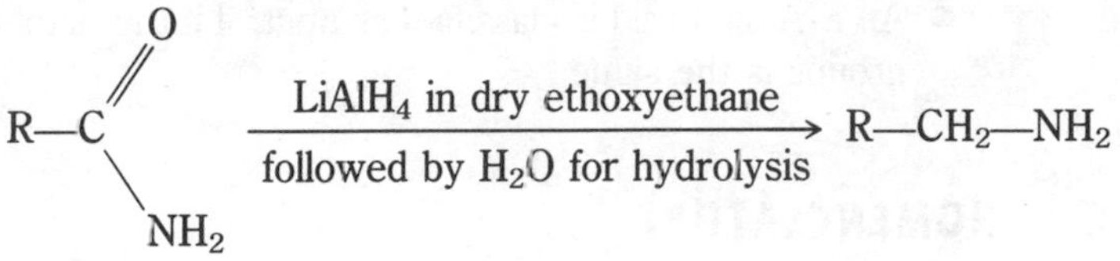

DEHYDRATION

Powerful dehydrating agents will convert amides into nitriles:

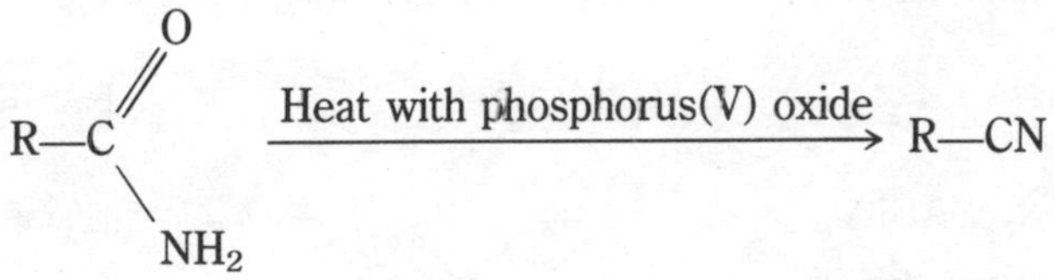

POLYAMIDES

- Polyamides are long chain polymers in which the units are held together by the amide linkage at regular intervals along the molecular chain

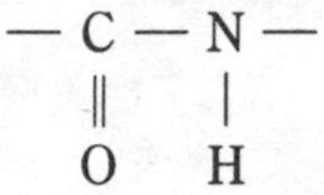

- Nylons are synthetic polyamides in which the number of C-atoms between each amide linkage may vary from one nylon to another

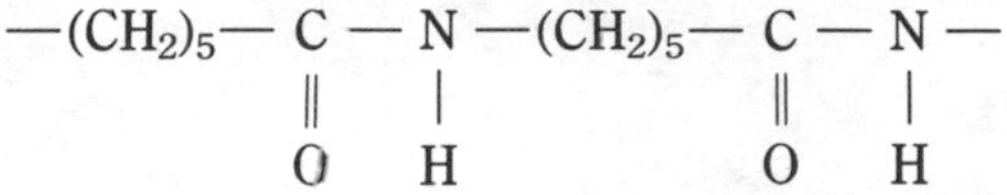

- Proteins are natural polyamides with only one —CHR— group between each amide (peptide) linkage but in which the R-group may be one of about twenty different acidic, basic or neutral groups

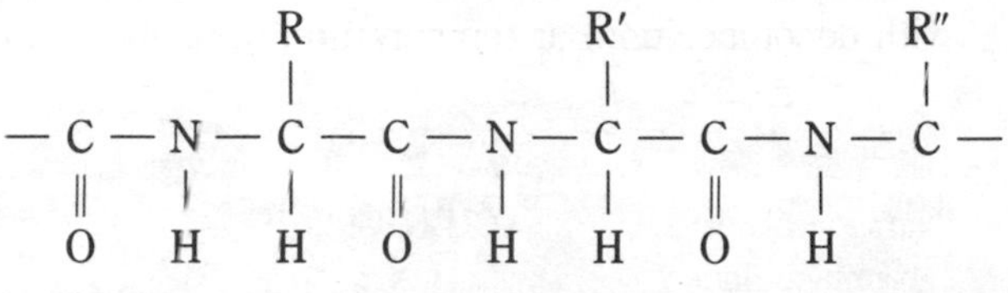

- Enzymes are protein molecules that act extremely efficiently as highly specific (bio)chemical catalysts

Nylons and proteins including silk and wool are extremely important polymers. Many form strong fibres. Hydrogen bonding involving the CO and NH groups between the neighbouring long polymer chains contributes to the strength of the fibres.

HYDROLYSIS OF PROTEINS

If you boil any protein (polyamide) with concentrated aqueous acid or alkali, hydrolysis will occur at the peptide (amide) linkages. The result of complete hydrolysis is a mixture of no more than twenty-five different α-amino-acids:

$$
\begin{array}{c}
\text{R} \quad\quad \text{R}' \quad\quad \text{R}'' \\
| \quad\quad\ | \quad\quad\ | \\
-\text{C}-\text{N}-\text{C}-\text{C}-\text{N}-\text{C}-\text{C}-\text{N}- \\
| \quad | \quad \| \quad | \quad | \quad \| \quad | \quad | \\
\text{H} \quad \text{H} \quad \text{O} \quad \text{H} \quad \text{H} \quad \text{O} \quad \text{H} \quad \text{H}
\end{array}
\rightarrow
$$

$$
\begin{array}{ccc}
\text{R} & \text{R}' & \text{R}'' \\
| & | & | \\
-\text{C}-\text{N}-\text{H} \quad\quad \text{HO}-\text{C}-\text{C}-\text{N}-\text{H} \quad\quad \text{HO}-\text{C}-\text{C}-\text{N}- \\
| \quad | \quad\quad\quad\quad \| \quad | \quad | \quad\quad\quad\quad \| \quad | \quad | \\
\text{H} \quad \text{H} \quad\quad\quad\quad \text{O} \quad \text{H} \quad \text{H} \quad\quad\quad\quad \text{O} \quad \text{H} \quad \text{H}
\end{array}
$$

α-amino-acid

AMINO-ACIDS

- α-Amino-acids have one carboxyl group and one amino group attached to the same (alpha α-) carbon atom

- α-Amino-acids are classified as acidic (or basic) if there are more (or less) $-CO_2H$ groups than $-NH_2$ groups

- An α-Amino-acid is classified as neutral if the number of $-CO_2H$ groups and $-NH_2$ groups is the same

NOMENCLATURE

The naturally occurring α-amino-acids that are the building blocks of proteins are not named systematically. Even if you have to make a special study of biochemistry, you would not have to remember the names and structures of all the α-amino-acids. However, I suggest you learn the four in Fig. 20.9.

NAME	FORMULA	NAME	FORMULA
glycine GLY aminoethanoic acid	$\begin{array}{c} NH_2 \\ | \\ H-C-H \\ | \\ CO_2H \end{array}$ neutral	alanine ALA 2-aminopropanoic acid	$\begin{array}{c} NH_2 \\ | \\ H-C-CH_3 \\ | \\ CO_2H \end{array}$ neutral
glutamic acid GLU	$\begin{array}{c} NH_2 \\ | \\ H-C-CH_2CO_2H \\ | \\ CO_2H \end{array}$ acidic	lysine LYS	$\begin{array}{c} NH_2 \\ | \\ H-C-(CH_2)_4NH_2 \\ | \\ CO_2H \end{array}$ basic

Fig. 20.9 Four important α-amino-acids

PHYSICAL PROPERTIES

The α-amino acids are white crystalline solids that are very soluble in water. In contrast to amines and carboxylic acids of comparable molar mass, α-amino-acids melt (usually with decomposition) at temperatures well above 200 °C: see Fig. 20.10.

Fig. 20.10 High melting point of amide acid caused by zwitterions

NAME	FORMULA	M_r	MELTING POINT/°C
aminoethanoic acid	$CH_2NH_2CO_2H$	75	262 (decomposes)
1-aminobutane	$C_4H_9NH_2$	73	−49
propanoic acid	$C_2H_5CO_2H$	74	−21

These properties are explained by the amino-acid existing as an inner salt or zwitterion (German: 'zwitter' = hybrid) formed by transfer of a proton from the $—CO_2H$ group to the $—NH_2$ group:

$$R—\overset{\overset{NH_2}{|}}{\underset{\underset{H}{|}}{C}}—CO_2H \rightarrow R—\overset{\overset{NH_3^+}{|}}{\underset{\underset{H}{|}}{C}}—CO_2^-$$

Except for glycine, an α-amino-acid can exist as enantiomers: see Fig. 20.11.

L(+) glutamic acid — letter ↑ specifies the form of the structure

D(−) glutamic acid — sign ↑ specifies the direction of rotation of the plane of polarised light

Fig. 20.11 Enantiomers of glutamic acid

These two isomers have identical physical properties except that one isomer will rotate the plane of polarised light to the left whilst the other will rotate the plane an identical amount to the right. The isomers also have identical chemical properties except for their reaction with another chiral molecule.

- The α-carbon atom is a chiral centre and is sometimes called assymetric because it has four different groups attached to it and is responsible for the existence of enantiomers and the optical activity of α-amino-acids
- The L-form of α-amino-acids (which may be (+) dextro- or (−) laevo-rotatory) is the building block of all proteins

NITRILES (CYANIDES)

"Important in synthesis. Link with nucleophilic addition to >C=O."

These toxic, dangerously flammable liquids are not strictly carboxylic derivatives but they are often included under that heading because, for example, they can be made by dehydration of amides:

$$R—CO—NH_2 \xrightarrow{-H_2O} R—CN$$

More importantly they are formed by heating alcoholic potassium cyanide with a halogeno-compound, thereby adding a carbon atom and precisely lengthening the chain: see Chapter 16.

PROPERTIES AND USES

They are important intermediates in organic syntheses. Nitriles are readily hydrolysed to carboxylic acids under acid conditions or its salt under alkaline conditions:

$$R—C{\equiv}N \xrightarrow[\text{aqueous reflux}]{H^+ \text{ (or } OH^-)} R—CO_2H \quad (\text{or } R—CO_2^-)$$

Nitriles can be reduced to primary amines:

$$R—C{\equiv}N \longrightarrow R—CH_2—NH_2$$

Propenonitrile (acrylonitrile) is an important monomer in the industrial production of poly(propenonitrile) – an addition polymer also known as Acrilan:

$$(n+4)\ \overset{H\quad\ CN}{\underset{H\quad\ H}{C{=}C}} \rightarrow —\overset{H}{\underset{H}{C}}—\overset{CN}{\underset{H}{C}}—\overset{H}{\underset{H}{C}}—\overset{CN}{\underset{H}{C}}—\left[\overset{H}{\underset{H}{C}}—\overset{CN}{\underset{H}{C}}\right]_n—\overset{H}{\underset{H}{C}}—\overset{CN}{\underset{H}{C}}—\overset{H}{\underset{H}{C}}—\overset{CN}{\underset{H}{C}}—$$

EXAMINATION QUESTIONS AND OUTLINE ANSWERS

Q1 The primary aliphatic amine 1-aminobutane reacts with nitrous acid to produce four products, **A**, **B**, **C** and **D**. Compounds **A** and **B** are unsaturated hydrocarbons and both have the molecular formula C_4H_8. On hydrogenation both are converted to the same saturated hydrocarbon, **E**. On reaction with hydrogen bromide both **A** and **B** yield the same product, **F**.

Compounds **C** and **D** both have the same molecular formula, $C_4H_{10}O$, and on mild dichromate oxidation give the products **G** and **H** respectively. Both **G** and **H** give a yellow/orange precipitate with 2,4-dinitrophenylhydrazine, but only **G** gives a red/brown precipitate with Fehling's solution.

    i) Identify, giving your reasoning, the compounds **A–H**. *(11)*
    ii) State which of them would exhibit geometrical isomerism. *(1)*
    iii) Suggest how the four products **A–D** arise from this reaction. *(2)*

(WJEC 1988)

Q2 a) Explain briefly what is meant by the term *nucleophile* and give equations to show ammonia taking part in: *(1)*
    i) a nucleophilic substitution reaction; *(2)*
    ii) an addition-elimination reaction. *(2)*
   b) Draw the mechanism for the reaction in a)i) and explain clearly which atom is acting as the *electrophile*. *(4)*
   c) Predict, giving your reasons:
    i) which of aminoethanoic acid and ethanamide will have the higher melting point; *(3)*
    ii) which of nitrobenzene and phenylamine will be the stronger base; *(4)*
    iii) which of ethylamine and ethanamide will be the better nucleophile. *(4)*

(ODLE 1988)

Outline answers

> **Q1**
>
> (i) A, B = but-1-ene and but-2-ene
> E = butane : F = 2-bromobutane
> C = butan-1-ol : G = butanal
> D = butan-2-ol : H = butanone
>
> (ii) but-2-ene
>
> (iii) $C_4H_9NH_2 \xrightarrow[H^+]{HNO_2} C_4H_9^+ + N_2(g) + 2H_2O$
>
> very reactive carbonation then reacts in various ways :- eg.
>
> $C_4H_9^+ \longrightarrow C_4H_8 + H^+$
>
> $C_4H_9^+ + OH^- \longrightarrow C_4H_9OH$
>
> **Q2**
>
> (a) } study the text (pp 275 and 300)
> (b) } and chapter 14

(c)(i)

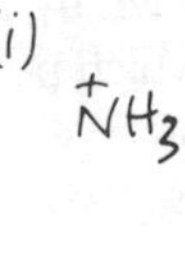

$$\overset{+}{N}H_3 - \overset{\overset{\displaystyle H}{|}}{\underset{\underset{\displaystyle H}{|}}{C}} - CO_2^-$$

aminoethanoic because ion-ion interactions possible.

(ii)

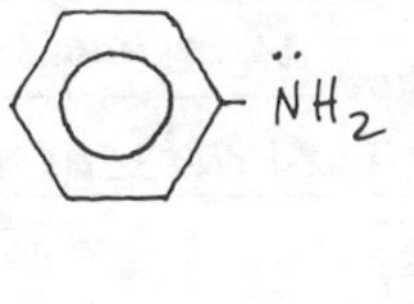

 $\ddot{N}H_2$

phenylamine because lone-pair on N atom but not in the $-NO_2$ group.

(iii)

$C_2H_5 \ddot{N}H_2$

ethylamine because lone-pair is not delocalized

EXAMINATION QUESTIONS AND TUTOR'S ANSWERS

Q3 a) For each of the classes of compounds given below, draw a full structural formula of one example and give the name of your chosen compound:

i) primary amine; (2)

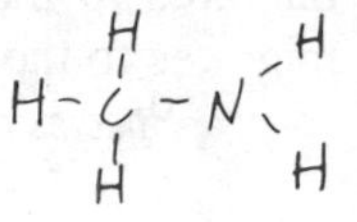

$$H - \overset{\overset{\displaystyle H}{|}}{\underset{\underset{\displaystyle H}{|}}{C}} - N \overset{\displaystyle H}{\underset{\displaystyle H}{\cdot}}$$ methylamine

ii) acid anhydride; (2)

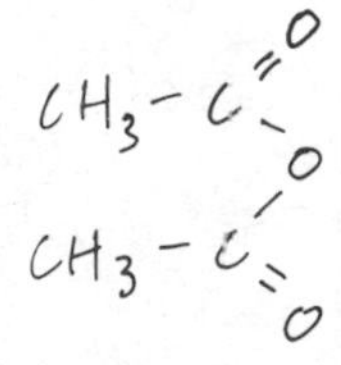

$$CH_3 - C \overset{\displaystyle =O}{\underset{\displaystyle O}{}}$$
$$CH_3 - C \overset{\displaystyle =O}{\underset{\displaystyle O}{}}$$

iii) aromatic diazonium salt. (2)

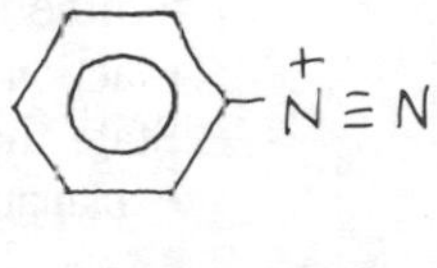

$$\text{—} \overset{+}{N} \equiv N$$

b) For each of the three compounds chosen in a), write a balanced equation in which your compound appears as a **product** and state the conditions under which each preparation takes place:

i) primary amine: (3)

equation $CH_3 Br + NH_3 \rightarrow CH_3 NH_2 + HBr$

conditions aqueous ammonia heated under pressure

ii) acid anhydride; (3)

equation $CH_3 CO_2^- Na^+_{(s)} + CH_3 COCl_{(l)} \rightarrow (CH_3 CO)_2 O + Na^+ Cl^-$

conditions heat sodium ethanoate with liquid ethanoyl chloride

iii) aromatic diazonium salt. (3)

equation $C_6H_5 NH_2 + HNO_2 + HCl \rightarrow C_6H_5 N_2 Cl + 2H_2O$

conditions Add ice-cold aqueous sodium nitrate to ice-cold solution of aniline in hydrochloric acid

(ODLE 1988)

Q4 a) State the type of isomerism shown by 2-aminopropanoic acid (*alanine*) and indicate the structural feature which permits the existence of two isomers. With the aid of diagrams, show the structural relationship between the isomers. Explain briefly how the two isomers can be distinguished.

Type of isomerism <u>optical</u>

Structural feature <u>an asymmetric carbon atom (four different groups attached) acts as a chiral centre.</u>

Isomer 1 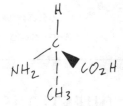 *Isomer 2*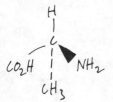

Explanation <u>They rotate the plane of polarized light in opposite directions.</u> (6)

b) When a solution of alanine is electrolysed at low pH, an organic species migrates to the cathode, whereas at high pH a different organic species migrates to the anode. Give the structure of each of the migrating species.

Species to cathode at low pH *Species to anode at high pH*

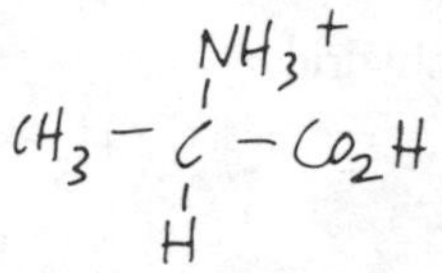

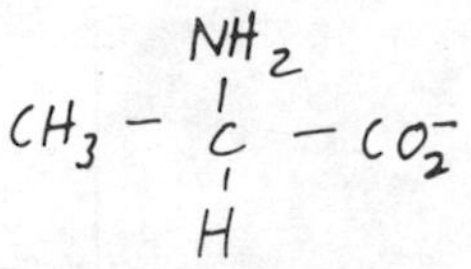

$$CH_3 - \underset{H}{\overset{NH_3^+}{C}} - CO_2H \qquad CH_3 - \underset{H}{\overset{NH_2}{C}} - CO_2^-$$

(2)

c) Synthetic alanine, prepared from propanoic acid by bromination followed by reaction with ammonia, differs from alanine obtained from natural sources. State the difference between natural and synthetic alanine and give a brief explanation as to why they differ.

Difference <u>It does not affect polarized light</u>

Explanation <u>Equal amounts of isomers 1 and 2 are formed.</u> (3)

(JMB 1988)

PRACTICE QUESTION

Q5 A popular question in organic chemistry asks you to compare a functional group in an alphatic and aromatic compound. You should prepare for these.

Describe the main reactions of amines, explaining any significant differences between the behaviour of alkylamines (such as butylamine) and arylamines (such as phenylamine). Quote formulae and equations wherever appropriate.

Show how the reactions of amines are used to produce at least two industrially important products, such as polymers and dyestuffs.

(NUFF 1988)

INDEX